D0998705

Agricultural Decision Making

Anthropological Contributions to
Rural Development

This is a volume in

STUDIES IN ANTHROPOLOGY

Under the consulting editorship of
E. A. Hammel, University of California, Berkeley

A complete list of titles appears at the end of this volume.

Agricultural Decision Making

Anthropological Contributions to Rural Development

Edited by

PEGGY F. BARLETT

Department of Anthropology
Emory University
Atlanta, Georgia

ACADEMIC PRESS

A Subsidiary of Harcourt Brace Jovanovich, Publishers

New York   London   Toronto   Sydney   San Francisco

COPYRIGHT © 1980, BY ACADEMIC PRESS, INC.
ALL RIGHTS RESERVED.
NO PART OF THIS PUBLICATION MAY BE REPRODUCED OR
TRANSMITTED IN ANY FORM OR BY ANY MEANS, ELECTRONIC
OR MECHANICAL, INCLUDING PHOTOCOPY, RECORDING, OR ANY
INFORMATION STORAGE AND RETRIEVAL SYSTEM, WITHOUT
PERMISSION IN WRITING FROM THE PUBLISHER.

ACADEMIC PRESS, INC.
111 Fifth Avenue, New York, New York 10003

United Kingdom Edition published by
ACADEMIC PRESS, INC. (LONDON) LTD.
24/28 Oval Road, London NW1 7DX

Library of Congress Cataloging in Publication Data
Main entry under title:

Agricultural decision making.

 (Studies in anthropology)
 Includes bibliographies and index.
 1. Economic anthropology––Addresses, essays, lectures. 2. Agriculture––Economic aspects––Decision
making––Addresses, essays, lectures. 3. Rural development––Addresses, essays, lectures. 4. Applied anthropology––Addresses, essays, lectures. I. Barlett,
Peggy.
GN448.A37 306'.3   80–17684
ISBN 0–12–078880–2

PRINTED IN THE UNITED STATES OF AMERICA

80 81 82 83 9 8 7 6 5 4 3 2 1

Contents

Chapter 1

Introduction: Development Issues and Economic Anthropology 1
PEGGY F. BARLETT

Chapter 5

The Attentive–Preattentive Distinction in Agricultural Decision Making 115

HUGH GLADWIN AND MICHAEL MURTAUGH

Chapter 6

Cost–Benefit Analysis: A Test of Alternative Methodologies 137

PEGGY F. BARLETT

Chapter 7

Risk and Uncertainty in Agricultural Decision Making 161

FRANK CANCIAN

Chapter 8

Forecasts, Decisions, and the Farmer's Response to Uncertain Environments 177

SUTTI ORTIZ

Chapter 9

Management Style: A Concept and a Method for the Analysis of Family-Operated Agricultural Enterprise 203

JOHN W. BENNETT

PART II

PATTERNS OF AGRICULTURAL DECISIONS

Chapter 10

Agricultural Business Choices in a Mexican Village 241

JAMES M. ACHESON

Chapter 11

Agrarian Reform and Economic Development: When Is a Landlord a Client and a Sharecropper His Patron? 265

KAJA FINKLER

Chapter 12

Stratification and Decision Making in the Use of New Agricultural Technology 289

BILLIE R. DEWALT AND KATHLEEN MUSANTE DEWALT

List of Contributors

Numbers in parentheses indicate the pages on which the authors' contributions begin.

JAMES M. ACHESON (241), Department of Anthropology, University of Maine, Orono, Maine 04038

PEGGY F. BARLETT (1, 137), Department of Anthropology, Emory University, Atlanta, Georgia 30322

JOHN W. BENNETT (203), Department of Anthropology, Washington University, St. Louis, Missouri 63130

SARA BERRY (321), African Studies Center, Boston University, Brookline, Massachusetts 02146

FRANK CANCIAN (161), School of Social Sciences, University of California, Irvine, California 92717

MICHAEL CHIBNIK (87), Department of Anthropology, University of Iowa, Iowa City, Iowa 52242

BILLIE R. DEWALT (289), Department of Anthropology, University of Kentucky, Lexington, Kentucky 40506

KATHLEEN MUSANTE DEWALT (289), Department of Behavioral Science, College of Medicine, University of Kentucky, Lexington, Kentucky 40536

KAJA FINKLER (265), Department of Sociology, Eastern Michigan University, Ypsilanti, Michigan 48197

HUGH GLADWIN (115), Department of Anthropology, University of California, Irvine, California 92717

CHRISTINA H. GLADWIN (45), Department of Anthropology, Northwestern University, Evanston, Illinois 60201

ALLAN HOBEN (337), Overseas Development Council, 1717 Massachusetts Avenue, Northwest, Washington, D.C. 20036, and African Studies Center, Boston University, Brookline, Massachusetts 02146

ALLEN JOHNSON (19), Department of Anthropology, University of California, Los Angeles, California 90024

MICHAEL MURTAUGH (115), School of Social Sciences, University of California, Irvine, California 92717

SUTTI ORTIZ (177), 2654 Exeter, Cleveland Heights, Ohio 44118

Preface

This book is about farmers' decisions—what goes into them and what effects they have, once made. The chapters address the following questions: How do farmers choose what to plant and how to plant it? What are the important variables that determine agricultural decisions? How are these variables best measured and how do they interact? How do farmers' own conceptions of the choice process match psychological, economic, and mathematical models? What are the limitations and uses of such models? What implications do recent findings in this area have for agricultural development programs?

The impacts of farmers' choices in agricultural production can be felt in diverse arenas. Individual farm decisions determine household profits and well-being, land use, capital requirements, and the adoption of new technology. They also affect such issues as prestige and leadership in the community and the long-term ecological stability of an area. These choices have implications for what products each nation has to export or to process and use domestically. Also affected are relationships between nations and such vital matters as their balance of payments and their ability to withstand poor weather, rising energy costs, or a rapidly growing population.

The agricultural decision research collected here focuses mainly on developing countries. In general, these nations are linked in the public attention to an impending global food crisis, increasing poverty, and political upheavals. To understand these international crises, however, we must look beneath generalizations about "food production," "transfer of technology," "rural instability," or "economic development" to see individual farmers, making choices. The purpose of this book is to show how anthropological understandings of farmers' choices and the reasons for them are useful, not only to explain rural development trends, but also to devise policies to bring about an improving standard of living for the majority of the world's peoples.

This book assembles a new generation of research in economic anthropology. Part I explores a series of theoretical and methodological questions concerning the use of formal models, cognitive versus statistical behavior models of decision making, attentive and preattentive aspects of agricultural decisions, measurement issues in evaluating the alternatives open to farmers, the role of risk and uncertainty and farmers' responses to them, and the tools to study agricultural decisions over the life-cycle of the household. While dealing with specific issues, each of these studies is based on detailed, long-term research in rural areas.

Part II is devoted to more comprehensive studies which explore the patterns of agricultural choices within one rural community. The impacts of nonagricultural alternatives on agricultural decisions, the causes and effects of traditional sharecropping agreements, and the importance of economic stratification and differential access to resources are discussed in this section.

Part III treats the implications of decision-making research for agricultural development policy and explores the decision-making context of aid programs. Whereas these final two chapters focus primarily on policy and development programs, many of the other contributions also take up these issues.

This volume is needed at this time for two reasons. First, the past decade has been enormously fruitful in refining anthropological research to understand and predict rural change. A compilation of these studies facilitates both the assessment of the advances made and the clarification of future research needs. These studies not only contribute to our understandings of rural change, but also contribute to a refinement of the discipline as well.

Second, there is a growing demand for the anthropological perspective among governmental and international agencies, and an increasing number of anthropologists are now employed in the development area. This volume serves to bridge the gap between academic researchers and development practitioners by presenting the theory, methods, and results of recent analyses in diverse development situations.

The next decades can be expected to bring ever more rapid change to the rural areas of the world: more population growth and pressure on food resources, more political upheavals and calls for economic and social justice, more ecological dislocations and increasing competition for scarce global resources. These trends must be understood in their local cultural context, however, and must be connected to the realities of life in developing countries. Influences from the international, national, regional, and local levels must be linked to the welfare of individual households and families. In this context, the decisions of agricultural producers will come to be of increasing importance, and growing apace will be our need to understand those choices. This volume is dedicated to furthering that understanding.

Warm thanks are due to Academic Press, which has supported this volume since its inception, and to Barbara Melvin who prepared the fine subject index. My special gratitude goes to all the contributors without whose hard work, cooperation, and encouragement this volume could not have been produced.

Chapter 1

Introduction: Development Issues and Economic Anthropology

PEGGY F. BARLETT

The 1970s has been a productive decade of research by economic anthropologists on agricultural development issues. This volume assembles the work of a diverse group of authors who have studied agricultural production, focusing on the decision processes of farmers. These chapters serve not only as a benchmark of research in economic anthropology but also as a demonstration of the important and practical contributions of anthropology to agricultural development programs and policy. This research transcends the substantivist–formalist controversy of the 1960s, integrating the strengths of both perspectives within its focus on agricultural decisions. Most of the authors here share the dual aims of this book, to advance the sophistication and accuracy of the discipline while bringing anthropological skills to bear on major world issues. These goals are complementary, for as we refine our theories, methods, and analyses, the value of anthropology for evaluating and guiding agricultural development increases as well.

Anthropology and the World Food Crisis

The dimensions of poverty and hunger have expanded with the size of human populations in recent years. "Possibly as many as 450 million to a billion persons in the world do not receive enough food" cites one

1

AGRICULTURAL DECISION MAKING:
Anthropological Contributions to
Rural Development

study (National Research Council 1977:1). Following any definition of the magnitude of the world food crisis comes an inevitable call for increased agricultural productivity: "Developing countries must double their food production by the end of the century [National Research Council 1977:2]." "Development plans are nowadays full of 'top priority for agriculture' [Lipton 1977:17]." Yet despite considerable efforts to improve the overall production of farmers in these countries, the results are mixed, and there is general agreement that poverty and malnutrition are increasing (deJanvry 1975; Griffin 1974; Lipton 1977; National Research Council 1977).

With the failure of the "development decade" of the sixties and the realization that increased agricultural productivity does not necessarily lead to an improved standard of living for the majority of rural families has come a greater concern with the distribution of income and with the overall welfare of farmers. This broader perspective is more congenial to the holistic approach of the anthropologist. The issues of poverty and hunger are now seen by many development experts to be affected by the complex institutions which structure rural life, and some have concluded that "agricultural change must take place within the context of institutional change [Griffin 1974:2]" (Bieri *et al.* 1972; Danda and Danda 1972; Stavenhagen 1977).

The institutions that affect farmers are diverse and complex, as well documented by the chapters presented here. This being the case, we can not expect to find "the simple anthropological parallel to dwarf wheat and miracle rice [Cancian 1977:2]." Our contribution lies not in defining yet another "key factor" that must be addressed, but rather in affirming the complexity and heterogeneity of farmers, their communities, their decisions, and of policies appropriate to their needs. Furthermore, "The contemporary rural community in a developing society is never isolated from the larger society [Castle 1977:3]." As also shown by a number of the chapters here, the external forces of world markets, developed countries, and international agencies can not be omitted from any attempt to understand local agricultural decisions or to propose solutions.

Overview of the Volume

The work collected in this volume shares several perspectives and explores some common themes. Following Wharton (1969:461), we seek to understand the production systems of peasant farmers—how they change, and what forces influence and inhibit change. We begin from the point of view that small farmers are neither irrational nor tradition-bound, and we assume that their agricultural patterns are the consequence of long- and short-term adaptations based on observation and experimentation. Determining first what agricultural decisions have been made, we can then pursue the impacts of these decisions: How do the choices of specific individuals and households affect the decision environment of others? How do agricultural choices influence other aspects of social and economic life? The focus of most of the studies here is on the decision process itself—what variables impact these choices and what layers of influences on decisions need to be understood.

A second theme of the research reported here is the issue of measurement. What tools from other disciplines and from traditional anthropological research prove useful in measuring choices and their determinants? What biases and shortcomings have been found in using such tools? All of the contributors to this volume have carried out long-term fieldwork in rural communities and have learned firsthand from farmers as well as from questionnaires and statistical analyses. What role do these qualitative understandings have, together with the quantitative data?

A third theme is the relationship of the decision maker's own perceptions to the practical process of food production. What are the folk categories, mental shortcuts, and preattentive processes that simplify agricultural decisions?

Finally, we are concerned to tie these research findings to their implications for development programs and policy. Some authors speak directly of specific agricultural development projects; others discuss more theoretical issues relating to policy. All of the chapters here, however, present information and analyses that relate directly to agricultural programs.

Some Issues and a New Role for Agriculture

Some recent issues in rural development point out the increasing complexity of development problems and proposed solutions. Agricultural development involves change in two dimensions: the kinds of crops grown and the way in which they are grown. In both dimensions, agronomic behaviors ramify into the entire social and political organization of rural life. For example, the shift from food-grain products to pasture in many areas of Central and South America has displaced many small farmers, turning farming communities into large ranches,

and setting off ecological repercussions whose effects are only beginning to be researched (Barlett 1977, n.d.; Holdridge and Tosi 1977;
Nations and Nigh 1978; Parsons 1976). Beef production is one of the
clearest changes in agriculture that involves complex world market
forces external to the peasant communities affected. Planned agricultural development projects also involve the introduction of new crops,
as farmers are urged to try citrus production, triticale, soybeans, vegetables, etc. Whereas many communities grow the same crops they have
grown for centuries, others are changing production decisions in response to the world market, to national price and market policies, and
to changing infrastructure. Plantation agriculture, although certainly
not new, has taken on a new character, as "luxury" fruits, vegetables,
and protein products are exported from the Third World to the developed nations. Criticism of this pattern argues that prime agricultural
land might better be used to feed the malnourished populations of the
exporting country (George 1977; Lappé 1975; Lappé and Collins 1977;
Lipton 1977). In contrast, many development planners see these exports
as essential to finance investments and to offset spiraling energy costs
and inflation. Although the magnitude of food exports today may be
unprecedented, the issue is not new. In the last century, a Russian
finance minister concluded, "We must export, though we die." As
Robert Cassen has said, however, he meant instead, "I shall export,
though you die [cited in Lipton 1977:359]." The distribution of development costs and benefits between different sectors of society, and
between different strata in a farming community, is a central issue in
development planning and one faced by many of the authors in this
volume.

Turning to the second major dimension of agricultural development,
decisions on agricultural techniques give rise to some of the same
issues. In many areas of the world, development projects have not
introduced new crops but have instead tried to increase the productivity of traditional crops. The most common single change is the
introduction of chemical fertilizers, often followed by insecticides, herbicides, machinery, and irrigation equipment. These changes in production methods destroy the one-way flow in which agricultural products enter into the cost of the production of all other things, as it was in
Ricardo's day, whereas "Agriculture does not utilize the products of
these other trades [Gudeman 1978:356]." Lester Brown notes that billions
of dollars are spent by U.S. farmers for "items purchased in the non-
farm or industrial sector," and "We can expect a steady rise in expenditures by farmers in the poor countries 'for the same sorts of inputs

[Brown 1970:59]." Many of these changes in cropping technique involve an increasing capital investment in agriculture, which usually affects small farmers differently from large farmers, again raising issues of the distribution of costs and benefits (Cochrane 1974; Hewitt 1973–1974).

Not only does this change in production techniques require the development of agricultural loan programs and extensive banking facilities in rural areas, but it also changes farmers' relationship with the world market. Dependence on trade and imports for production inputs is a new dimension for agriculture and one whose consequences we are only beginning to perceive. Historically, agricultural areas have been particularly resilient because their production units can continue to function though the central polity may collapse (Greenwood 1974). Although the collapse of product markets, and resulting boom and bust cycles, is well known and its consequences understood, we have yet to see how "developed" agriculture in Third World countries can respond to an extended collapse of trade and thus of access to production inputs. The increased vulnerability of rural areas to political and economic vicissitudes is thus an important aspect of agricultural development.

The Historical Perspective on Intensification and Development

Whether talking of new crops or new techniques, agricultural development implies increased productivity of land units. This increase has historically been achieved by labor intensification (Barlett 1976; Boserup 1965; Dumond 1965; Netting 1974), as shown by the remarkable productivity of irrigation agriculture in such countries as China, Japan, and Java. Much of the current world emphasis on agricultural development, however, stresses capital intensification. Anthropologists have long been interested in the coordination, both local and regional, required by labor intensification, but the kinds of economic and political control imposed by competing corporate giants and international lending agencies have very different characteristics.

Furthermore, as rural communities are more closely integrated with the world market economy, they must necessarily respond more directly to prices and market trends. The basis of such prices and trends, however, is relatively short-term supply and demand pressures. Many traditional agricultural systems, on the other hand, are adaptations to long-term ecological and economic forces (Price 1978; Williams 1977). Thus the transition from "traditional" to more "developed" farming practices is a transition to a different framework for decision making, one which may not be adapted to survival in the bad years, but rather to profit maximization in the average years (Johnson 1971; Lipton 1968; Wharton 1971).

In contrast to the industrial agriculture model which stems from the experience of the United States, many researchers are drawing attention to the greater efficiency of small farms, both in productivity per land unit and in the use of scarce resources (Lipton 1977; Loomis 1976; Rochin 1977). Especially as government planners in developing countries move away from optimism about the capacity of industry to absorb labor, the attractiveness of labor-intensive agricultural strategies increases. The anthropologist's understandings of farming communities and especially small-farmer production methods and decisions become essential inputs into these new programs to encourage labor-intensive development.

Thus, the changes implied by agricultural development in many Third World countries raise the following general questions:

1. What is the impact of new crop choices and new markets on the food supplies and nutritional standards of different rural sectors of these countries?
2. What effect does the addition of purchased inputs have on the agricultural community and its subsectors and their relationship with the nation and the world market?
3. Do either of these types of agricultural change tend to freeze or exacerbate inequalities in rural areas? Do they tend to reinforce or weaken the agrarian institutions that have led to the gap between the rich and the poor?
4. What will be the ramifications of increased agricultural productivity via capital intensive agribusiness?
5. How will agricultural adaptations be affected by replacing long-term decision criteria with more short-term considerations?
6. How can labor intensive techniques and perspectives be integrated into agricultural development plans and programs?

These questions can not be answered in simple, global terms. They lead us to look at diverse local areas, to seek solutions within specific cultural contexts, and to understand the emerging patterns in light of the decisions made both by farmers and by planners. Through an analysis of these decisions will come the clarification of the major determinants of these changes, and thus the avenues by which the changes can be affected. It is to these ends that these chapters have been collected.

Economic Anthropology and Research Perspectives

Economic anthropology in the 1960s was occupied primarily by debates between formalists, who argued that formal economic concepts derived from Western market economies can be applied to all cultures, and substantivists, who argued that such concepts as rationality and maximization are culture-bound and distort the reality of non-Western, nonmarket economies. Formalists conceptualize the economy as a rational choice process based on the allocation of scarce resources to alternative ends. Individuals are the units of economic analysis and are assumed to act on the basis of self-interest. Substantivists, on the other hand, define the economy as the process of material-means provisioning for society and focus on the institutions that structure this process.

As the decade neared its end, there began to emerge a growing body of research that focused on rural change and agricultural development, and that labeled itself neither substantivist nor formalist. Many of the issues of the previous controversy are not pertinent to this problem-oriented research, since most of it was carried out within economies that are partially, if not wholly, market oriented and since the social and institutional environments are included as important parts of any "formal" economic analysis. Not only are "substantive" perspectives joined with formal analysis, but also the rigidity and accuracy of many formal economic concepts and assumptions have been questioned. Classical economic terms such as rationality, preferences, maximization, and utility have been explored, operationalized, and then criticized, as they have been adapted to fit local realities.

Most of the researchers in this volume reflect the perspective that all human behavior involves choices—some easy, some agonizing, some constrained, and some unconscious. As Goodfellow notes, even "custom" does not remove choice from highly constrained situations: A Bantu woman customarily cooks and works in the fields each day, but she must still decide how much time to allocate between cooking and fieldwork. Though the Bantu say a fixed number of cattle are required for certain payments, "All evidence points to the fact that a goat may be substituted for a cow [Goodfellow 1950:93]." Distinctions between "primitives," "peasants," and "farmers" are now seen as stemming from different cultural adaptations based on different values of the same variables. Ortiz concludes, "Peasant societies give us more dramatic examples, perhaps, but they are not totally different from market economies [Ortiz 1967:194]." Cancian also points out the consequences of more rigid distinctions: "The question whether Zinacantecos are economic maximizers or prisoners of tradition . . . is a bogus question that leads to scientifically incorrect and politically dangerous descriptions of peasant societies [Cancian 1972:189]."

Thus, we have moved away, over this past decade, from simple questions with yes or no answers and are now pursuing issues by seeking to determine *when* farmers behave as one would predict from strict maximization formulae and when they do not. Further, we attempt to define what variables, institutional or otherwise, will account for these differences. We have avoided an emphasis on "structural typology" (Strickon and Greenfield 1972:7) in favor of a more actor-oriented analysis, which stresses the diversity of behavior within a local context of institutions, customs, and conditions. Our emphasis has moved beyond the consideration of achieved versus ascribed characteristics of persons to a study of relationships. This focus is infused with the dynamic perception of relationships as *negotiated*, of agricultural choices as fluid and responsive to the decision-making environment. Such a perspective leads not only to a measurement of the diversity of behavior within rural communities, but also to a clarification of and sometimes to the measurement of the variables that interact to produce the behavior outcome. "The strategies and tactics employed by the various actors have to be viewed as the outcome of the interplay of a number of variables rather than the result of simple structural demands or psychological states [Strickon and Greenfield 1972:14]."

This concern with a range of variables and the complexity of agricultural development issues should not imply that all questions can only be answered with a laundry list of "important factors." Strickon and Greenfield note that access to resources (material or nonmaterial) is "perhaps" the key variable (1972:15). Acheson, Bennett, Cancian, DeWalt and DeWalt, in this volume, show the importance of access to land and wealth in understanding agricultural decisions. Barlett and Finkler explore the role of family labor resources, and Finkler notes the crucial role of capital and water resources. Yet in each case, there are complex circumstances that affect resource distribution and the diversity of decision outcomes that emerge. These in-depth studies give conclusive proof that simple solutions and worldwide verities are doomed to disappointment. Local realities will always distort the mathematical curves, and no simple model can hold up past class, kinship, ecological, and governmental differences. Yet, as can be seen in the following, in many communities, and in development agencies, the same variables reemerge to play important roles in affecting agricultural decisions.

Issues in Agricultural Decision Research

The decision making process of farmers involves a range of factors that are taken into account. Each farmer usually makes choices within the context of the household and is influenced by the household's needs and goals as well as by the resources available to the household. These resources include not only land, water, labor, etc., but also social resources such as information about agricultural methods or credit and any influence or political power necessary in many areas to successful agricultural production. Finkler describes how some families with political clout obtain access to irrigation water more easily and, thus, are in a position to benefit from sharecropping arrangements. These families have access to capital and credit as well and can thus choose production strategies that other farmers can not.

The range of agricultural options open to farmers is usually determined by the interaction of the natural environment with the larger social environment. Thus, government programs to provide better roads and housing are shown by Chibnik to affect labor allocation decisions in Belize. Market forces together with ecological variations in the Guatemalan mountainside are cited by C. Gladwin, H. Gladwin, and Murtaugh as structuring decisions to plant corn on the hillsides and vegetables on the valley floor. Bennett stresses the "multidimensional social system" in which decisions are made, and DeWalt and DeWalt show that attention to one "key" variable will not predict all agricultural innovations. Berry and Cancian point out that the decision making environment includes the decisions made by others—the social environment of decisions. Seemingly irrational decisions to conform to tradition or to spend resources in "social investments" are described by Berry as clearly rational when seen in their total context.

Although peasant villages are often seen as uniformly "poor" by many outsiders, the internal wealth and status differences within these communities are emphasized by a number of the authors here. DeWalt and DeWalt trace different patterns of innovation among the different socioeconomic strata in one Mexican *ejido*. Cancian elaborates the role of rank differences in wealth for decisions made under uncertainty and risk. Acheson shows that differences in household resources will result in different assessments of "good" business investments. These authors illustrate the importance of taking stratification into account in order to understand variation in choices and outcomes.

Within this range of variables being studied, the contributors to this volume differ on the extent to which their analyses reflect the actual mental process of the farmer. Chibnik's "statistical behavior approach" is explicit in revealing some relationships between variables that farmers themselves do not perceive. Chibnik demonstrates that the village a man lives in, and the character of that village, are more important than the size of his family in determining how he allocates his labor. Both variables are said by informants to be a part of the decision on whether to engage in agriculture or to seek wage labor, but the decision makers themselves can not say whether one factor or another takes precedence. Chibnik thus uses informants' statements and perceptions but finds his own analysis a better predictor of behavior. Cancian, Barlett, Acheson, and Finkler, as well as Chibnik, carry out their analyses without attempting to link them to the cognition of the farmer. The actual decision process is seen to be a black box; individual variation is expected; and group patterns become the focus of research.

C. Gladwin discusses a theory of "real-life decisions," which proposes stages and procedures that directly correspond to farmers' own understandings. She outlines the difference between Stage I, which narrows the wide range of options open to decision makers to the few that will be seriously considered, and Stage II, in which the final decision is made. The criteria of choice in Stage I involve "elimination by aspects"; in Stage II, each option is ordered on the basis of one important aspect, and the top option must then pass all the constraints of the decision situation in order to be chosen.

Ortiz also explores the statements of farmers about their decisions to determine their response to uncertainty, their ability to plan, and to predict the future. Her research shows that farmers do not base their decisions on forecasts of future events or on prospects ranked according to their likelihood but, instead, on the range of recently experienced events. H. Gladwin and Murtaugh further explore the cognitive process by proposing that decision making takes place in two modes: the attentive and the preattentive. This distinction explains some aspects of current decisions and also suggests how past decisions are integrated into behavior and choice patterns. Gladwin and Murtaugh illuminate why some farmers can not always discuss the reasons for their decisions, as these choices and aspects of choices may be preattentive. Both of these last two chapters describe the information storage process, as observed in agricultural decisions, and explore its relevance for understanding real-life behavior. The attentive–preattentive distinction further affects how we as researchers generate the variables that we include in our analyses.

A number of the authors here use and criticize methods and tools from economics. Cancian, Johnson, and Barlett draw attention to the tendency of some microeconomic techniques to be normative rather than descriptive, and illustrate the importance of testing these ideas against real-world data. Johnson's use of linear programming, to discover whether the Machiguenga obtain a balanced diet in the most efficient way possible, explores the limitations of formal analysis. He sounds a cautious note on the ability of models to predict complex decisions accurately, drawing attention to the reductionism involved in operationalizing such models and in assuming that humans make decisions like computers do. Formal models operationalize our perceptions of reality, and these perceptions will reflect certain values and thus are inevitably liable to inaccuracies and bias.

Barlett presents three methods of replicating the process of weighing the costs and benefits of agricultural options and demonstrates that Chayanovian calculations, based on returns to household labor, involve fewer arbitrary assumptions about opportunity costs and thus are less open to researcher bias and distortions. She measures the extent to which these methods accurately predict farmers' decisions and shows that farm households in Paso weigh their needs and resources in ways that are not appropriate to calculations based on capitalist firms.

Chibnik and Ortiz discuss and criticize the use of the economic concepts of maximization, choice, and efficiency, and join Johnson and Barlett in calling for qualitative, ethnographic interpretations, in combination with quantitative measures. Ortiz discusses farmers' responsiveness to price incentives and shows that a long list of qualitative variables must be taken into account to make the economists' price-response model work. She stresses, however, that psychological, cultural, and institutional variables should not be taken into account in order to "rationalize the poor performance of the mathematical models," but rather must be "introduced at the relevant point in the formal description."

Acheson uses the economists' internal rate of return and net present value calculations to assess business investments, but while he finds differences between these methods and the *ganancia* calculations used by people in Cuanajo (a calculation that is very similar to the Chayanovian method preferred by Barlett), he finds the former more accurate. Finkler likewise contrasts the economists' production function with farmers' traditional distribution of benefits to the different factors of production. Her calculations show that sharecroppers contributing land to the bargain always receive a lower share than the value of what they have contributed. Although these last two authors are less critical of economists' methods than the former, they both nevertheless flesh out these economic calculations with qualitative data, as advocated in the preceding.

Cancian explores the distinction between risk and uncertainty for farmers of different economic rank and predicts that poor farmers are more willing to innovate under uncertainty because they have less to lose; the rich farmers are more likely to innovate later, when uncertainty is less. These hypotheses are tested by comparing behavior outcomes of innovation decisions. This methodology complements Ortiz's work on a similar topic, whose data are drawn more from informants' statements. Berry suggests that risk may be seen as a variant of uncertainty because there is no such thing as certain knowledge of the future. Thus, she holds that uncertainty avoidance on the part of rich farmers can be reinterpreted as profit maximization because they can afford to let poorer farmers test out for them the costs of a new crop or technique.

Several chapters in this volume take the study of agricultural decisions into new dimensions. Bennett's work is based on 12 years of study of the same region in Canada and studies the history and "life cycle" of the entire farm enterprise, looking not at individual decisions but at the cumulative outcomes of many such decisions as they emerge in an overall management style. Whereas his starting point is folk categories of management style, he compares these categories to objective measures and thereby illuminates the components of different styles. He concludes that household needs and resources are important determinants of decisions made within each phase of the enterprise cycle.

Acheson's work illustrates the complexity of understanding farming decisions, given that households have different resource combinations and have other nonagricultural investment options to consider as well. The links between agricultural and nonagricultural decisions are also explored by Chibnik and Finkler. The four combinations of household business choices that Acheson describes illustrate both the internal diversity of choices within the community and the importance of access to land and to other household resources.

Acheson's analysis provides a detailed example of Berry's emphasis on identifying the constraints on rural decisions. Whereas nearly all the authors explore this topic to one degree or another, Hoben's discussion of decision makers inside development agencies such as AID clarifies the constraints that act to keep resource allocations divergent from policy goals. He outlines the institutional structure within which projects are designed, defended, and implemented, and notes that both the decision criteria of project personnel and the bureaucracy's rewards to their performance are not oriented toward effectiveness in increasing rural welfare or farmers' participation in development projects, though both these goals are part of official AID policy. Hoben's analysis clarifies the role of ideology among donor decision makers, but also provides a good example of Berry's point that attitudes and preferences are linked to the actors' circumstances.

Berry also stresses that development policymakers operate in a complex decision making environment and need more than just information and money to design policies aimed at addressing rural poverty and stagnation. She notes that political policy is affected by constraints from politicians' sources of power and patronage and that no decisions can be separated from their social interconnections. Hoben's chapter provides confirmation for these points.

Several authors discuss the political implications of their research. Johnson notes that faith in formal models tends to put the researcher or policymaker in the position of the "enlightened" and justifies actions against the will of the local population. Cancian warns against the false sense of security derived from such models. H. Gladwin and Murtaugh discuss the ways such "outsider" understandings of local realities may leave out essential preattentive factors. One solution is to allow the farmers themselves to determine the usefulness of innovation or policies; the other is to train policymakers and agronomists to probe more successfully for the preattentive considerations in farmers' traditional decisions.

Barlett describes how certain assumptions about the cost of family labor can make a farm look like it is losing money, when, in fact, the family chooses only to invest extra labor in some crops when that labor has no alternative uses. Some government policies have been proposed, based on the former calculations, that would decide that such farms are "inefficient" and thus will not receive credit, technical support, or scarce inputs. Such policies will tend to increase inequalities between farmers and may adversely affect overall productivity and community stability in many rural areas. Berry notes the same point for agricultural programs that benefit farmers who can withstand risk; those households who avoid risk may often be responding to their poverty and vulnerable position, and such programs will only strengthen the local elites. Cancian notes that these risk-avoiders may also be high-ranking individuals who seek to protect their position, but both authors agree that policies aimed at benefiting poorer farmers must take account of rank and resource access to be successful.

Finkler's analysis of increasing stratification in an *ejido*, whose land resources are supposedly equally distributed, shows that inequalities in the distribution of *any* resources needed for agricultural production may result in advantages to certain households and subsequently to increased inequalities in wealth and power. Three aspects of Finkler's analysis are shared by most of the contributions in this volume, though the theories, methods, and data presented here are diverse. Decision making is shown to involve a complex interaction of variables; access to resources emerges as a particularly important variable; and both these points have significant policy implications for rural development programs that seek to affect agricultural decisions.

References

Barlett, Peggy F.
  1976 Labor Efficiency and the Mechanism of Agricultural Evolution. Journal of An-
       thropological Research 32(2):124–140.
  1977 The Structure of Decision Making in Paso. American Ethnologist 4(2)285–308.
  n.d. Agricultural Choice and Change: Economic Decisions and Agricultural Evolution
       in a Costa Rican Community.
Bieri, Jurg, Alain de Janvry, and Andrew Schmitz
  1972 Agricultural Technology and the Distribution of Welfare Gains. American Jour-
       nal of Agricultural Economics 54:801–808.
Boserup, Ester
  1965 The Conditions of Agricultural Growth. Chicago: Aldine.
Brown, Lester R.
  1970 Seeds of Change. New York: Praeger.
Cancian, Frank
  1972 Change and Uncertainty in a Peasant Economy. Stanford: Stanford University
       Press.
  1977 Can Anthropology Help Agricultural Development? Culture and Agriculture No.
       2:1–8. Bulletin of the Anthropological Study Group on Agrarian Systems.
Castle, Emery N.
  1977 A Framework for Rural Development. Culture and Agriculture No. 3:1–6. Bulle-
       tin of the Anthropological Study Group on Agrarian Systems.
Cochrane, Willard W.
  1974 Agricultural Development Planning: Economic Concepts, Administrative Proce-
       dure, and Political Process. New York: Praeger.
Danda, Ajit K., and Danda, Dipali G.
  1972 Adoption of Agricultural Innovations in a West Bengal Village. Man In India
       54(4):303–319.
deJanvry, Alain
  1975 The Political Economy of Rural Development in Latin America: An Interpreta-
       tion. American Journal of Agricultural Economics 57(3):490–499.
Dumond, D. E.
  1965 Population Growth and Cultural Change. Southwestern Journal of Anthropology
       21:302–324.

George, Susan
  1977   How the Other Half Dies. New York: Allenheld, Osmun.
Goodfellow, D.M.
  1950   Principles of Economic Sociology. London: Routledge.
Greenwood, Davydd
  1974   Political Economy and Adaptive Processes: A Framework for the Study of
         Peasant-States. Peasant Studies Newsletter 3(3):1–10.
Griffin, Keith
  1974   The Political Economy of Agrarian Change: An Essay on the Green Revolution.
         Cambridge: Harvard University Press.
Gudeman, Stephen
  1978   Anthropological Economies: The Question of Distribution. Annual Review of
         Anthropology 7:347–377.
Hewitt de Alcantara, Cynthia
  1973– The "Green Revolution" as History: The Mexican Experience. Development and
  1974   Change 5(2):25–44.
Holdridge, Leslie R. and Joseph A. Tosi, Jr.
  1977   Report on the Ecological Adaptability of Selected Plants for Small Farm Produc-
         tion in Six Regions of Costa Rica. Mimeograph. Tropical Science Center. San
         Jose, Costa Rica.
Johnson, Allen
  1971   Security and Risk-Taking Among Poor Peasants. In Studies in Economic An-
         thropology. George Dalton, ed. p.144–151. Washington, D.C.: American An-
         thropological Association. Anthropology Studies No. 7.
Lappé, Frances Moore
  1975   Fantasies of Famine. Harpers 250:1497.
Lappé, Frances Moore and Joseph Collins
  1977   Food First: Beyond the Myth of Scarcity. New York: Houghton Mifflin. With
         Cary Fowler.
Lipton, Michael
  1968   The Theory of the Optimizing Peasant. Journal of Development Studies 4(3):
         327–351.
  1977   Why Poor People Stay Poor. Cambridge: Harvard University Press.
Loomis, Robert S.
  1976   Agricultural Systems. Scientific American 235(3):99–104.
National Research Council, Commission on International Relations
  1977   World Food and Nutrition Study. Washington, D.C.: National Academy of Sci-
         ences.
Nations, James D. and Ronald B. Nigh
  1978   Cattle, Cash, Food and Forest. Culture and Agriculture No. 6:1–5.
Netting, Robert Mc.C.
  1974   Agrarian Ecology. Annual Review of Anthropology 3:21–56.
Ortiz, Sutti
  1967   The Structure of Decision-Making Among Indians of Colombia. In Themes in
         Economic Anthropology. Raymond Firth, ed. p.191–228. A.S.A. Monograph No.6.
Parsons, James J.
  1976   Forest to Pasture: Development or Destruction? Revista de Biología Tropical 24
         (Supplement 1):121–138.
Price, Barbara J.
  1978   Demystification, Enriddlement, and Aztec Cannibalism: A Materialist Rejoinder
         to Harner. American Ethnologist 5(1):98–115.

Rochin, Refugio I.
  1977   Rural Poverty and the Problem of Increasing Food Production on Small Farms:
         The Case of Colombia. Western Journal of Agricultural Economics 1(1):181–186.
Stavenhagen, Rodolfo
  1977   El Campesinado y las Estratégias del Desarollo Rural. Cuadernos del CES 19.
         Centro de Estudios Sociológicos. El Colegio de Mexico. Mexico City.
Strickon, Arnold and Sidney M. Greenfield, eds.
  1972   Structure and Process in Latin America: Patronage, Clientage, and Power Sys-
         tems. Albuquerque: University of New Mexico Press.
Wharton, Clifton R.
  1971   Risk, Uncertainty, and the Subsistence Farmer: Technological Innovation and
         Resistance to Change in the Context of Survival. *In* Studies in Economic An-
         thropology. George Dalton, ed. p.152–179. Washington, D.C.: American An-
         thropological Association. Anthropology Studies No.7.
Williams, Glyn
  1977   Differential Risk Strategies as Cultural Style Among Farmers in the Lower
         Chubut Valley, Patagonia. American Ethnologist 4(1):65–83.

PART I

THEORETICAL ISSUES AND METHODOLOGICAL PERSPECTIVES

Chapter 2

The Limits of Formalism in Agricultural Decision Research

ALLEN JOHNSON

An often noted difference between economics and anthropology is the preference of the economist for formal prescriptive models and the anthropologist for broad-ranging ethnographic descriptions. A review of the major journals in each field bears this out. Economists' journals are dominated by arguments in mathematical–deductive form that usually are neither tested nor even exemplified by empirical data, whereas anthropologists' journals are dominated by ethnographic descriptions of individual cases strung together by loosely phrased theoretical commentaries. However, there is also a substantial minority of the opposite preference in each field: Economists have their ethnographers and historians, and anthropologists have their mathematicians.

Rarest of all, in both fields, is the presentation of formal arguments or models tested by empirical data. Even though most scholars know that an ideal of scientific method is the use of rigorous theory to develop expectations (predictions, retrodictions) about the world that can be compared with directly observed events so that discrepancies between model and data can be used to improve the model and to gather new and better data, the tendency to polarize to one extreme or the other— the formal–prescriptive or the empirical–descriptive—apparently remains too powerful to resist.

In both camps, there is a certain lack of understanding about what formal models can and can not accomplish. Those who favor formalism

19

AGRICULTURAL DECISION MAKING:
Anthropological Contributions to
Rural Development

tend to overlook the consistent failures of their models to predict interesting real-world events, holding instead to a kind of "persistent
optimism" (Dreyfus 1972) that the failure is due either to a temporary
imperfection in the model or to shortcomings in the data. Those who
favor ethnography, on the other hand, overlook the benefits that the
rigor of formal models brings to our reasoning, and may in practice
overemphasize the chaos and mystery of human economic behavior.
The middle ground between these extremes may be avoided in part
because it is an uncomfortable place to be: For the formalist, because it
exposes the inevitable empirical narrowness of the model, and for the
ethnographer, because it exposes confusion in the theory that lies
behind the description.

I take it for granted that formal model building based on rigorous
deductive reasoning is a powerful aid in the analysis of economic
behavior, in nonindustrial as well as in industrial settings. But I am
concerned about the tendency of model builders to become absorbed in
the interiors of their models and to lose apparent interest in their
outward, empirical usefulness. In Simon's (1977) terms, formalists show
a marked preference for "well-structured problems" over "ill-structured
problems," even though it is the "ill-structured" form of a theoretical
problem that is closest to the real-world problem we are trying to solve.

In this chapter, I will explore the implications of the well-structured–
ill-structured dichotomy by applying a formal model to an aspect of
agricultural decision making in a subsistence economy, and then examining discrepancies between the model and observed behavior. I will
argue that the formal model is by itself virtually uninterpretable without reference to an ethnographic context that can be provided only by
participant observation over a long term of field research. This is no
momentary obstacle to complete formalization of a problem but is,
rather, an inherent limitation on the extent to which formal models can
account for observed outcomes of agricultural decisions.

The polarization into two camps has turned the middle ground into a
sort of depopulated buffer zone across which opponents regard one
another with a palpable antipathy. For example, Georgescu–Roegen
(1979:318–325), whose "institutional" economics, with its holistic
breadth and dialectical mode of reasoning, has been criticized by formalists as "vague and impressionistic," responds by criticizing his
critics for their "arithmomania." Spirited debate is always welcome:
Still, we should bear in mind that this is not a contest one side is likely
to win. And the atmosphere is such that those who choose to occupy the
middle ground are apt to be regarded by both groups as having "left
camp," either to be disregarded or else treated as part of the opposition.

Another form of avoiding the middle ground is to deny that opposing positions exist. Simon (1977:137–153), for example, argues that there is no essential difference between normative and descriptive economics. He sees normative economics as a set of descriptive statements transformed into imperatives. Thus, if descriptive economics asserts "firms behave so as to maximize profits, normative theory asserts maximize profit! [p. 138]." But Simon leaves out altogether the descriptive issue of whether and under what circumstances real firms maximize profits or otherwise behave according to formal models.

To the degree that this middle ground remains barren, the study of agricultural decision making suffers. Agricultural decision research has substantive practical relevance owing to the precarious state of the world's food supply and the major dislocations wrought by modern changes in technology and markets. Even if it is true that the majority of these changes are impersonal, historical, and beyond the control of any group or agency, there still remain arenas in which governments and international firms and foundations exert powerful influences on local agrarian populations. The quality of the understanding on which these agencies act has a direct bearing on the well-being of the people on whom they act, and on the "success" of their programs. It is of particular importance whether or not our understanding includes the ability to anticipate the reactions of local farmers to new opportunities and constraints.

The polarization of agricultural decision studies creates two traditions that are each weaker in isolation than they are together. In the next section, I briefly review the success of formal models in predicting food-related decisions in non-Western economies. This is followed by a formal analysis (linear programing) of food-production decisions in a subsistence economy that just is beginning to experience economic modernization. Next I explore discrepancies between the predictions of the model and the observed economic behavior, in some detail, to illuminate what the formal model accomplishes and what it does not, and why. Finally, I return to the formal–descriptive polarization to argue the need for research and policies specifically aimed at combining both approaches in full partnership.

Predicting Observed Outcomes from Formal Models in Economic Anthropology

The capacity of formal decision models to predict or retrodict observed economic behavior in nonindustrial economies, although undeniable, has been limited severely, especially when contrasted with optimistic forecasts of success by enthusiasts (e.g., Buchler and Nutini 1969:2–7; White 1973:386–387). Anthropologists have had no success, for example, with the theory of games, which attempts to formalize optimizing strategies for situations of competition between two or more "players." Early game-theory papers by anthropologists (Buchler and Nutini 1969; Moore 1957) were programmatic, speculative, and presented no quantitative data, either in the cells of a game matrix or as empirical tests of game-theory solutions.

Other game-theory studies, where quantitative data were presented, were fundamentally flawed. Gould (1963) presented a quantitative game matrix from which he calculated an optimizing strategy in Tanzanian farmers' choice of crops, but then he notes, almost in passing (p. 294), that he manufactured the data matrix only for the purpose of illustrating his argument about the *potential* uses of game theory. Davenport's (1960) analysis of Jamaican fishing strategies was based on the unacceptable assumption that the opponent, Nature, was purposely and strategically shifting ocean currents to minimize the fishermen's catch. Thus, although the model does predict fishermen's actual behavior, this result is meaningless. When the model is restated correctly with the reasonable assumption that nature is indifferent rather than malevolent, the game collapses into a simple cost–benefits analysis that utterly fails to predict actual fishermen's behavior (Read and Read 1970).

Elsewhere (Johnson 1978:141–157), I have reviewed a number of cases where economic anthropologists have compared quantitative predictions with empirical data. The formal models from which predictions are derived include (among others) a computer simulation of decision processes, a profit function based on input–output analysis of production, and a "central place" analysis of a peasant marketing system. In most of these cases, a small but significant portion of the data is accounted for by the model. But in each of these cases it becomes necessary to draw on qualitative ethnographic considerations to explain why the majority of the data fail to conform to predictions of the model (cf. Boyd 1975:253–254).

Of special relevance to the example I will give is a quantified linear-programing model that predicts agricultural behavior in a nonindustrial economy. White (1973:395–399) analyzes agricultural strategies among the Kapauku Papuans, using data collected by Pospisil (1963). White's linear-programing model predicts the amounts of labor and land that the Kapauku should use to maximize their calorie food production;

observed Kapauku behavior conforms *exactly* to White's prediction. This result, however, is partially a logical outcome of how the model was constructed, rather than an empirical verification. White (1973) set the constraints on the model so that the allowable solutions could either be equal to or less than the empirically observed amounts of labor and land the Kapauku use, *but they could not exceed either amount.* When a solution was sought in the form of a strategy that would maximize food production (calories), logically it employed the maximum amounts of land and labor available to it (i.e., those that the Kapauku actually use).

White's model did allow for the possibility that calories could be maximized with less than the observed labor and land, and consequently the formal analysis suggests that there is no way for the Kapauku to achieve the same level of calorie production by using less land or labor than they do. But the model does not account for the observed fact that the Kapauku actually use less land and labor than they have available.

An interesting prediction of White's model concerns the mix of three crops (sweet potatoes, sugar cane, and root crops) the Kapauku should plant. This prediction is quite different from observed behavior: The Kapauku actually plant 49% of the land in sweet potatoes, 32% in sugar cane, and 19% in root crops, whereas the linear-programing solution is to plant 85% in sweet potatoes, none in sugar cane, and 15% in root crops. White explains the discrepancy as due to a "balance of different crops [being] sought [p. 398]," but this, of course, is one of those qualitative, post hoc explanations I identified as the rule when formal models fall short of predicting observed behavior. Still, as in other cases, White's model is close enough to the mark that it predicts some of the data (e.g., the low proportion of root crops) and can be used as a springboard for further, more informed exploration of the data.

In short, it is fair to say that rigorous deductive models have not yet been accurate in predicting or otherwise accounting for observed economic behavior in nonindustrial settings. Formal model builders would agree, no doubt, that this is to be expected: Deductive models rarely predict all the observed cases (and even when they do, we still allow that nature may present us with an exception tomorrow). In an area of knowledge as fragmentary and uneven as non-Western economies, we probably should be delighted at even the partial success of a model. But I am not concerned here simply with the models' failures to predict. Rather, I am concerned with the discernable tendency of model-builders to go only as far as their models permit and to stop there. This reflects an implicit stance that everything of interest in the universe of economic decision making ultimately is formalizable, and that other modes of understanding are inferior or unimportant and should be eliminated. We turn now to an examination of this assumption, in the context of a specific example.

Example: A "Well-Structured Problem"

The formal economist has a preference for "well-structured problems." In Simon's (1977:238–239; 305–307) terms, a well-structured problem is tied intimately to a developed system of deductive reasoning, and has the following properties:

1. A set of axioms or previously proved theorems.
2. A sequence of expressions derived mechanically from the axioms or theorems in Property 1 by means of legitimate operations.
3. A solution, which can be the last expression in the sequence 2, if the problem is purely formal; or else, if the solution involves acting upon the world in some way, there must also be a body of knowledge that will "reflect with complete accuracy . . . the laws (laws of nature) that govern the external world [p. 306; parentheses in original]."

Linear programing is generally conceded to be a well-structured problem-solving technique of proven usefulness, both by the proponents (e.g., Simon 1977:143) and the opponents (e.g., Georgescu–Roegen 1979:320) of formal analysis. The present example is a variety of the "optimum diet" problem, which in its general form asks what combination of available foods provides an adequate diet at the lowest cost. Because specific foods differ in both nutritional value and price, the task of linear programing is to find the amounts of each food that, taken together, will meet the nutritional requirements of a population while minimizing cost. Obviously, cheaper foods will be substituted for more expensive foods whenever possible, and expensive foods will be included only in the smallest amounts necessary to meet the minimum requirements for scarce nutrients.

This problem-solving technique seems especially promising in regard to subsistence farmers, who could be expected to allocate their scarce labor supplies with the goal of providing a good diet for the least cost. If we can operationalize "good diet" to mean "fully nutritious" and "least cost" to mean "least caloric energy [work] expended in food production," then linear programing will provide a solution that predicts the optimum strategy for achieving these ends.

Data for the present example were collected in a community of horticultural Indians in the tropical rainforest of the Peruvian Amazon.[1] The Machiguenga are a recently contacted group of Arawakan-speaking Indians who traditionally live in single-family houses, or in clusters of two or three houses, separated from each other by large expanses of tropical forest. Recently, such clusters have been attracted into the sphere of influence of small "school communities," where they settle within about an hour's walk of the school to obtain medicines, small quantities of trade goods (e.g., machetes, aluminum pots), and an opportunity for their children to attend primary school taught in their native language. Ethno-historical evidence indicates that the gradually increasing local population densities (average density is .8 persons per square mile), in interaction with the greater availability of steel tools, have led the Machiguenga to plant bigger gardens and to spend less time obtaining wild foods such as game and fish than they used to (Johnson 1977). Keeping in mind these changes in the conditions of food production, what mix of production strategies would provide a nutritious diet for the least work?

In seeking a solution to a problem like this with a formal technique, usually it is necessary to try different formulations of the problem. The linear-programing routine (here, in the form of a computer program) requires input of an array of foods and their nutritional contents, a cost vector (the labor cost in calories of work expended per kilogram of each food), and a vector of nutritional requirements (see Tables 2.1 and 2.2, and Footnote 4). The computer then searches through the available foods for combinations that meet the nutritional requirements, minimizing for cost. But the required input may be presented to the problem-solving routine in a variety of forms, each of which will lead to a unique solution.

The simplest form of the present problem groups Machiguenga foods under three production strategies:

1. Game and other wild foods obtained by hunting and gathering along forest trails.
2. Fish obtained by various techniques from local rivers and streams.
3. Domesticated crops cultivated in slash-and-burn gardens.

These are the three main categories of food production acknowledged by the Machiguenga, and each is regarded as a distinct strategy with no significant overlaps between them. Each provides a "food basket" or aggregate of foods, and each basket differs from the other two in cost

[1] For discussion of methods see Johnson 1975, Johnson and Behrens 1980, and Montgomery and Johnson 1976.

TABLE 2.1
Labor Costs and Nutrient Content of Machiguenga Food Strategies

	Hunting–gathering	Fishing	Gardening	Total required per house[a] per year
Labor cost per kilogram (kcal)	1151	739	80	
Nutrient content per kilogram				
Calories	880	1010	1330	4,540,965
Protein (gm)	55	179	15	60,980
Calcium (mg)	590	200	360	2,044,000
Phosphorous (mg)	730	1800	490	2,044,000
Iron (mg)	15	7	13	37,230
Vitamin A (r.e.)	1050	0	190	1,700,900
Thiamine (mg)	.9	.3	.7	2,190
Riboflavin (mg)	.8	.8	.4	2,590
Niacin (mg)	15	30	6	28,760
Vitamin C (mg)	350	0	210	124,100

Source: Johnson and Behrens (1980).
[a] Average household = 7.4 persons.

and nutritional value. The nutritional value of a particular basket can be calculated by averaging the quantities of a given nutrient provided by each food in the basket, in proportion to its share by weight, in the whole basket. Inspecting Table 2.1, we can see that the garden basket is clearly the least expensive, but of somewhat limited nutritional value compared to the expensive hunting and gathering basket, which is the most balanced source of nutrients. Fish are intermediate in cost, but relatively unbalanced in nutrients.

The first run of the computer was set to find a solution that met minimum requirements of nutrients without further constraints. It is not surprising, given the great cost differentials between strategies, that the solution given in this run was to devote 100% of labor time to gardens. Since the garden basket in aggregate has the typical nutritional value of a root crop, this would be analogous to telling an American family to eat nothing but inexpensive potatoes, since, if they were to eat plenty, they would get enough of every required nutrient at a low cost. Obviously, this solution is unacceptable because the human body is unable simply to sift through the nutrients in potatoes, discarding the overabundant calories while retaining the scarce other nutrients.

Hence, the first solution presented here (Solution 1 in Table 2.2) is

TABLE 2.2
Observed Behavior and Linear Programming Solutions[a] Compared

| | Percentage of minimum requirement provided by strategy | | | | | | | | | |
	Calories	Protein	Calcium	Phosphorus	Iron	Vitamin A	Thiamine	Riboflavin	Niacin	Vitamin C
Solution 1 (Labor = 2,816,600 kcal/household/year)										
hunting–gathering 95%										
fishing 0%										
gardening 5%	100	232	100	128	159	164	155	101	160	972
Solution 2 (Labor = 278,100 kcal/household/year)										
hunting–gathering 0%										
fishing 0%										
gardening 100%										
maize 10%										
arrowroot 53%										
plantain 20%										
pigeon peas 17%	100	166	100	177	125	100	247	100	125	357
Observed behavior (Labor = 990,000 kcal/household/year)										
hunting–gathering 23%										
fishing 22%										
gardening 55%	210	248	128	196	251	88	230	120	183	1208

[a] Solutions 1 and 2 were computed by the L–P routine of the SOUPAC library of statistical computer programs.

under the constraint not to exceed the food energy requirement (calories), although all other nutrients in the model are free to exceed the minimum requirements. This constraint literally reversed the solution by calling for 95% of labor to be invested in hunting and gathering and only 5% in gardens; fish, again, are excluded altogether. This solution is about as far from observed behavior as it can be: In fact, the Machiguenga devote about one-fourth of their labor to hunting and gathering, another one-fourth to fishing, and the remaining one-half to gardens.

This extreme and unrealistic solution reflects the rigidity of the model in this form: The solution is forced to the hunting–gathering strategy by the poor nutritional balance of the garden and fish food-baskets. Although it fails to predict actual behavior, Solution 1 does help us to understand why the hunting–gathering strategy, despite its great cost, might be attractive to subsistence cultivators because of the nutritional diversity it offers.

The rigidity in the model to this point comes from considering the diet problem for subsistence farmers as one of production, not consumption. Simply because the Machiguenga produce food in certain quantities, it does not follow that they consume everything they produce, or that they consume everything in proportion as it is produced. Field observations show that the Machiguenga are exceedingly careful not to waste foods obtained in the forest or the rivers, but this is not the case for garden foods, where some foods are indeed "wasted." In particular, manioc, which is rich in calories but poor in other nutrients, is produced in great quantities. It is *the* staple food of the Machiguenga, and its overproduction is understood best as a security mechanism in a society where links of interdependence among households are comparatively weak and fragmentary. The food energy stored in the manioc roots in the garden acts as a kind of reserve fund in case of emergency. The Machiguenga eat foods selectively from their gardens, so that when a garden is abandoned manioc and occasionally other root crops are left over. In fact, one of the Machiguenga expressions for an abandoned garden is *ashi shintori,* meaning 'it [the manioc] belongs to peccary.'[2]

The only obstacle to treating each food separately is that a unique cost

---

[2] I often am asked whether this could be a deliberate strategy to attract peccary with manioc "bait." There is no doubt that it does work out this way in fact, for peccary (and other game) are bagged in gardens from time to time. But, when I put this question to a Machiguenga informant, he laughed and said, "Oh no, it is too much work for too little return. The peccaries are just pests in the gardens that we have to live with." I conclude that "extra" manioc is intended primarily as a security hedge, although it certainly does no harm to the meat supply.

cannot be assigned in the same way as prices are set for foods in a market place. A Machiguenga hunting trip, for example, is not labor expended in obtaining a single food but is completely opportunistic. The hunter usually has a single species of game in mind, depending on the season and information from neighbors about sightings of game, but he takes a variety of arrows intended for different categories of game; in short, he will hunt whatever he encounters. The hunting trip is also an occasion for opportunistic gathering of fruits, nuts, palm hearts, and insect foods. Thus the labor of a hunting–gathering trip produces an average of several foods, just as garden labor produces an aggregate of food crops. Still, there is nothing to prevent the Machiguenga from varying the proportions of crops in their gardens, and evidence is clear that they do.[3]

Hence, another form of the problem treats each food as separate, assigning to it an average cost proportional to its share of its particular basket, but allowing the linear-programing routine to sort through 18 foods and to select any combination.[4] The solution to this second form of the problem is to devote 100% of labor to garden foods, divided as follows among specific crops: 10% to maize, 53% to cocoyam, 20% to plantains, and 17% to pigeon peas (Solution 2, Table 2.2). With just these four garden foods, a balanced diet is obtained without need for hunting and gathering, or fishing.

Comparing Solutions 1 and 2 to the observed behavior of the Machiguenga (Table 2.2), we see that the observed behavior falls between the two solutions. The first solution is much more expensive than the observed pattern and is roughly analogous to a gourmet diet of exotic foods. The second solution is much less expensive than the observed pattern and is analogous to "good plain fare," nutritious but monotonous. Neither solution is close to predicting the observed pattern of Machiguenga food production. The two solutions do, however, effectively bracket the observed pattern and help us to understand it. The first solution may be said to establish the plausibility of expending

---

[3] That the Machiguenga do experiment with crops is clear. They plant experimental gardens (as do subsistence farmers typically—see Johnson 1972) where they cultivate new crop varieties obtained from neighbors as well as from distant regions.

[4] The 18 foods are, hunting and gathering (5): game, grubs and other insects, palm hearts, fruits, and nuts; fishing (1): various fresh-water fish, averaged; gardening (12): maize, manioc, cocoyam, sweet potato, sugar cane, papaya, pineapple, plantain, guava, beans, pigeon peas, banana.

Actually, if we count separate species of game, insects, fruits, nuts, and fish, the number of foods surely exceeds 100, and this does not take into account the many varieties of each garden crop distinguished by the Machiguenga.

effort in the expensive hunting and gathering strategy. The second helps us to understand how the garden can provide a substantially balanced diet despite the heavy emphasis on raising poorly balanced root crops. In fact, the actual balance of the 12 foods in the garden complex is not unrelated to the predictions of the model. If we group manioc and yams together with cocoyam into a category, "root crops," group plantains with bananas, and group beans with pigeon peas as "legumes," the comparison between model and observed behavior looks like this:

| Crop category | Crop proportions by weight (Solution 2) | |
	Model	Observed
maize	10%	5%
root crops	53% (cocoyam)	76% (cocoyam, manioc, yams)
banana/plantain	20% (plantain)	7% (banana, plantain)
legumes	17% (pigeon peas)	0.2% (pigeon peas, beans)
other	—	12% (sugar cane, papaya, pineapple, guava)

The main discrepancies here are the "overproduction" of roots (previously explained in terms of "security"), and the "underproduction" of legumes (protein foods with which the model would replace Machiguenga fish and game). We also may note that Solution 2 leaves sugar cane and fruits out of the picture, neglecting their possible role as sources of sugar for quick energy (an important function of these crops among the Machiguenga). Apart from this, there is an approximate fit between model and observed reality that is quite suggestive.

Neither solution has anything to say about fish, however. In order to explore and interpret the various inconsistencies between formal model and empirical reality, and to examine the role formal models can play in our understanding of this and other problems in agricultural decision making, we now turn to Simon's (1977) concept of the "ill-structured problem."

The Linear-Programing Example as an "Ill-Structured Problem"

In general, the problems presented to problem solvers by the world are best regarded as Ill-Structured Problems. They become Well-Structured Problems only in the process of being *prepared* for problem solvers. It is not exaggerating much to say that there are no Well-

Structured Problems, only Ill-Structured Problems that have been formalized for problem solvers [Simon 1977:309; emphasis added].

Behind the linear-programing diet problem lies an original form of the
problem that, though general and imprecise, is nonetheless the essence
of the question we are really asking: Given that the Machiguenga
produce their own food through their own efforts, what is the most
satisfying diet they can obtain, taking into account the dissatisfaction
entailed in producing it?

Preparing this problem for a well-structured problem solver like
linear programing requires two steps:

1. *Reduction* of the original problem to a small subset of attributes
   permitted by the formal problem-solving routine.
2. *Assumptions* about the data needed to fill the gaps between our
   imperfect knowledge of the real world and the perfect knowledge
   demanded by the routine. Actually, these are simply successive
   stages in moving the problem from an intuitive to an objective
   form.

Reduction of the problem is necessary because any formal problem-
solving routine is highly restricted in the kinds of information it can
use. I (like any ethnographer after long-term fieldwork among one
people), can list readily many factors that play a role in the Machiguen-
gas' calculations about food production. Nutrition is certainly one of
these. The Machiguenga recognize that different foods have different
properties, and that a balance must be struck between them. The late
afternoon meal—their main meal of the day—ideally includes a starchy
staple (manioc, cocoyam), a vegetable (palm heart, leafy garden vegetables), and a small portion of meat (fish, game; red meats are regarded as
potent, to be eaten in small quantities).

Also relevant to their nutritional attitudes is a strong emphasis on
food diversity, seen both in their conversations about food and in the
large variety of foods we observed in their homes. The Machiguenga,
however, are more likely to discuss this matter in terms of "taste," an
area in which they have an ample vocabulary for describing the flavors
and textures of food. Many of the foods in their diet are of excellent
quality; the fish, in particular, are equal in flavor and texture to our
salmon or trout, and when roasted or smoked are without peer. But the
Michiguenga also show a high regard for all the foods in their diet, and
rarely express distaste for any particular food (except in the case of
personal food taboos, a subject too complex to be taken up here). Thus,
they would eat with apparent relish not only sweet ripe fruits, but also
green ones that to me seemed dry and tasteless.

The dissatisfaction entailed in productive work is also complicated. Participant observation has convinced me that the Machiguenga enjoy a certain amount of work, and if they were somehow forced not to work they would feel impoverished. But they also prefer certain kinds of work over others. Garden work, on the whole, is least preferred; the monotonous routines in the tropical sun wear them down. Fishing is the most sociable kind of work; usually, it is undertaken by whole families or groups of families, and is characterized throughout by a spirit of joviality and ribaldry. Hunting and gathering, although in terms of pure energy expenditure by far the most demanding labor, is clearly the most preferred. Forest trails are steep and difficult, but the forest is cool, attractive, and continuously surprising, and the foods obtained there are among the most desired. Sometimes the trip into the forest has the quality of a picnic, and it would be difficult for the anthropologist to see it as work at all, were not detailed advance planning and animated discussion of alternative routes evidence of an underlying seriousness.

In this light, the linear-programing form of the problem appears almost absurd in its simplicity: a labor cost vector (measured in calories of energy expended), a vector of minimum nutritional requirements, and an array of foods and their nutrient contents. It must be stressed that this is not by choice but by necessity: There simply exist no methods for including all the considerations that we have good reason to believe enter into people's efforts to balance the satisfaction from food with the cost of obtaining it.

But even in this reduced form, we still do not have a well-structured problem. The main features of a well-structured problem are met in this example by linear programing itself: that is, axioms and mechanical procedures for producing a solution. It is the caveat that Simon adds, almost as an afterthought, that really opens Pandora's box: the need to reflect "with complete accuracy" the relevant attributes of the external world. For our knowledge of that world is far from perfect and recedes in quality with our distance from the scene we are trying to understand. In preparing the data for a well-structured solution, we make innumerable assumptions that in effect "gloss over" the incompleteness of the data. How rare it is to see the presentation of a formal model compared to an empirical data set, then followed by a detailed list of all the assumptions and small "fudgings" that were needed to complete the data analysis! If this were required (as something approximating it is in the laboratory sciences), the list would take up more space than the analysis. Instead, we find a certain impatience with these data questions, that just "get in the way" of clear thinking and lead to "vague and impressionistic" analyses.

To illustrate, I choose three of the numerous questions that can be asked about the present data, and give examples of the kinds of assumptions that are inevitable if a formal model is to be tested by empirical data.

1. What is a nutritious diet? As our knowledge of human needs goes, we have a rather good body of reliable knowledge on diet and nutrition. For the model, recommendations made by respected international panels of nutritionists were followed (FAO 1970, 1973; NAS 1974). Yet these recommendations are open to many criticisms. For example:

(*a*) The minimum requirements are set high enough that persons with very unusual needs for particular nutrients can be satisfied. The National Academy of Sciences notes that its estimate of calcium requirements could probably be halved without serious health consequences. Inflated values are given for certain other nutrients as well. That this inflation is by 100%, rather than by some more precise figure like 82% or 117%, indicates the level of accuracy that is available to us. Also, apropos the discussion, a political aspect of the high intake recommendations has been identified by some observers (Clark 1970:10–15).

(*b*) There are complex interaction effects between particular nutrients. A well-known example is that both maize and beans contain incomplete proteins, but in combination each complements the other to constitute a good source of protein. As it happens, the second linear-programing solution, combining maize and pigeon peas, might accidentally lead to a similar complementarity, but the food composition data do not contain this information (IN-CAP 1961) and this complementarity of proteins is in no way taken into account by the linear-programing procedure in seeking a solution. There are many other, even less well-understood, interaction effects (e.g., that high rates of protein consumption also require higher rates of calcium) that cannot be included in the well-structured form of the problem.

2. What is the true picture of Machiguenga food production? The data reported here are reliable data, in terms of the state of the art in this kind of research; all are based on representative sampling and have been quantified according to generally accepted procedures. But other questions remain. For example, most of the data were collected over a 13-month period of fieldwork in 1972–1973. How can those data claim to

be representative of an average year? I can report that the Machiguenga considered it an unexceptional year, but this is certainly "soft" data. Furthermore, the data were collected on a sample of only 105 adults and children, because these were the only people accessible in reasonable walking time from our field residence. What of the isolated forest settlements inaccessible except for short visits after difficult treks, or other communities living at different elevations and under different ecological circumstances? Again, we are at the limits of present knowledge, for either our tool kit of methods is inadequate, or research funds and personnel are insufficient ever to answer these questions satisfactorily.[5]

3. Another set of uncertainties enters when we try to mesh our knowledge of specific populations with our knowledge of food and nutrition in general. For example, the analysis of the nutritional content of Machiguenga foods is based on published analyses of typical foods from Latin America. There is no guarantee that the maize or plantain samples analyzed in published studies are identical to the local varieties grown by the Machiguenga. Likewise, we have no way of knowing if our nutritional theory concerning a healthy diet, which is based largely on studies of populations in industrialized societies, is applicable to an Amazon Indian population.

"Persistent Optimism" and the Limits of Formalism

Given these two transformations from an ill-structured into a well-structured problem—that the problem must be reduced to a fraction of its original form, and that the empirical data are far from meeting the formal requirements for perfect information—it is not reasonable to expect formal decision models to predict more than certain limited features of any relatively comprehensive description of economic behavior. This does not mean that the *idea* that people make decisions in terms of their own perceived self-interest, is wrong. We cannot follow Dalton (1969:67–68) in denying that traditional agriculturalists make choices. But it is not enough to say that decision theory *in principle* applies to all human behavior, because *in practice* it cannot; or rather, for the majority of cross-cultural instances, formal decision analyses do

---

[5] These data shortcomings are not simply true of anthropology. Economic data are also full of lacunae and sampling problems. For example, despite the current energy "crisis," there is no reliable knowledge concerning how much petroleum is now in storage in the United States. Yet model builders in both fields tend to gloss over these *inherent* problems rather than confront them and accept the implications.

not account for the data without the ad hoc, intuitive, soft considerations that formal models are explicitly supposed to avoid.

At this point I am sure that objections will have begun to mount in many readers' minds. Is not this simply a temporary situation owing to the newness of our efforts? Does not the great success of formal economics at predicting economic behavior in western economies imply that success in other areas is likely as soon as we catch up to economists in sophistication? What is the point of these discouraging words if we have nothing better to offer? I would reply that the situation is more fundamental than temporary, that the success of economics is limited even in western economies, and that we do have something better to offer.

The idea that failure to predict is a temporary situation, to be expected in a new field, reflects the persistent optimism that is common among model builders. Simon, for example, despite his inventive excursions outside the most narrow formalisms of economics, in the end returns repeatedly to this optimistic position (Simon 1977:151–152, 174, 241, 319, 324). He is consistent in arguing that there is no fundamental difference between mathematical–deductive decision theory and the adjustments required by his concepts of the "good enough solution" and the "ill-structured problem." His view is that every ill-structured problem can be resolved into a finite set of well-structured problems, each of which separately can lead through suitably mechanical procedures to unambiguous solutions. These rigorous solutions, then, *somehow* will add up to a sort of macrosolution to the ill-structured problem. He walks repeatedly to the edge of the known world, so to speak, looks down into the chasm beyond, and then hurries back to the touchstone of formal economics. In my view, this form of belief falls in the domain of faith rather than science (Johnson 1978:12–21).

There is nothing wrong with this per se, as long as it is recognized for what it is. But this particular belief may be questioned on two counts. First, there is the troublesome inability of formal models to predict observed outcomes, as noted earlier in this chapter. Model builders typically do not test their models with new empirical research, tending rather to illustrate them with simplified, or even imaginary, examples. This has led Dreyfus (1972:188) to suspect many formalists of choosing highly simplified areas of experience where problems *can* be solved with formal procedures, rather than letting prior theoretical commitments determine which problems should be attempted. Buchler and Nutini (1969:2–3), in fact, explicitly favor this selective approach: "If it is unrealistic to assume that the universe conforms to mathematical principles, then we must search for those aspects of the universe which admit of mathematical formulations." This seems to put the cart before the horse, to say that we are to be mathematicians first and problem-oriented social scientists only when the mathematical conditions permit.[6]

Second, behind this persistent optimism lies a highly dubious psychological assumption that "The mind can be viewed as a device operating on bits of information according to formal rules. . . . [It is] a third-person process in which the involvement of the "processor" plays no essential role [Dreyfus 1972:68]." The assumption is that when people anywhere make decisions, their problem-solving procedures are identical in all the significant ways to the mathematico–deductive routines we program computers to follow. Specifically, holistic processes of pattern recognition and evaluation by integrated "selves" that bear traditions and have emotions as well as logic are effectively eliminated from the theory.

There is a natural tendency for devotees of a unified theoretical framework like that of formal economics to exaggerate the scope and power of the theory. Homans (1967) does this when he treats the principles of rational choice as the unique general theory for all the social sciences—all others being regarded as special-purpose theories for limited problem areas. In the present case, Homans would say that the optimizing decision model in the form of linear programing is the embodiment of the general theory, whereas the specific model of ecological adaptation in the form of nutritional requirements and energy costs reflects a special purpose theory useful for applying this particular case of the general theory.

But this can be easily turned around to make adaptation the general theory: Human beings must survive to be able to make decisions, just as they must make decisions to survive. It really makes no sense, for example, to ask whether the Machiguenga are primarily decision makers seeking satisfaction or organisms adapting to an environment, since both are equally true and in no way contradictory. And if an adaptation perspective leads us to consider decision processes that cannot be formalized in the foreseeable future, like the desire for dietary diversity

[6] At this point, in an earlier draft of this chapter, two readers wondered why Behrens and I had not tried to bring the model and data closer together by "guestimating" more parameters, like the "security value" of manioc, or the "negative utility" of agricultural work versus hunting–gathering work. The point is precisely this: Such guestimates are not the genuine quantification demanded by the model. To present such a "fudged" correspondence between model and data would be misleading at best and at worst would violate the strict canons of science concerning what it means to test a model.

or the subjective enjoyment of work, then these processes should not be ignored, but treated qualitatively and, in a word, ethnographically.

The Irreducible Need for Ethnography: Political Considerations

But, can our plea for a middle ground between formalist and holistic–qualitative approaches be justified in view of the demonstrated success of formalist principles via constant testing of economic theory in daily practice in Western economies? First, we must note that doubts about the adequacy of formalist theory even in Western economies are crystallizing as old models fail to account for recent behavior affecting unemployment, inflation, and energy: "The malaise of which economists now suffer is not imaginary . . . ; the respect that economists inspire among business circles is not what it used to be [Paul Samuelson, quoted in Georgescu–Roegen 1979:319]." Second, when our interest shifts to behavior during economic change in non-Western societies, the balance has to shift in favor of holistic ethnographic research, because economic circumstances in these societies are too unfamiliar for existing theories to be expected to apply without modification.[7]

The decision whether to rely on our ready-made formal theories or to turn to new empirical descriptions for guidance is not simply a scientific question, but also a political one. The implicit perspective of formal economic theory is that all the necessary information concerning the economic motivation of the actors already exists fully in the theory. The rigorous deductive form of the theory ensures us accurate and reliable solutions to such problems as resource allocation, factor mix, marketing strategy, and so on. The ability to predict the actual behavior of real farmers, storekeepers, and middlemen can be defined as a nonissue; if observed behavior does not conform to the models, this is just taken as evidence that the behavior falls short of "optimum" (or, perhaps, that our empirical descriptions are faulty).

What makes this a political assumption is that it joins formal economists with change agents as allies convinced of their superior

[7] Most readers will be familiar with Burling's (1962) article, "Maximization Theories and the Study of Economic Anthropology," but I would urge a fresh look at the section (pp. 805–809) where he discusses the limitation of economics in practice to items and processes that carry a price tag.

knowledge and engaged in an effort to bring enlightenment to the less informed. It is also a political fact that "ignorant" populations are often poorly controlled by the central governments trying to change them, and that the likely outcome of "successful" change is increased central political and economic control. The assumption that our models are more correct than those of the actors they refer to can be taken as a license to introduce economic change even against the wishes of local populations, on the grounds that the people "do not know what is good for them."

The criticisms of formal modeling at this level are equally political. One asserts the right of local populations to self-determination, without undue external pressure or misleading claims about the benefits of change (Bodley 1975:168–169). Another stresses the existence of great local variation in environmental and social circumstances that make it impossible for outsiders to have a sufficient understanding of the situation to safeguard their theoretically reasonable models against prescribing empirically absurd strategies (Cancian 1977). In either case, the political message is that the people whose lives are directly affected by the change should determine its course.

How this issue is viewed by governments and other agencies of economic development determines how the role of the anthropologist is evaluated. For example, a colleague of mine, a physician and specialist in health-care delivery, was asked to visit hospitals and clinics in several developing countries as a consultant to a firm of economists evaluating government health-care programs. At the end of his 2-week tour of duty, the physician asked why they had not included an anthropologist, since there were many cultural and social factors to be taken into account in planning health-care improvements. They answered that they would have liked to, but that they needed to complete such a study in about six weeks, from initial research to final report and recommendations; they said that an anthropologist would need a year or two, and then still might not have the information they wanted.

This last comment is an exaggeration, since anthropologists who are familiar with a region can draw upon previous experience in conjunction with a relatively short term of additional research to offer unique and valuable insights. The experienced ethnographer can visit a new community in a region with which he or she is already familiar, and in a rather short time develop reliable forecasts or "anticipations" on specific questions like "What form of credit do the people want?" "Will they accept cooperatives?" "Will they try high-yield hybrid seed?" The answers often will be imprecise compared to quantitative predictions derived from profit-maximization models, but they will stand a greater chance of being plausible and leading to feasible policy; and that also is a kind of precision.

At a deeper level, what seems to be at stake is a fundamental difference in approach to evaluating the effectiveness of research on economic change. The economist, confronted by practical realities in the form of limited budgets and political pressure to reach decisions rapidly and to keep the "facts" simple and numerical, naturally prefers to operate within the framework of clear-cut models of decision making under certainty, rather than deal with the ambiguities of how people perceive their own needs, the complicated and "folksy" notions they have about their environment, and a more humanized, qualitative sense of what is "good for them." It is in these last two senses that anthropologists evaluate the effectiveness of their research on economic change.

The machine-like procedures of formal decision theory assume machine-like people making decisions in the real world. But the fact that people are not machine-like is a major reason that formal models have trouble predicting their behavior. The ethnographer, however, is not a machine. In the course of ethnographic fieldwork, there develop not mere impressions, but solidly based intuitions drawing upon the same ability to weigh large numbers of factors and arrive at good enough solutions as the people being studied. I want to stress that these intuitions are neither "vague" nor "impressionistic," but reflect the experience and accumulated cultural wisdom of the people themselves.

In our present example, Solution 2 misleadingly suggests that the Machiguenga could provide a suitable diet by farming about half as much land as they now do; they need neither forests nor streams. This could be said to justify the current government's plans to legalize Indian control over small plots of land, opening up the remainder of their common habitat to legal settlement by impoverished farmers from the overpopulated highlands. But the Machiguenga now obtain most of their materials for tools, houses, matting, utensils, and their prized diversity of diet, in the forest. To obtain these under the proposed changes, they would have to work for wages and raise cash crops. This would integrate them into the Peruvian economy, and increase the flow of cash (which in government statistics will count as economic development).

Questions about whether this is "fair" to the Machiguenga (who, incidentally, control their own population quite effectively and do not need to expand their territory), or about how to evaluate the change in "quality of life" that accompanies a rise in "standard of living," are generally neglected and may even appear naive in the light of political reality. The scientist and the moralist may have little to do with what actually happens in economic change, which is often enough an implacable process of historical change that sweeps everyone before it. But in those areas of economic change where there is some control, in the form of public policy decisions and the expenditure of development funds, and where there is a commitment to local participation in the change process, there is no alternative but to include, as an equal partner to the formal theoretical analysis, an ethnographic investigation capable of fleshing out the analysis and rooting it in the local soil.[8]

Conclusion

In his review of mathematical "optimization" studies in anthropology, White (1973) writes: "Anthropology to date has contributed only in a limited degree to the empirical refining of the theories of optimization, even though anthropological fieldwork provides the best possible natural laboratories for such work [p. 387]." Both parts are true: The contribution has been limited, and the possibilities are great. It may be, however, that more than "refining" is needed. The formalism and effectiveness of economics are not due to power inhering in the theory as such, but to a particular operationalization in the form of "price theory," that is culturally and historically limited in its application.

In this chapter I have deliberately understated the benefits of formalist model-building for an understanding of problems in agricultural decision making. Anyone who has done formal modeling will see no paradox when I say it is at once a highly organizing experience and an imaginatively stimulating one. It promotes clarity of thought, detailed attention to measurement, and orderly exploration of alternative possibilities. I have emphasized instead the reasons for avoiding a reductionistic adherence to formalism and the associated tendency to depreciate alternative approaches, particularly when this reductionism becomes an implicit part of the political decision process.

The unique powers of a formal analysis are purchased at the expense of flexibility, particularly in the ability to anticipate the creativity, or "novelty" (Georgescu–Roegen 1979:321) of human solutions to eco-

[8] Cancian (1978) has pointed out that the professional anthropologist is too expensive for local agricultural communities. But anthropologists can take a regional approach, gaining direct experience in a few specific communities and generalizing through briefer visits to numerous other communities in a region.

nomic problems. Agricultural decision making is full of surprising behavior. Several of the chapters in this volume attest to the mystification (and frustration) of economists or government officials when confronted with the behavior of particular farmers. In fact, what is actually surprising is that anyone would think that an abstract theory, operationalized with reference to an industrial firm or similar limited frame, could prescribe behavior for farmers who have lived in an environment their entire lives, observed countless details about its soils, crops, weather, labor supply, market prices, and government intervention, and have integrated these experiences with cultural "rules of thumb" into a total understanding that all our research methods in combination can hardly fathom.

But if the formal and abstract technology of our discipline cannot by itself generate the understanding we seek, we have another aid of inestimable value in the person of the ethnographer, who develops the ability to apprehend the whole context of decision making by living in the locale long enough to overcome ethnocentric bias and sense the rationality by which decisions are made locally. This understanding can be made more careful and rigorous by being formalized wherever possible in models of proven effectiveness, but it can never fully be reduced to or comprehended by them.

Acknowledgement

I thank Peggy Barlett, Frank Cancian, Michael Chibnik, Nicholas Georgescu-Roegen, Daniel Gross, Orna Johnson, David Richards, and Douglas White for their helpful comments.

References

Bodley, John
　1975　Victims of Progress. Menlo Park, CA: Cummings.
Boyd, David
　1975　Crops, Kiaps, & Currency: Flexible Behavioral Strategies Among the Ilakia Awa of Papua New Guinea. Ph.D. Dissertation, Anthropology. UCLA.
Buchler, Ira, and Hugo Nutini, Eds.
　1969　Game Theory in the Behavioral Sciences. Pittsburgh: University of Pittsburgh Press.
Burling, Robbins
　1962　Maximization Theories and the Study of Economic Anthropology. American Anthropologist 64:802–821.

Cancian, Frank
    1977   Can Anthropology Help Agricultural Development? Culture and Agriculture, No.
            2 (March 1977).
    1978   Practical Generalizations in Development Anthropology. Paper presented at the
            Tenth International Congress of Anthropological and Ethnological Sciences, New
            Delhi.
Clark, Colin
    1970   Starvation or Plenty? London: Secker and Warburg.
Dalton, George
    1969   Theoretical Issues in Economic Anthropology. Current Anthropology 10:63–101.
Davenport, William
    1960   Jamaican Fishing: A Game Theory Analysis. Yale University Publications in
            Anthropology 59:3–11.
Dreyfus, Hubert
    1972   What Computers Can't Do: A Critique of Artificial Reason. New York: Harper.
FAO
    1970   Requirements of Ascorbic Acid, Vitamin D, Vitamin B_{12}, Folate, and Iron. Food
            and Agriculture Organization, Nutrition Meetings Report No. 47. Rome.
    1973   Energy and Protein Requirements. Food and Agriculture Organization, Nutri-
            tion Meetings Report No. 52. Rome.
Georgescu–Roegen, Nicholas
    1979   Methods in Economic Science. Journal of Economic Issues 13:317–328.
Gould, Peter
    1963   Man Against His Environment: A Game Theoretic Framework. Annals of the
            Association of American Geographers 53:290–297.
Homans, George
    1967   The Nature of Social Science. New York: Harcourt Brace, and World.
INCAP
    1961   Tabla de Composición de Alimentos para Uso en América Latina. Guatemala:
            Instituto de Nutrición de Centro América y Panama.
Johnson, Allen
    1972   Individuality and Experimentation in Traditional Agriculture. Human Ecology
            1:149–159.
    1975   Time Allocation in a Machiguenga Community. Ethnology 14:301–310
    1977   The Energy Costs of Technology in a Changing Environment: A Machiguenga
            Case. In Material Culture. H. Lechtman and R. Merrill, Eds. pp. 155–157. St.
            Paul: West.
    1978   Quantification in Cultural Anthropology. Stanford: Stanford University Press.
Johnson, Allen, and Clifford Behrens
    1980   Machiguenga Food Production Decisions: A Linear Programming Analysis. Un-
            published Manuscript, University of California, Los Angeles.
Montgomery, Edward, and Allen Johnson
    1976   Machiguenga Energy Expenditure. Ecology of Food and Nutrition 6:97–105.
Moore, Omar Khayam
    1957   Divination: A New Perspective. American Anthropologist 59:69–74.
NAS
    1974   Recommended Dietary Allowances. Washington, D.C.: Food and Nutrition
            Board, National Academy of Sciences.
Pospisil, Leopold
    1963   Kapauku Papuan Economy. New Haven: Yale University Publications in An-
            thropology.

Read, Dwight, and Katherine Read
  1970 Critique of Davenport's Game Theory Analysis. American Anthropologist 72:351–355.
Simon, Herbert
  1977 Models of Discovery. Boston: Reidel.
White, Douglas
  1973 Mathematical Anthropology. *In* Handbook of Social and Cultural Anthropology. J. Honigmann, ed. pp. 369–446. Chicago: Rand McNally.

Chapter 3

A Theory of Real-Life Choice: Applications to Agricultural Decisions

CHRISTINA H. GLADWIN

Introduction

Reviews of the recent literature on agricultural decision making show that there is no scarcity of theories about how farmers make these decisions (Anderson 1979; Anderson, Dillon, and Hardaker 1977; Dillon 1971; Roumasset et al. 1979). Indeed, the topic of farmers' risk-aversiveness is quite fashionable (Anderson 1974; Boussard and Petit 1967; Cancian 1972; Dillon and Scandizzo 1978; Johnson 1976; Moscardi 1979; Moscardi and deJanvry 1977; Ortiz 1979; Roumasset 1976). Most of the currently popular theories, however, suffer from the fact that they do not take into account the simplifying procedure or heuristics that people use in real life to make their decision-making process *easier* (Quinn 1971, 1975, 1978). In fact, people seem to need simple rules of thumb when making everyday decisions (Cyert and March 1963; Hall and Hitch 1951). They must maintain fairly consistent and communicable strategies for dealing with a highly varied environment (Simon 1959; H. Gladwin and Murtaugh, this volume) and must deal with constraints on their cognitive information-processing capabilities (Slovic, Fischhoff, and Lichtenstein 1977).

The theory of choice presented here incorporates some of the simplifying procedures people use in making everyday real-life decisions. Because it does, it differs from the economists' normal assumption that

AGRICULTURAL DECISION MAKING:
Anthropological Contributions to
Rural Development

decision makers can rank order all the available alternatives on preference or indifference. Instead, it posits a psychologically more realistic two-stage model of the choice process that may be represented by a decision tree, a decision table, or a set of decision rules.

The chapter is organized so that the theory is outlined and aspects are defined in the first section. In the second section, the choice process of Stage 1 of the theory is described, and agricultural examples are given of Stage 1 decisions made by farmers in Lauderdale County, Alabama, and the *Altiplano* or Western Highlands of Guatemala. In the third section, the algorithm or cognitive steps of Stage 2 are presented. In the fourth section, decision-tree models of several agricultural decisions made in Alabama, Guatemala, and Puebla, Mexico, are presented. The reader should note that some mathematical symbols are used to formalize the theory in the second and third sections. The reader who is intimidated by mathematical language can ignore the symbols and still understand the choice process, given the written explanation and numerous examples.

Outline of the Theory

The theory assumes, following Lancaster (1966, 1971) and Tversky (1972), that an *alternative* is a set of characteristics or aspects. An *aspect* is an attribute or dimension or factor or feature of an alternative. For example, the aspects of a car (in a hypothetical choice between alternative used cars to buy) could be its cost, its miles per gallon, or its appearance. Furthermore, the theory assumes that all aspects are discrete. Thus, when decision makers use a continuous quantitative dimension such as cost, they either treat it as a constraint (for example, is cost greater than or equal to $4000?) or they categorize it in such a way that only an ordering, or a semiordering with "just noticeable differences" (Luce 1956), of the alternatives on the aspect cost will influence the choice (e.g., cost of Car 1 is greater than the cost of Car 2). An algebraic representation of the choice process follows from this assumption.

The theory hopes to explain how people in everyday life make choices among a large number of objects. For example, a car buyer, when trying to decide between used cars, might pick up a newspaper with several hundred cars listed and arrive at three used cars that he or she is interested in, after 15 minutes of scanning the page. The theory says that the car buyer has gone through the first of two stages in the choice process, Stage 1 being identical to Tversky's (1972) choice process of elimination by aspects. The car buyer has eliminated (rapidly, often unconsciously) all cars that have some unwanted aspect: The buyer does not want a truck, van, or two-door sedan; he or she does not want a car more than 4 years old, and does not want a European car or a convertible. Thus, the buyer is left with a feasible subset of alternatives that can be physically examined and decided about in a more detailed way. The latter decision process is Stage 2.

Once the alternatives are narrowed to a feasible subset, the "hard-core" decision process occurs in Stage 2. People typically go through Stage 1 quickly and think that the "real" decision process is Stage 2. In that stage, decision makers choose among the remaining alternatives by considering the aspects or attributes or factors of each alternative. They may simplify the decision process further by eliminating some aspects on which the alternatives have equivalent values. For example, if two alternative cars are both under $4000, the car buyer eliminates the aspect "Is cost under $4000?" and concentrates on other, more relevant decision factors.

After elimination of irrelevant aspects, the decision maker picks one of the aspects to order, partially or fully, the alternatives on. (For example, the gasoline economy of Car 1 may be poorer than that of Car 2.) Then, the buyer considers the constraints that are imposed on the alternatives from the environment, social system, or context and passes the ordered alternatives through the (unordered) constraints. If the alternative ordered first does not pass all its constraints, the alternative ordered second ("the second-best") is allowed a chance at passing all its constraints. If no alternative passes all its constraints, another strategy is employed. Stage 2 or the hard-core choice process is therefore essentially an algebraic version of *maximization subject to constraints,* a choice principle found in any microeconomics text (Henderson and Quandt 1971) or linear programming text (Wagner 1969). However, the hard-core choice process is algebraic, since there is an *ordering* of alternatives on an aspect rather than maximization of a continuous function. This algebraic process may be represented by an algorithm, a decision table or tree, or a set of decision rules.

Aspects

As in Lancaster (1966, 1971), an alternative is considered to be a set of characteristics or aspects, but with the important difference that "an aspect can represent values along some fixed quantitative or qualitative dimensions (e.g., price, quality, comfort) or they can be arbitrary features of the alternatives that do not fit into any simple dimensional structure [Tversky 1972:285]." For example, if a farmer is deciding on the set of crops to plant this year, the aspects of the alternative crops could be relative yields and profitability of the crop on the farmer's fields, relative riskiness of the crop, the extent of the farmer's knowledge about the crop, and the amount of land, time, labor, capital, and/or credit that the farmer has available to grow the crop in question. Each alternative crop has a value on each of these aspects.

However, whereas Lancaster assumes "that all characteristics are quantitative and objectively measureable so that the assertion that b_{ij} is the quantity of the ith characteristic possessed by a unit amount of the jth good has universal and [in principle] empirical meaning . . . [Lancaster 1971:15]," in the present theory, it is assumed that when decision makers use a continuous quantitative dimension or aspect, they discretize it or categorize it into a surprisingly small number of categories. This assumption may be unusual in economics, but it is not unusual in psychology, dating from evidence presented by Miller (1956).[1] Therefore, in the choice processes of Stages 1 and 2, a continuous aspect will be treated either as a constraint (e.g., "Do you have the capital or credit to buy inputs for potatoes this year?") or can be categorized in such a way that only the ordering of the alternatives on an aspect (e.g., the gross returns of potatoes are greater than the gross returns of corn) influences the choice. As will become obvious later, an algebraic representation of the choice process follows from this assumption.

Stage 1

When confronted with a large number of alternatives, decision makers narrow the set to a feasible subset that satisfies certain minimal conditions. For example, given eight to ten different possible crops to

[1] This evidence is moving the field of psychological decision making away from the normative models of decision making still popular in the field of economic decision making (Anderson, Dillon, and Hardaker 1977), to models that describe choice in terms of information-processing phenomena: "A psychological Rip Van Winkle who dozed off . . . would be startled by the widespread change of attitude exemplified by statements such as 'In his evaluation of evidence, man is apparently not a conservative Bayesian: he is not a Bayesian at all.' [Kahneman & Tversky 1972:450]," or " 'man's cognitive capacities are not adequate for the tasks which confront him,' [Hammond 1974:4]," or " 'people systematically violate the principles of rational decision making when judging probabilities, making predictions, or otherwise attempting to cope with probabilistic tasks (Slovic, Fischhoff, and Lichtenstein 1976:169)' [Slovic, Fischhoff, and Lichtenstein 1977:3–7]."

plant, a farmer in the *Altiplano* of Guatemala will eliminate some of them rapidly or unconsciously or *preattentively* (Tversky 1972; H. Gladwin and Murtaugh, this volume).[2] He or she might eliminate vegetables because of a lack of irrigation, or potatoes because of a lack of knowledge of how to plant them or apply pesticides. The farmer might not even think of growing coffee because the land is at an unsuitably high altitude. The choice will be finally narrowed to a "feasible" subset of crops requiring a more detailed decision in Stage 2.

Stage 1 is essentially Tversky's (1972) elimination-by-aspects theory, a choice process in which "an aspect is selected . . . and all the alternatives that do not include the selected aspect are eliminated. If a selected aspect is included in all the available alternatives, no alternative is eliminated and a new aspect is selected [p. 285]." It should be noted that any shared aspect (e.g., cost) that can be formulated as a constraint (e.g., is cost of the crop greater than $20/ha?) can be a selected aspect that is not shared by all alternatives, and is therefore capable of eliminating some of them.

Stage 1 of a Cropping Decision

For a decision at time *t*, the choice process of Stage 1 can be represented by a tree. For example, Stage 1 of the cropping decision made by farmers in the *Altiplano* of Guatemala is represented by the tree in Figure 3.1. For a farmer in the *Altiplano*, there are eight possible crops or systems of crops to plant: corn in association with two kinds of beans (corn + *frijol* + *haba*), wheat, potatoes, vegetables, fruit trees, a monocrop of beans (*frijol*), a monocrop of bush beans (*haba*), and coffee or coffee in association with avocado (coffee + avocado).[3] In some subregions of the *Altiplano* with appropriate climatic conditions, it is also possible to rotate or plant two or three different crops in the same field in one year. In this case, the system of crops is denoted here as wheat + vegetables, or vegetables + potatoes + vegetables, or potatoes + potatoes. The symbol "+" therefore denotes "planted in association with" or "in rotation with."

---

[2] The *Altiplano* of western Guatemala is an area of roughly 22,000 km² or 8400 miles, and is considered here to include the departments (or states) of Chimaltenango, Sololá, Totonicapán, Quezaltenango, San Marcos, Huehuetenango, and El Quiché.

[3] Theoretically, this set of "possible" crops could be much larger, and for farmers in the *Altiplano*, include crops grown in other parts of Guatemala (e.g., cotton, banana, tobacco, and sugar). The larger the set of possible crops, however, the longer the time needed for interviews with farmers. To shorten interviewing time, it was therefore decided to consider as alternatives only those crops that had *some* possibility of passing the altitude, soils, and water constraints in the *Altiplano*.

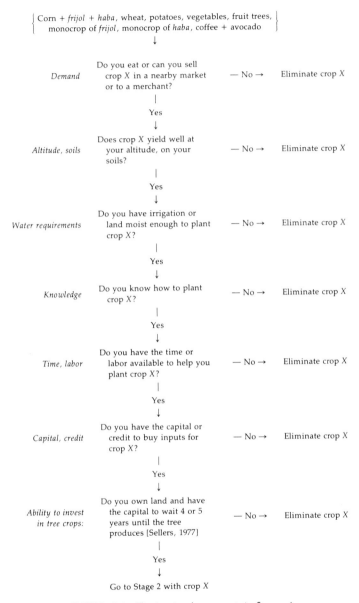

FIGURE 3.1. Elimination by aspects in Stage 1.

There are also six minimal conditions or constraints that a specific crop must satisfy in order to pass Stage 1:

1. *Demand:* The farmer must either have a consumption need for the crop or be able to sell it at a nearby market or to a trader.
2. *Altitude, soils:* The crop must have "good" yields on the farmer's own fields, that is, on those soils and at that altitude. ("Good" yields are arbitrarily defined by the farmers in the region.)
3. *Water requirements:* The farmer must have either irrigation or land moist enough to plant the crop or system of crops. (To plant two or three crops of vegetables each year in the *Altiplano*, the farmer must have irrigation on some fields.)
4. *Knowledge:* The farmer must know enough to plant the crop. Clearly, how much knowledge is "enough" depends on a subjective judgment made by each farmer.
5. *Time or labor:* The farmer must personally have adequate time or have labor (family and/or hired labor) available to plant the crop.
6. *Capital or credit:* The farmer must have the capital or credit to obtain the necessary inputs (seed, fertilizer, insecticides, labor) to plant the crop.

For a farmer to plant fruit trees as a monocrop, he must also satisfy a seventh minimal condition:

7. *Ability to invest:* The farmer must own the land and have the capital to be able to plant fruit trees and then wait 4 or 5 years until the trees produce fruit (Sellers 1977). Clearly, a farmer who owns only a little or no land will not be able to invest in fruit trees as a monocrop.

To predict the individual farmer's crop choice, data about each possible crop are put down the tree in Figure 3.1 (i.e., each farmer is asked a series of six questions about each crop in the "possible" set at the top of Figure 3.1). If a farmer answers yes to *all* six questions or criteria for a crop (e.g., wheat), then the model in Stage 1 "sends" the farmer to Stage 2 with that crop. If a farmer answers no to one question or criterion for a crop, then that crop is eliminated from the set of feasible crops that are sent to the more detailed decision process of Stage 2. In the latter case, the model predicts that the farmer will not grow that crop. The reader should note that only one no is needed to eliminate a crop; six (or seven in the case of tree crops) yeses are needed for a crop to pass Stage 1 and go on to Stage 2. The outcome or output of Stage 1 is thus a subset of feasible crops for each farmer (e.g., {corn + *frijol* + *haba*, wheat, potatoes}).

Stage 1 of a Fertilizer Decision

Another agricultural example of elimination-by-aspects comes from the decision of the *kind* or type of chemical fertilizer to apply at planting in Lauderdale County, Alabama. A farmer in that area has five different kinds of fertilizer to choose among: granulated or bagged fertilizers, bulk blends, powders, liquids, and suspensions. These are shown in the offered set at the top of Figure 3.2. In this model, it is hypothesized that the farmer eliminates (rapidly, maybe unconsciously) powdered fertilizer because it is difficult to spread on the field. Four types of fertilizer are then submitted to an availability or supply constraint. Since bulk blends and clear liquids are not readily available in Lauderdale County, the farmer takes two alternatives, suspensions and granular combinations, to Stage 2 to make a more detailed choice.

Stage 1 Formalized

In general, if one adopts the set notation of Tversky (1972), the choice process at time t can be represented by a tree as shown in Figure 3.3 and by a mechanism for selecting the aspects. In that notation, if x is an alternative or member of an offered set A (i.e., $x \in A$), then for each alternative x, there exists a nonempty set of aspects $x' = \{\alpha, \beta, \gamma, \ldots\}$. The aspects then define subsets of A: $A_\alpha = \{x \mid x \in A$ and $\alpha \in x'\} =$ those alternatives of A that have the aspect α. Similarly, $A_\beta = \{x \mid x \in A$ and $\beta \in x'\} =$ those alternatives of A that have the aspect β, and $A_\gamma = \{x \mid x \in A$ and $\gamma \in x'\}$ those alternatives of A that have the aspect

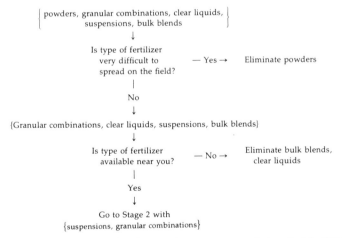

FIGURE 3.2. Stage 1 in the decision between types of chemical fertilizers.

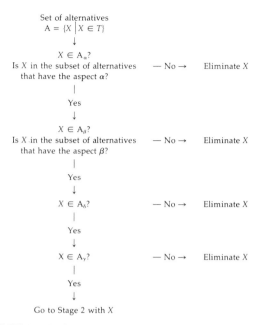

Set of alternatives
$A = \{X \mid X \in T\}$

$\downarrow$

$X \in A_\alpha$?
Is X in the subset of alternatives   — No →   Eliminate X
that have the aspect α?

|

Yes

$\downarrow$

$X \in A_\beta$?
Is X in the subset of alternatives   — No →   Eliminate X
that have the aspect β?

|

Yes

$\downarrow$

$X \in A_\delta$?   — No →   Eliminate X

|

Yes

$\downarrow$

$X \in A_\gamma$?   — No →   Eliminate X

|

Yes

$\downarrow$

Go to Stage 2 with X

FIGURE 3.3. Stage 1: Elimination by aspects at time $\underline{t}$.

γ (p. 286). The tree in Figure 3.3 states that, of all the alternatives $x \in A$, the alternatives that do not have the aspect α are eliminated; those that do not have the aspect β are eliminated; and so on until all the alternatives x in set A have been exhausted. The subset remaining after the elimination process $(A_\alpha \cap A_\beta \cap \cdots \cap A_\gamma)$ then passes to Stage 2.

Selection of Aspects

Given the elimination by aspects process, one must specify the mechanism whereby the aspects that eliminate are selected.[4] In Tversky's (1972) theory, the selection of aspects is probabilistic: "An aspect is selected with probability proportional to its weight . . . [p. 285]." The weight of an aspect is its subjective weight or value or utility; therefore, as in Lancaster, a utility function on the set of *aspects* is assumed. "Let *u*

---

[4] Here I am referring to the cognitive process by which aspects are selected by the decision maker. Given an actual model of a real-life decision, one can also ask how the *specific* aspects that appear in the model are selected by the model-builder. The latter question is the topic of a chapter in itself, and has been addressed by this author in two previous papers (C. Gladwin 1979a, 1979b).

be a scale that assigns to each aspect a positive number representing its utility or value." Then,

$$Pr(\text{aspect } \alpha \text{ is selected}) = \frac{u(\alpha)}{\Sigma_{\beta \in A' - A^\circ} u(\beta)},$$

where α and β are aspects that belong to at least one alternative in the offered set, but do not belong to all the alternatives. (Here, $\beta \in A' - A^\circ$; $A' = \{\alpha \mid \alpha \in x' \text{ for } some \; x \in A\}$, A' is the set of aspects that belongs to at least one alternative in A; and $A^\circ = \{\alpha \mid \alpha \in x' \text{ for } all \; x \in A\}$, A° is the set of aspects that belongs to all the alternatives in A, Tversky 1972:286.)

Because the selection of aspects is probabilistic in Tversky, the aspect selected to eliminate alternatives may vary in repeated decisions, and therefore the alternatives that are not eliminated, the choices, may vary in repeated decisions. The theory thus explains the observed inconsistencies in choice behavior: At different points in time, people make different choices.[5] Note, however, that the utility or subjective value of the *aspects* does not vary over time. An aspect may carry more subjective value than other aspects but still not be chosen to eliminate alternatives *all* the time (i.e., with probability 1). It follows that, although the aspects selected vary over time, the sequence of aspects selected in any one decision at one moment of time build one path to one chosen subset. "Any . . . sequence of aspects can be regarded as a particular state of mind which leads to a unique choice. In light of this interpretation, the choice mechanism at any given moment of time is entirely deterministic; the probabilities merely reflect the fact that at different moments in time different states of mind (leading to different choices) may prevail [Tversky 1972:296]."

Although Tversky's selection-of-aspects process is intuitively plausible and theoretically appealing, the question remains as to whether people select aspects probabilistically or use a deterministic procedure that produces different choices in repeated decisions. Elsewhere, I present an example of a deterministic choice procedure that produces different *choices* by the same decision maker in repeated decisions (C. Gladwin 1975). Similarly, a deterministic choice procedure could select different *aspects* over time. In the next section, decision rules are used to select one choice from among several possible alternatives. It is just as

[5] He also shows that this theory reduces to that of Luce (1959) and Restle (1961) as a special case, and satisfies Debreu's (1960) objection to Luce's choice theorem (Coombs *et al.* 1970:150).

plausible for decision rules (a deterministic choice process) to select *aspects*, which then are used in other decision rules to select the alternatives. In this case, the choice process will have rules behind the rules, or reasons behind the reasons for the choice (for a fuller explanation, see C. Gladwin 1977:38–54). The deterministic selection of aspects would be an alternative process to Tversky's probabilistic selection.

Tversky's choice process stops when one alternative remains after elimination of all others. One criticism of the present theory is that it does not specify the mechanism by which Stage 1 ends and Stage 2 begins. H. Gladwin and Murtaugh suggest that Stage 1 occurs unconsciously or *preattentively* in routine decisions (choices from menus in restaurants, choices of movies or TV shows, choices of restaurants). In these kinds of choices, the decision maker eliminates alternatives while his attentive consciousness is focused on something else—for example, a conversation. "The boundary between stage 1 and stage 2, then, is the boundary between preattention and conscious choice. However, the decision maker may look at the menu and come up with the one obvious choice. In this case, there is really no stage 2 and the whole process is handled preattentively [H. Gladwin and Murtaugh 1975:18]." Alternatively, Zulauf suggests that, even when only one alternative remains after elimination by aspects, the decision maker may want to take a harder look at it in Stage 2, to decide whether to choose it or nothing (C. Zulauf, personal communication). In this case, however, one may again assume there are two alternatives, "it" and "nothing."

Stage 2

The conscious or hard-core decision process occurs in Stage 2, a six step process.

Step 1

Aspects that are included in at least one alternative are mentally listed or considered. Given a difficult decision, however, many people resort to taking out a sheet of paper and drawing a line down the middle. This procedure works best with only two alternatives, and, incidentally, was the method recommended by Ben Franklin for arriving at a decision. They then list the aspects of the choice. For example, Table 3.1 is a list of the aspects that, it is hypothesized, influence the choice between liquid fertilizers (suspensions) and granulated fertilizers made by farmers in

TABLE 3.1
Factors Influencing the Choice between Granulated Fertilizers and Suspensions in
Lauderdale County, Alabama

Aspects	Liquid fertilizers or suspensions		Granular combinations
1. Cost per ton of nutrient	Liquid applied by dealer	≈	Granular applied by dealer
	Liquid applied by dealer	>	Granular applied by farmer
2. Distribution or coverage of field	Uniform	≈	Uniform if broadcast
	Uniform	<	If fertilizer is applied in a row, under the seed
3. Can other chemicals (e.g., preemergent herbicide, minor elements) be mixed with the fertilizer?	Yes	>	Yes: minor elements No: herbicides
4. Is specialized equipment required for application?	Can custom-hire	≈	Own or custom-hire
5. Number of trips across the field at planting	Two trips: Trip 1 for fertilizer and herbicides mixed; Trip 2 for seed		Two or three trips: Trip 1 for herbicides; Trip 2 for seed and fertilizer if applied in a row, or just seed if broadcast; Trip 3 for just fertilizer if broadcast.
6. Extent of labor input needed if farmer applies	Laborer attaches a hose to the nurse tank and transfers material to applicator tank	<	Laborer must pick sacks up and fill hopper
7. Can I get the N–P–K ratio I want?	Yes	≈	Yes

Lauderdale County, Alabama.[6] In Stage 2 of that decision, farmers consider factors or aspects such as cost per ton of nutrient in each type

[6] The listing and elimination of aspects is usually mental and rarely done on paper. In order to present a clear example of Step 2, however, Table 3.1 was formulated and may appear too rational a step to occur in a real-life decision process (D. Kronenfeld, personal communication). The decision criteria in this example were elicited from farmers and fertilizer suppliers in Lauderdale County, Alabama, and fertilizer experts in the National Fertilizer Development Center, Tennessee Valley Authority, and the International Fertilizer Development Center, Muscle Shoals, Alabama.

of fertilizer, distribution or coverage of the field provided by each type, the ability to mix other chemicals with the fertilizer, the need for specialized equipment to apply the fertilizer, the number of trips across the field required at planting with the fertilizer, the amount of labor needed to apply the fertilizer, and finally, whether the farmer can get the desired N–P–K ratio with that type of fertilizer.

Step 2

To simplify the decision process even further, some aspects may be eliminated or not considered by the decision maker. The following is a list of strategies or heuristics used by decision makers to eliminate aspects. The list is not intended to be exhaustive.

(a) If an aspect is of little or no subjective worth to the decision maker, that aspect is eliminated (or not even considered or listed in Step 1).

(b) If all alternatives have equal or equivalent values, on an aspect (in the sense of less than a just noticeable difference, Luce 1956), that aspect is eliminated.

(c) If two aspects are of equal or equivalent importance, and the order of the alternatives on one aspect is the opposite of the order of the alternatives on the other (e.g. $\alpha_{x_1} > \alpha_{x_2} > \alpha_{x_3}$ and $\beta_{x_3} > \beta_{x_2} > \beta_{x_1}$), then both aspects are eliminated (Zulauf, personal communication).

(d) If one aspect affects the decision process only through another aspect, and does not have a separate effect, the two aspects are considered as one aspect.

In the example in Table 3.1, the farmer has listed seven subjectively important aspects of granular fertilizers and suspensions (Step 2a), and then has eliminated the aspects "Is specialized equipment required for application?" and "Can I get the N–P–K ratio I want?" because both kinds of fertilizers have equivalent values on those aspects (Step 2b). That is, the farmer can own or custom hire the equipment needed to apply either kind of fertilizer, and, in Lauderdale County, Alabama, can get the desired N–P–K ratio with either bagged fertilizers or suspensions. Furthermore, the farmer considers the two aspects "Can other chemicals be mixed with the fertilizer?" and "number of trips across the field" to be really one factor to be considered, although both aspects may be listed mentally (Step 2d). Clearly, the farmer can cut down the number of trips across the field (and therefore save gas and decrease the compactedness of the field), if preemergent herbicides, fertilizer, and trace elements can be mixed and applied at the same time. These two aspects therefore are considered to be one aspect of this particular decision process. Thus, four aspects remain: cost per ton of nutrient, distribution or coverage of the field, ability of other chemicals to be mixed with the fertilizer, and the extent of labor input if the farmer applies the fertilizer personally.

Step 3a

From the subset of aspects not eliminated, the decision maker chooses or selects *one* aspect on which alternatives are ordered. There are two plausible ways to choose the "ordering" aspect:

1. The decision maker selects the aspect with the greatest utility or subjective worth:

$$\text{select } \alpha \text{ such that } u(\alpha) \geq u(\beta) \qquad \text{for all } \alpha, \beta \in B, \qquad (1)$$

where B is the subset of aspects ($B \subset A$) which remain after elimination of aspects in Step 2.

2. The decision maker chooses the aspect α by means of a choice function "not built up from orderings," $C_t(\alpha, \beta)$:

$$C_t(\alpha, \beta)_{\beta \in B} = \alpha \qquad \text{if } \alpha \, P \, \beta \text{ for every } \beta \in B, \qquad (2)$$

where P is not an order, that is P is asymmetric, but not transitive or connected (Arrow 1951:21).

Although the use of a choice function "not built up from orderings" will not be completely defined in this chapter (see C. Gladwin 1977:28–37) for a more complete treatment), the reader should have an intuitive "feel" for the difference between the two choice mechanisms expressed in 1 and 2. If the decision maker uses a utility scale or function over the set of aspects $\alpha, \beta \in B$, as in 1, then he can rank-order the aspects on subjective worth or importance. Thus the farmer, in the example in Table 3.1, can say whether "number of trips across the field" is more important to him than the aspect "distribution or coverage of field." The choice function in 2 says that the farmer does not *ordinarily* rank-order aspects while making decisions, but simply knows from his experience or production rules that both aspects, "number of trips" and "coverage of the field," must be considered. One of his production rules tells him to choose the aspect α as the ordering aspect. Step 3a also signifies that the selection of aspects at this stage in the choice process is deterministic, and not probabilistic as in Stage 1. This means that, *given the same set of aspects B after Step 2*, the decision maker will always select the *same* ordering aspect α in repeated decisions.

Step 3b

1. If the alternatives are mutually exclusive, the decision maker orders the alternatives on the "ordering" aspect α:

$$\alpha_{x_1} > \alpha_{x_2} > \alpha_{x_3}$$

This order may also be a semiorder:

$$\alpha_{x_1} >>_\delta \alpha_{x_2} >>_\delta \alpha_{x_3};$$

where

$$\alpha_{x_1} >>_\delta \alpha_{x_2}$$

if and only if $\alpha_{x_1} > \alpha_{x_2} + \delta$ and δ is a "just noticeable difference" (Luce, 1956).

For example, the farmer in Lauderdale County uses the aspect "cost per ton of nutrient" on which to order or rank the alternative kinds of fertilizer. As stated in Table 3.1, a comparison of granulated fertilizers and suspensions on cost depends on whether the farmer or the dealer applies granulated fertilizers. (It is assumed that the dealer applies suspensions.) If the farmer owns equipment and personally applies granular, the cost of granular is less than that of suspensions. If, however, the dealer applies both kinds of fertilizer, the cost is almost the same.

2. If the alternatives are *not* mutually exclusive, then the decision maker *partially* orders the alternatives on the ordering aspect α:

$$\alpha_{x_1} > \alpha_{x_2}$$
$$\alpha_{x_3} > \alpha_{x_2}.$$

This order may also be a semiorder:

$$\alpha_{x_1} >>_\delta \alpha_{x_2} \quad \text{and} \quad \alpha_{x_3} >>_\delta \alpha_{x_2}.$$

Examples of real-life decisions in which alternatives are not mutually exclusive choices are numerous. A woman buying clothes will often buy two blouses if she cannot decide which one she likes best, and if she has enough money. A department chairman who cannot rank two candidates for "assistant professor" will sometimes try to search for the funds to hire them both. A further agricultural example is provided by the cropping decision of farmers in the *Altiplano* of Guatemala. In Stage 2 of that decision, which will be presented in Figure 3.6, a farmer compares the profitability of wheat to that of corn and beans, and the profitability of potatoes to that of corn and beans. Normally, wheat and potatoes are not rank-ordered on profitability, however, because the farmer will try to plant both wheat and potatoes if there is enough land to do so. In this example then, wheat, potatoes, and corn (and beans) are not necessarily mutually exclusive alternatives (i.e., a farmer may plant all three crops if there is enough land to do so).

Step 4: Constraints

For each of the remaining aspects, the decision maker or the environment (social–economic system or context) imposes or formulates a minimum condition or requirement that has to be satisfied by the chosen alternative. In fact, some constraints may already be formulated or exogenously given or imposed on the decision makers by their or others' scarce resources or previous decisions. The decision maker is then not conscious of formulating constraints out of some aspects: they are just *there*, and represent the powerlessness of the individual against the environment. Moreover, some alternatives may have to pass constraints that other alternatives do not. Farmers who are choosing whether to plant wheat or potatoes as their cash crop in the *Altiplano*, for example, are aware that they have to plant wheat in fields close enough to a road, so that after harvest they can carry the wheat to the thresher (*trilladora*), which is stationed on the road. Potatoes, as the alternative cash crop, do not face this constraint (the need to be planted in fields near to the road), since they need no processing but are ready to sell right after the harvest. Thus the constraints imposed on the alternatives by the environment or social system surrounding the decision maker may not be the same, and one may speak of an alternative passing or not passing all "its" constraints.

As in Stage 1, constraints can be formulated either from quantitative aspects or from qualitative features of the alternatives that do not fit into any simple dimensional structure. If an aspect of an alternative is quantitative or continuous, the decision maker selects a threshold S_j for each of its aspects j remaining after elimination of aspects in Step 2 and selection of the ordering aspect α in Step 3, such that an alternative X_i can be chosen if and only if:

$$X_{ij} \geq S_j, \qquad \text{for all of its aspects } j \in B - \alpha,$$

where X_{ij} denotes the value of X_i on the aspect j.

If an aspect $j \in B - \alpha$ is an arbitrary feature of the alternative, then the constraint is of the form $X_i \in A_j$, where A_j is the set of alternatives that have the aspect j. The alternative X_i must have the feature j if it is to pass the constraint.

For example, a farmer in Lauderdale County, Alabama, formulates the constraint "Do you have the labor (family or hired) available to apply fertilizer yourself this year?" from the aspect "labor input needed." (Other constraints used by the farmer can be seen in Figure 3.3 in the next section.)

Step 5

The decision maker passes the ordered alternatives through the constraints, which are not necessarily ordered. An alternative must pass *all* of its constraints to be chosen. If no alternative passes all of its constraints, the decision maker goes to Step 6.

Since there is an ordering of alternatives on an aspect, and a passing of alternatives through constraints, the choice process presented here is an algebraic version of maximization subject to constraints, the choice principle that serves as the underlying assumption of most microeconomic theory (Henderson and Quandt 1971). Since the choice process in Steps 3–5 is algebraic, it may be represented by a decision tree, table, flowchart, or by a set of decision rules. Agricultural examples of Step 5 will therefore be presented in "tree" form in the next section.

Step 6

A decision maker may follow one of several plausible strategies if no alternative passes all the constraints. (Again, the list of strategies is not intended to be exhaustive.)

(a) The decision maker eliminates the ordering aspect α, and returns to Step 3a to select another ordering aspect β; either the aspect with the next highest subjective worth or utility, or an aspect chosen by another production rule. The alternatives are ordered on that aspect. Then:

The highest ranking alternative on that aspect is chosen, or Steps 4 and 5 are repeated.

(b) The decision maker keeps the ordering of the alternatives on the initial ordering aspect α and returns to Step 4. Either the threshold S_j in the constraint $(X_{ij} \geq S_j)$ formulated from the aspect j is lowered, or that constraint is eliminated, and Step 5 is begun. If still no alternative passes, the decision maker returns to Step 4 and lowers the threshold on another constraint. Either the production rules determine which constraints to relax, or the decision is made to relax the least important constraints. This process is continued until at least one alternative passes.

However, upon relaxation of one or more constraints, both alternatives may have passed the set of constraints. In this case, both alternatives may be chosen if the decision maker cannot decide between them.

(c) The decision maker keeps the ordering of the alternatives on the aspect α and simply chooses the highest ranking alternative on that aspect. The reader should note that this strategy *resembles* what economists call "a trade-off": The alternative chosen is better on aspect α but worse on aspect β, so that the decision maker is said to "trade-off" some of aspect α for some of aspect β. Actually, the decision maker decides that aspect α is more important than aspect β (or the production rules decide) and the alternative ranked highest on aspect α is chosen.

(d) The decision maker decides *not to decide* at time t, and searches for new alternatives or waits to see if an alternative can now pass the constraints it failed earlier. Thus at time $t + 1$, he or she may change the feasible subset proceeding from Stage 1, and start again in Stage 2. An example is President Carter's decision to postpone the decision on the neutron bomb.

Decision Trees

After the listing and elimination of aspects, the choice process in Stage 2 also may be represented by a decision tree. The use of trees as models of behavior is not new in the social sciences. The phrase markers of transformational grammar are trees (Chomsky and Miller 1963). Trees of "embedded sets" are structural models in psycholinguistics (Boyd and Wexler 1973). Trees are found in the multidimensional scaling literature (Degerman, 1972), and in computer simulations of human problem solving (Miller, Galanter, and Pribram 1960; Newell, Shaw, and Simon 1958). A lexicographic ordering model, which is common in the decision-making literature, is just one kind of tree (Debreu 1954; Roumasset 1976). Trees can be found in the normative decision-making literature (Raiffa 1968), and in the anthropological decision-making literature (Barlett, 1977; C. Gladwin, 1975, 1976, 1977, 1979a, 1979b; H. Gladwin, 1971, 1975; Quinn 1978).

Indeed, decision trees are a simple way to represent visually the logical relationships between alternatives, decision criteria, or constraints, and outcomes that are specified in Steps 3–6 of the algorithm presented above. Since those cognitive steps already have been described, this section will explain decision trees by presenting agricultural examples of different kinds of trees.

Two Mutually Exclusive Alternatives and Multiple Constraints

The most common type of decision tree in Stage 2 is one with two alternatives and two or more constraints. The alternatives are normally mutually exclusive: The decision maker is constrained to choose only one of two good alternatives (e.g., the woman with a budget constraint can buy only one blouse). A tree of this type will have both right and left branches, with constraints as nodes of the tree.

An example of this kind of tree is shown in Figure 3.4, representing the choice between suspensions and granulated fertilizers made by a hypothetical farmer in Lauderdale County, Alabama. In that tree, read from top to bottom, farmers minimize (i.e., order the alternatives on) cost per ton of nutrient supplied by the fertilizer. If the farmer owns the equipment to apply granulated fertilizers personally, the cost of granular fertilizer will be less than the cost of suspensions, and he or she will go down the left-hand branch of the tree. If a dealer applies both kinds of fertilizer, the costs are about equal, and the farmer goes down the right-hand branch of the tree.

On the left-hand branch of the tree, the farmer faces first a labor constraint: Even though the equipment is available to apply granular fertilizer, the labor to keep filling the hopper may not be available. If not, the farmer passes to the right-hand branch of the tree. If the labor (family or hired) is available, granular and liquid fertilizers are then compared on the aspect "coverage of the field." In general, the coverage will be more uniform with granulated fertilizer if it is row-applied (i.e., applied alongside the row of plants). In this case, the model predicts that the farmer will buy fertilizer in granular form, since the cost is less, labor is available, the coverage of the field is more uniform, and the same number of trips will have to be made (2) across the field with both granulated fertilizers and suspensions. If the granular fertilizer is to be broadcast, however, the coverage of the field is about equal with either granular or liquid fertilizer, and granular fertilizer loses that advantage. Moreover, the number of trips across the field increases when granulars are broadcast, if other chemicals (e.g., preemergent herbicides, minor

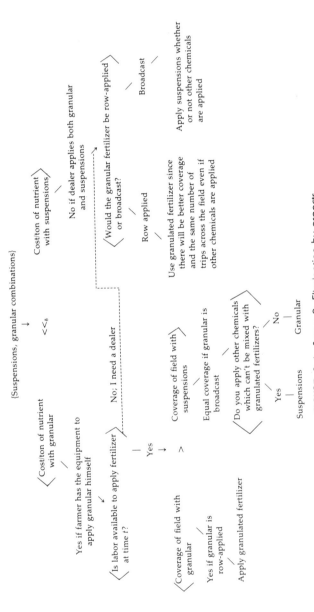

FIGURE 3.4. Stage 2: Elimination by aspects.

elements) also are applied at planting. Thus the farmer chooses "suspensions" if he or she applies other chemicals at planting to minimize the number of trips across the field (and the costs of gas and compactedness of the soil). However, granular fertilizer is chosen if other chemicals are not applied with the fertilizer, since the number of trips across the field in that case would be the same and therefore not a criterion that "cuts."

On the right-hand branch of the tree, a dealer applies both kinds of fertilizer for the farmer, so that cost per ton of nutrient is about the same for both kinds. In that case, if the granular fertilizer is row-applied, the model predicts that the farmer will apply granular fertilizer since the coverage of the field will be more uniform, and the number of trips across the field will be the same even if other chemicals are applied. In case the granular fertilizer is *broadcast,* the model predicts that the farmer will buy suspensions for two reasons. First, if other chemicals are applied, the number of trips across the field will be minimized by applying suspensions. Second, if other chemicals are not applied at planting, granulated and liquid fertilizers are about the same: They have approximately the same uniform coverage of the field, roughly the same costs, and require an equal number of trips across the field. In the interests of parsimony therefore, the criterion "Do you apply other chemicals that can't be mixed with granular fertilizers?" can be omitted on the right-hand branch of the tree, since it does not really cut the sample of farmers into a subset who apply suspensions and a subset who apply granular fertilizers. When this criterion or constraint is omitted, the model predicts only one outcome, "Apply suspensions," on the extreme right-hand branch of the tree.

The tree in Figure 3.4 thus illustrates a common feature of tree models of real-life decisions, which is that some of the constraints are included on some paths but omitted on others. That is, some of the constraints cut on some paths, but do not cut on others. For example, the labor and other chemicals constraints appear or cut on the left-hand branch of the tree, but not on the right-hand branch. The labor constraint is irrelevant and therefore omitted on the right-hand branch, since all the farmers going down that path would have a dealer apply the fertilizer. Similarly, the other chemicals constraint is irrelevant or does not cut on the right-hand path for the reasons outlined. A decision tree model that is properly specified, therefore, does not include all the possible or logical combinations of constraints or conditions that it theoretically might include. For parsimony, irrelevant or redundant constraints on a path are omitted. Unfortunately, time did not allow a testing of the decision trees in Figures 3.2 and 3.4 with a separate, second set of decision makers.

Multiple, Mutually Exclusive Alternatives and One Constraint

If there is one constraint and multiple (e.g., $X \geq 2$) alternatives, Stage 2 can be represented by a right-branching tree. An example of this kind of tree is given by a model of Fante fish sellers' choice of market (C. Gladwin 1975:86–90). In this type of decision tree, the alternatives are ordered on an aspect (e.g., the riskiness of the market). The alternative ordered first (e.g., the least risky market) then attempts to pass through the only constraint: "Is the probability that the market be at least "somewhat good" greater than or equal to .5?" If the alternative ordered first passes the constraint, it is chosen. If not, the alternatives ordered second and third (and so on) attempt to pass the constraint. If all alternatives fail, another strategy from Step 6 is employed.

A Fante fish seller will thus go to the least risky market if she thinks that probably it will be profitable. If not, the next risky market is examined to see if that probably will be profitable. If not, the most risky market is looked at. In short, the fish seller is minimizing risk subject to a profitability constraint. This particular decision model predicted 90% of the actual choices made between three alternative markets during one fishing season (C. Gladwin 1975:108–177).

Adoption Decisions: One Alternative and Multiple Constraints

If there is one new alternative, which can be accepted or rejected (such as a new technology, a new job, a new spouse, a new house), and multiple constraints, Stage 2 can be represented by a left-branching tree. Adoption–decision trees are examples of this kind of tree, since the decision is whether to accept or reject an innovation, such as a new variety of seed, fertilizer, machine, or a new farming practice. The decision is really between two alternatives (Adopt the innovation–Do not adopt; Marry–Do not marry; Take the job–Do not take the job; Buy the new house–Do not buy). Constraints demand that the new alternative or innovation perform better than the status quo on at least one dimension; and equally as well on a number of dimensions, criteria, or factors. Thus the decision criteria constrain the innovation rather than the status quo, since it is assumed that the status quo passes the same constraints, because the decision maker has been following this strategy up to the present time.

Examples of adoption–decision trees can be seen in a previous study of farmers' decisions to adopt or not to adopt the recommendations of the Plan Puebla, a rural development project in Puebla, Mexico that aimed to increase yields of corn on rainfed farms. Since the recommendations of the project were different for different agroclimatic production zones, the study focused on the decision processes of a sample of farmers in one village of one zone. In 1973–1974, the recommendations for village farmers were to get credit for fertilizer, to increase plant population, to increase the number and change the timing of fertilizer applications, and to use a recommended level of fertilizer per hectare (ha). Since farmers seemed to decide to adopt each of the first three recommendations or farming practices on their own merits or independently of the others, rather than to adopt a technological package, separate models of the first three adoption decisions were developed. The results of testing the decision models were in general as expected: The models predicted 82–97% of actual adoption choices made by village farmers during one cropping season. For brevity, the reader is referred to C. Gladwin (1976, 1977, 1979a, 1979b) for a description of the models and results.

The Decision to Adopt Urea

An example of a tree model of an adoption decision that is currently being made by farmers in the *Altiplano* of Guatemala can be seen in Figure 3.5, the decision of whether or not to apply urea, a nitrogen fertilizer, on corn and/or wheat. This is a totally different decision from the one to apply *any* kind of chemical fertilizer, since the latter decision was made by farmers in the *Altiplano* 10–20 years ago. (The latter decision therefore would be very hard to study *now* since one needs a subsample of adopters and a subsample of nonadopters to elicit or observe decision criteria. Moreover, without a big sample of nonadopters, it is very difficult to test the model to see whether the elicited criteria cut). However, interviews with farmers and agronomists in ICTA, the national Institute of Agricultural Science and Technology, showed that farmers could improve their corn and wheat yields if they changed the *type* of chemical fertilizer they applied.

Traditionally, farmers in the *Altiplano* apply bagged mixtures of nitrogen and phosophorus (20–20–0 or 16–20–0) on corn, after planting at the time of the first rains, and on wheat, at planting. ICTA recommends an *additional* application of nitrogen fertilizer or urea (46–0–0) ten days

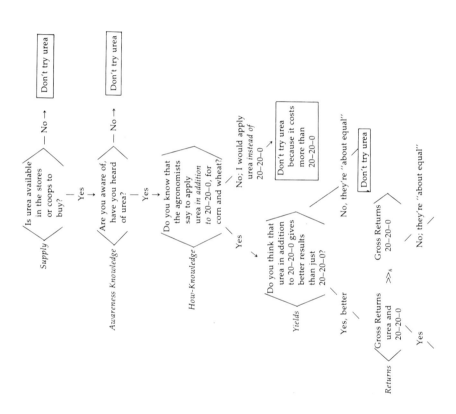

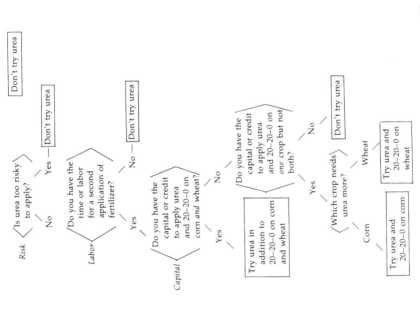

FIGURE 3.5. The decision to apply urea in addition to 20-20-0 for corn and wheat.

before flowering (*candeleo*) for corn and 40 days after planting for wheat. Since few farmers apply urea *in addition to* the first application of 20–20–0, the important question becomes, "Why not? What are the factors limiting adoption of urea?"

The decision tree in Figure 3.5 pinpoints the constraints or factors limiting adoption (here defined as initial trial) of urea for a farmer who plants corn and wheat. All the constraints concern urea, the innovation, or compare the recommended combination of fertilizers, urea and 20–20–0, to the traditional application of just 20–20–0. In general, the recommendation of urea and 20–20–0 must perform better than the status quo on at least one criterion, and urea must pass all constraints.

1. *Supply:* Urea must be available in the store, bank, or cooperative where the farmer obtains fertilizer.
2. *Awareness knowledge:* The farmer must have heard of or be aware of urea (Byrnes 1977).
3. *How-knowledge:* The farmer must know that the agronomists recommend urea *in addition to* 20–20–0, and not just urea. (This criterion assumes that urea costs noticeably more than 20–20–0, so that a farmer who thinks urea and 20–20–0 are substitutes will not apply urea.)
4. *Yields:* The farmer must expect better yields with the combination of 20–20–0 and urea.
5. *Gross returns:* Given costs and yields, the farmer must think that returns (yields minus cash costs) will increase with adoption of urea. The combination of urea and 20–20–0 must perform better than the status quo on at least this criterion.
6. *Risk:* The farmer must think that urea is not too risky, or must be willing to take an added risk.
7. *Labor or time:* The farmer must have the time or the family or hired labor for a second application of chemical fertilizer.
8. *Capital or credit:* The farmer must have the capital or credit to apply urea and 20–20–0 on both corn and wheat. If the capital for one crop but not two is available, then the farmer should allocate scarce capital to the crop that needs it more. (If a farmer only grows corn, these allocation criteria are not necessary. If the farmer has a different crop mix (e.g., corn and potatoes or corn and vegetables or corn and coffee), the crop which "needs urea more" will of course vary.)

This example illustrates some important features of adoption–decision trees. Clearly, some of the criteria have a sequential order. For example, there has to be a supply of urea in the region before farmers can decide whether or not it is profitable to buy. Likewise, farmers have to be aware of urea and know what the recommendation really is before they look at its profitability or riskiness. Indeed, these criteria may be considered Stage 1 constraints. However, the order of criteria in the tree is unimportant (in the sense that it does not affect the predicted outcome) because *all criteria* have to be passed on the left-hand path leading to the outcome "Try urea in addition to 20–20–0." If the farmer fails to pass *one* constraint (i.e., if the answer is no to questions about supply or knowledge or profitability or capital or labor, or if the answer is yes to the risk question, then the tree model predicts nonadoption. In this way, the decision tree pinpoints the most important factors limiting adoption, since a simple count of the number of farmers down each path leading to the outcome "Do not try it" shows which paths are more important than others. Because the order of criteria on the tree is unimportant, the ordering aspect, in this example "gross returns," does not have to be the topmost criterion in the tree. Moreover, the order of criteria in the tree may be switched, to see if the main factors limiting adoption change when the order of criteria is changed (C. Gladwin 1977:281–284).

Alternatives That Are Not Mutually Exclusive, with Multiple Constraints

If the alternatives are not mutually exclusive, then the decision maker *partially* orders the alternatives on the ordering aspect. The crop choice made by farmers in the *Altiplano* of Guatemala is an example of this kind of decision. As explained in the first section of the chapter, for each farmer, there corresponds a subset of feasible crops, which have passed in Stage 1, six or seven minimal conditions or constraints: a consumption or market constraint, altitude, soil, and water constraints, knowledge, time or labor availability, and capital or credit constraints. Given this subset of feasible crops, a farmer passes to Stage 2 in Figure 3.6, which *allocates* the farmer's available land (owned or rented) to the crops that pass Stage 1 constraints. Thus Stage 2 asks, "*Hay terreno?*" or 'Is there enough land to plant crop X'? The single question or criterion becomes a string of criteria or a flowchart, however, because the crops that pass Stage 1 are not necessarily mutually exclusive alternatives. That is, if a farmer has enough land to plant the consumption crop corn and one or more cash crops (e.g., wheat, potatoes), that farmer will do so. (Diversification is used either to decrease risk or to avoid making a difficult decision, or both.) If the farmer does not own or operate much land, however, the crops that pass Stage 1 *compete* for the little land

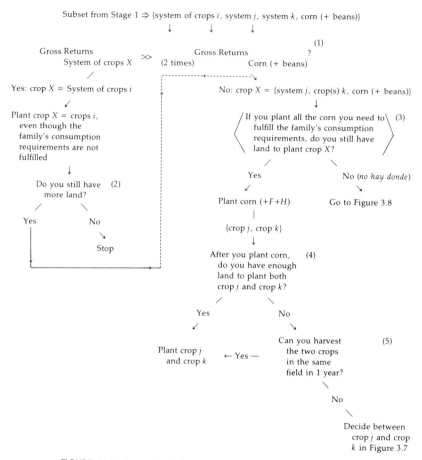

FIGURE 3.6. Stage 2 of the cropping decision in the Altiplano.

available, and the decision (and decision model) becomes more complicated. In the most general terms, Stage 2 of the model proposes that farmers in the *Altiplano* give first priority to crops or systems of crops[7] that are at least twice as profitable as corn, the main consumption crop. Second, they plant as much corn as is necessary to fulfill the family's

---

[7] A system of crops is a set of crops that are interplanted or multicropped (Hildebrand 1976). In the *Altiplano*, corn is intercropped with beans (*frijol* and *haba*), so is written corn (+ beans). A system of crops is also defined here as a set of crops that is harvested on the same field of land in one year (e.g., a first harvest of wheat and a second harvest of peas, or two harvests per year of potatoes, or three harvests per year of vegetables).

consumption requirements between harvests. Third, if farmers have still more land, they then plant a crop or system of crops that is not twice as profitable as corn, but may be equally as or a little more or less profitable than corn (+ beans).

THE PROFITABILITY CRITERION

At the top of the flowchart in Figure 3.6 is the subset of crops that pass all the criteria of Stage 1. This subset varies from farmer to farmer, even within homogeneous subregions of the *Altiplano*. For generality and purposes of explanation, it is assumed here that three systems of crops and corn (+ beans) have passed Stage 1. The first criterion at the top of Figure 3.6, the profitability criterion, asks if the profitability or gross returns (value of the production of a crop minus the cash costs of production) of each crop or system of crops is at least *twice* as great as the gross returns of corn (+ beans). Hence the *ganancia* or 'profitability' of wheat is compared to that of corn (+ beans); the profitability of two or three crops of vegetables is also compared to that of corn (+ beans). However, the order on the aspect "profitability" is a partial and not a full order: The profitability of one crop of wheat is not compared to that of two or three crops of vegetables or any other cash crop. *All* of the alternatives in the feasible set at the top of Figure 3.6 are *not* rank-ordered on the aspect "profitability" (Step 3b, 2). Instead, each alternative is compared with, or ordered with respect to, the consumption crop corn (+ beans), since—as the farmers testify—*Maíz es principal* or 'Corn is first'.

Given this partial ordering, the profitability criterion asks if a crop or system of crops in the feasible subset is at least twice as profitable as corn (+ beans) (i.e., *si tiene doble la ganancia*). The crop(s) may be three or four times as profitable as corn, but must be at *least* twice as profitable. In this flowchart, only crop *i* is considered by the farmer to be twice as profitable; crops *j* and *k*, and of course corn (+ beans), are not. As a result, only crop *i* goes down the left-hand path of the tree; the rest of the crops in the feasible set go down the right-hand path to Criterion 3.

Examples of crops that are at least twice as profitable as corn (+ beans) are many in the *Altiplano*: two to three crops of vegetables and one crop of potatoes in Almolonga, Quezaltenango; two crops of vegetables *or* one crop of vegetables and one crop of corn (+ beans) in Santa Rita–San Ramon in San Marcos; a rotation of wheat and a vegetable, or 'bush beans' (*frijol de suelo*) and potatoes, or two crops of vegetables in Tecpán, Chimaltenango; two crops of potatoes in *la zona de paperos* 'Concepción Chiquirichapa' in Quezaltenango; coffee or coffee + avocado in San Lucas Toliman, Sololá; a monocrop of fruit trees in Chichicastenango, El Quiché.

These systems of crops go down the left-hand branch of the tree to the outcome, "Plant crop X even though the family's consumption requirements of corn (+ beans) are *not* fulfilled." A farmer with a crop that is twice as profitable as corn should therefore plant the profitable crop even if this takes some land *out* of corn, with the result that the family's consumption requirements for corn are not met for that year—a risky thing for a farmer in the *Altiplano* to do. Clearly, the only reason the risk can be taken is that replacing corn and beans with a crop that is really worth "the trouble" is profitable. Since the cash crop is at least twice as profitable as corn, the farmer is sure that that crop can be grown and sold and corn can be bought in the market place, even if there is a shortage of corn with a resulting high price in the market. With crops that are less than twice as profitable as corn (e.g., wheat or one crop of potatoes), the farmer does not have this guarantee.

If the farmer still has (operates) more land after planting crop i, Criterion 2 in the model sends him or her to the right-hand branch of the tree. If not, the model stops. For example, in Almolonga, farmers grow three or four crops of vegetables per year on the one-half to two *cuerdas*[8] of the irrigated flat land they have. They do not grow corn on that land, in spite of the fact that they cannot then meet their family's consumption needs for corn. If they do not also own or rent hilly, nonirrigated land, the model predicts they will *not* grow corn at all. Similarly, farmers in San Lucas Toliman grow coffee (or coffee + avocado) on what low altitude land they *own*. If they do not also own or rent land too high for coffee, they will *not* plant corn.

THE CONSUMPTION CRITERION

On the right-hand branch of the tree in Figure 3.6, Criterion 3 asks if the farmer has more land left to plant a cash crop(s), and enough corn to fulfill the family's consumption requirements for a year (the period between harvests of corn in the *Altiplano*). The quantity of corn that is required for consumption depends, clearly, on the size of the family and the productivity of the land and corn variety that the farmer plants. If the answer is no, *no alcanza el maíz todo el año*, then the farmer is sent to Figure 3.8, p. 79, and is asked the string of questions in Figure

---

[8] A *cuerda* is the unit of measurement for land, and its size varies across the *Altiplano*; in Totonicapán, 1 *cuerda* = .0406 hectares (ha); in Almolonga and the Xela Valley, 1 *cuerda* = .0441 ha; in San Pedro Jocopilas, Quiché, 1 *cuerda* = .0635 ha; and in Tecpán, 1 *cuerda* = .1128 ha.

3.8, to determine if a cash crop *and* corn (+ beans), or just corn (+ beans) should be planted.

If the answer is yes to Criterion 3, *sí, alcanza el maíz todo el año*, the model predicts that corn should be planted first and then one, two, or more cash crops. The latter decision is easy if only one cash crop is left at this stage in the decision process, for then the farmer will certainly plant that cash crop—and corn (+ beans). Probably half of the farmers who plant corn and wheat in Totonicapán and Quezaltenango fit into this category, i.e., only one cash crop, wheat, fulfills all the conditions of Stage 1. Since wheat is not usually twice as profitable as corn (+ beans), farmers go down the right-hand path of the tree with the subset: {corn, wheat}. Since they have enough land, they first plant enough corn to fulfill the family's consumption requirements, and then plant wheat.

DIVERSIFICATION CRITERIA

The decision is more complicated if *two* or more cash crops are left in the feasible subset {crop *j*, crop *k*} at this point. These farmers have to decide whether or not to diversify. The diversification criteria (4 and 5) are the simplest criteria imaginable. If farmers have enough land to diversify, i.e., plant both crop *j* and crop *k* (e.g., wheat and potatoes), the model predicts that they will plant both crops. If they do not have enough land to plant both crops on separate fields, but they can rotate the crops on the same field within one year, the model also predicts that they will plant both crops. If they do not have enough land and cannot rotate the crops within the year, then they are sent to the subdecision or subroutine in Figure 3.7 to decide between the two cash crops.

Submodels or Subroutines in Stage 2: The Decision between Two Cash Crops

The model in Figure 3.7 is the subdecision between two cash crops (e.g., wheat and potatoes), *when the farmer does not have enough land to plant them both*. The first three criteria look at (logically) possible combinations of orderings of the cash crops on *ganancia* or 'profitability' and risk. The criteria do *not* look at the relative riskiness of the crops without also looking at their relative profitability.

The top node in Figure 3.7 asks if the most profitable crop (*j*) is also the least risky. If so, the model predicts that the farmer plant that crop. This is an obvious outcome, and may be the case when *Altiplano* farmers are deciding whether to plant irrigated potatoes or irrigated vegetables, a relatively more profitable, less risky crop.

The next node, Criterion 2, asks if the more profitable crop is also the

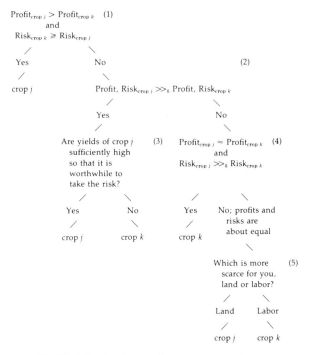

FIGURE 3.7. The decision between two cash crops.

riskier crop (i.e., if the order of the two crops on profitability agrees with the order of the two crops on risk). In the *Altiplano,* this is usually the case for wheat and potatoes, potatoes being the higher-valued, riskier crop (in terms of production *and* marketing).

If yes, the more profitable crop is also riskier, Criterion 3 asks if the farmer is willing to *take* the risk of the riskier, higher valued crop. Criterion 3 hypothesizes that a farmer is willing to take this risk if the yields of the higher-valued crop are sufficiently high that it is worth the farmer's while to take the risk. For example, in a region with potato yields of 2500–3000 pounds per *cuerda* (lb/cda), farmers would be willing to take the risk of potatoes; whereas in a region with potato yields of 1500–2000 lb/cda, farmers would not take the risk of potatoes but would plant wheat.

If a farmer does not perceive the higher-valued crop to be also riskier, then the farmer goes down the right-hand path of the tree to Criterion 4. Criterion 4 asks if the crops are equally profitable, and if one is clearly more risky than the other. If this is the case, the model predicts that the farmer will plant the less risky crop, since there is no reason to take the risk of the riskier crop. This would be the case between wheat and potatoes, when the farmer thinks they have about equal returns.

If the farmer perceives no difference on returns or risk between the two cash crops, the model hypothesizes that another criterion be selected to decide between them (an example of Step 6a, 1 of the algorithm presented earlier). Criterion 5 proposes that the farmer plant the crop that uses *less* of the *scarcer* resource. For example, if the farmer in the Xela Valley has more land available than time or labor, then wheat should be planted, given that wheat and potatoes are considered to be about equal on risk and profitability.[9] However, if the farmer has relatively more available labor than land, then potatoes should be planted (Barlett 1977).

Implications for Diversification of Crops

Diversification of crops is usually handled in an economic model (e.g., a portfolio-selection theory model) in this way: Farmers choose a set or "portfolio" of crops by trading off a little profitability for less risk (Markowitz 1959). In this model, however, diversification of crops is not dependent primarily on the profitability and riskiness of the alternative crops as would be the case in a PST model, but is dependent primarily on whether or not the farmer has enough land to diversity cash crops (Criterion 4, 5 in Figure 3.6). Thus the diversification rule is: "Diversify if you have 'enough' land to do so." How much land is enough depends on the subjective estimate of the farmer, who must take into account the size of his or her fields, their location, and the economies of scale associated with each crop. If the farmer decides there is not enough land for both cash crops, this model compares the cash crops on profitability and riskiness in Figure 3.7. However, the consumption crop corn is not treated as another cash crop, and so never is compared with the other crops in Figure 3.7. Corn in this model is therefore treated as something very different and special, in agreement with the farmers' reverence for *El Santo Maız*.

---

[9] The neoclassical economist would immediately point out that the profit-maximizing crop does use less of the farmer's more scarce resource. Therefore, Criterion 5 is implicitly included in the profitability evaluations of Criteria 1–3. Although this is true, farmers in the *Altiplano* hardly ever take out pencil and paper and *hacer la cuenta*, 'do accounts' of their profits from crops. Instead, they roughly estimate which crop is more profitable, which uses more inputs, and which is more susceptible to losses. Apart from such a rough, subjective estimate of profitability, a farmer may think of the inputs (land or labor) he has available to produce the crop. Although related, Criterion 5 can thus be considered a different criterion from Criteria 1, 2, and 4.

FIGURE 3.8: THE CASH CROP(S) AND CORN COMPETE
FOR THE SAME LAND

The consumption requirement on the right-hand branch of the tree in
Figure 3.6 asks if the farmer has enough land to plant a cash crop(s) and,
in addition, enough corn to fulfill the family's consumption requirements for the period between harvests (one year in the *Altiplano*). If the
farmer says no, the string of questions in Figure 3.8 is asked, to see if
there might be extenuating circumstances that would lead the farmer to
take some land out of corn to put into a cash crop(s); even though doing
so would mean having to buy some corn in the market for home
consumption. The decision for the farmer in Figure 3.8 is thus between
a crop mix of a cash crop(s) and corn, versus a crop mix of just corn (+
beans).

In Figure 3.8, there are four criteria that lead the farmer to grow a cash
crop and corn (+ beans), even though that would not fulfill the family's
consumption requirements for corn. The first two criteria ask if the
farmer can somehow find or "squeeze out" some land for a cash crop,
without significantly decreasing the area planted to corn. Criterion 1
says that the farmer will plant crop X and corn if it is possible to
multicrop or interplant cash crop X and corn, in such a way that yields
of corn per *cuerda* do not diminish substantially. ICTA's experiments
with multiple cropping in the *Altiplano*, called *surcos dobles* or *relevos*,
have shown that this is possible (Hildebrand, *et al.* 1977:6–25). Criterion 2 says that the farmer will plant crop X and corn if land can be
rented for the cash crop, and owned land is devoted to corn. This
constraint is really a subdecision: A farmer will rent land if crucial
inputs are available: land to rent; time to search for the owner before
planting time; and capital to pay for the land; and other inputs, such as
seed, fertilizer, and labor. In addition, the farmer must think that
renting the land would prove profitable.

Criterion 3 asks if there are agroclimatic, socioeconomic conditions
that limit the production of corn to only a portion of the farmer's land.
For example, the farmer might think that corn does not produce well on
all the land. Alternatively, corn may be planted only on the fields
around the house, to discourage the theft of green corn in the field—by
people and birds. On the other hand, rotation of corn with a cash crop
(e.g., wheat) might be thought to increase production of corn. For any
of these reasons, Criterion 3 predicts that the farmer will plant a cash
crop in addition to corn.

Criterion 4 is the simplest reason why a farmer would plant a cash
crop in addition to corn (+ beans): farmers in a cash economy need
cash. Criterion 4 simply asks if a farmer plants a crop to have cash in
case of need. It is expected that farmers who have off-farm labor and can

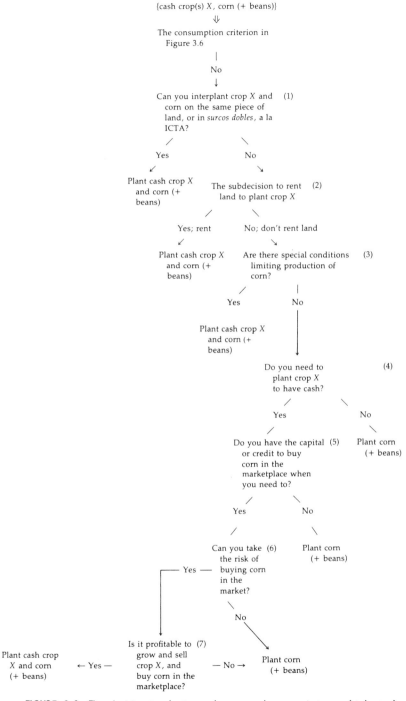

{cash crop(s) X, corn (+ beans)}
⇓

The consumption criterion in
    Figure 3.6

No

Can you interplant crop X and   (1)
    corn on the same piece of
    land, or in *surcos dobles*, a la
    ICTA?

Yes   No

Plant cash crop X
and corn (+   The subdecision to rent   (2)
beans)   land to plant crop X

Yes; rent   No; don't rent land

Plant cash crop X   Are there special conditions   (3)
and corn (+   limiting production of
beans)   corn?

Yes   No

Plant cash crop X
and corn (+
beans)

                          Do you need to   (4)
                          plant crop X
                          to have cash?

                          Yes   No

                          Do you have the capital (5)   Plant corn
                          or credit to buy   (+ beans)
                          corn in the
                          marketplace when
                          you need to?

                          Yes   No

                          Can you take (6)   Plant corn
Yes ——   the risk of   (+ beans)
                          buying corn
                          in the
                          market?

                          No

Plant cash crop   Is it profitable to (7)
X and corn   ← Yes —   grow and sell   — No →   Plant corn
(+ beans)   crop X, and   (+ beans)
                          buy corn in the
                          marketplace?

FIGURE 3.8. The decision to plant a cash crop and corn or just corn (+ beans).

get cash from other sources would say no to this criterion. Likewise, farmers who grow a cash crop only *por el gasto,* 'for consumption needs' would answer no to Criterion 4.

Although most farmers in the *Altiplano* are in constant need of cash, so that Criterion 4 is *almost* a truism, there are three additional constraints acting to *dis*courage them from growing a cash crop and *en*courage them to plant all the land to corn (+ beans), to *try* to fulfill the family's consumption requirements. The first constraint, Criterion 5, proposes that a farmer plant just corn (+ beans) if the capital is not always available to buy corn in the market place *cuando no alcanza,* when 'the family runs out of corn'. For farmers with severe capital constraints, planting and storing a year's supply of corn is insurance against a later shortage of capital. The second constraint, Criterion 6, hypothesizes that a farmer plant only corn (and no cash crop) if it is too risky to buy corn in the market. Due to the lack of a price-support system for corn, there is considerable fluctuation in the price of corn in the market. (Wheat, as opposed to corn, has a fixed price.) Thus farmers who do not secure their consumption needs for corn (on average, 2500 lb/year for a family of six) at harvest time may later find that others besides themselves are demanding corn in the market, raising its price. Moreover, during a serious corn shortage, farmers may not find any corn to buy. Thus farmers who remember previous shortages of corn do not want to risk another such episode, and so plant the land they have to just corn (+ beans).

The third constraint, Criterion 6, proposes that a farmer must think it is worthwhile to plant and sell a cash crop, and then buy corn in the market, before some of his corn land will be planted to a cash crop. Clearly, selling a cash crop and buying corn is 'worth the trouble' (*vale la pena*) or profitable if the farmer earns 50–75% more with the cash crop. If only 30% more with the cash crop can be earned it may not be worthwhile. In the latter case, the model predicts that the farmer will plant just corn (+ beans).

In conclusion, factors such as the availability of multiple-cropping technology, the availability of rentable land, the presence of agro-socioeconomic conditions that limit the area planted to corn, and the need for a cash crop in a cash economy—all act to encourage the farmer to plant a cash crop, even though by doing so, enough corn cannot be produced to fulfill the family's consumption requirements for corn. On the other hand, factors such as lack of capital, the risk inherent in trusting family consumption to the vagaries of the market place, and the perceived nonprofitability of growing and selling some kinds of cash crops (e.g., wheat) act to discourage the farmer from planting a cash crop on corn land.

This model, presented in Figures 3.1, 3.6, 3.7, and 3.8, predicted 90%

of the crop-mix decisions made by 130 farmers in different subregions of the *Altiplano*. For a more detailed summary of the results, the reader is referred to C. Gladwin 1980.

Conclusion

A two-stage theory of choice, incorporating some of the simplifying procedures people use in making everyday, real-life decisions, has been presented. It was hypothesized that decision makers treat an alternative as a set of discrete aspects. In the first stage of the choice process, they eliminate, often preattentively, all alternatives containing some aspect they do not want (H. Gladwin and Murtaugh, Chapter 5, this volume; Tversky 1972). In the second, hard-core stage of the decision process, they eliminate irrelevant aspects, order the alternatives on one important aspect, and pass the ordered alternatives through unordered constraints. Stage 2 of the theory is thus an algebraic version of maximization subject to constraints, and may be represented by an algorithm, decision tree or table, or set of decision rules.

As examples of the theory, models of several agricultural decisions made by farmers in the Third World and the United States have been presented. Decisions modeled include: farmers' choice of crops in the *Altiplano* of Guatemala; farmers' decisions of the type or kind of chemical fertilizer to apply in Lauderdale County, Alabama, and the *Altiplano* of Guatemala; marketing decisions of fish sellers in Cape Coast, Ghana; and adoption decisions of farmers in Puebla, Mexico, and the *Altiplano*. The predictability of these hierarchical models has been surprisingly high: The models tested predict 85–95% of the actual choice data used to test the model. The ability of decision trees to predict actual decisions is only remarkable, however, because most studies of agricultural decision making do not *test* the predictability of the model against actual choices of individuals (see Anderson, 1974; Benito 1976; Moscardi and deJanvry 1977). Thus, although simple in appearance, decision-tree models have shown great operational potential in predicting the actual choices made by individual farmers.

In addition to theoretical simplicity and descriptive accuracy, decision-tree models also have practical significance for agricultural development planners. By pinpointing the main constraints or factors limiting farmers' choices (e.g., factors limiting adoption of new technology), decision-tree models can make policy recommendations. Decision-tree models can answer questions such as: (a) Why the farmers did not adopt the recommended new technology; (b) as land size decreases, will farmers take some land out of the consumption crop and switch to higher-valued cash crops? Why? In which instances?; and (c) As labor costs increase, will farmers change the kind of chemical fertilizer they apply? By showing policy planners *how* farmers are making their agricultural decisions, decision-tree models can recommend effective policy changes, such as: (a) improvement of the supply of urea to consumer cooperatives in the region; (b) to allow farmers to diversify their crop mix, improve yields of the consumption crop in the region; and (c) if the fertilizer is row-applied, farmers will not change the kind of fertilizer they apply—they still need extension help with that kind of fertilizer. In short, because decision-tree models *predict*, policy planners can have some confidence that they know what is happening at the farm level, and can be effective at the farm level.

Acknowledgements

This chapter was initially part of the author's doctoral dissertation at the Food Research Institute, Stanford University. I am therefore grateful to advisors and colleagues at Stanford, especially Carl Zulauf. The agricultural examples were developed during fieldwork in Puebla, Mexico, during 1973–1974; in Lauderdale County, Alabama, in 1977; and in the *Altiplano* of Western Guatemala in 1977–1979. The latter research was supported by a Rockefeller postdoctoral fellowship and the International Fertilizer Development Center (IFDC) at Muscle Shoals, Alabama. Invaluable help in the *Altiplano* was given by colleagues at the Institute of Agricultural Science and Technology (ICTA) in Guatemala. Naturally, the views expressed in this chapter are mine and do not necessarily reflect those of IFDC, ICTA, or the Rockefeller Foundation. I appreciate the advice and support given by these institutions and colleagues, especially M. Chinchilla, K. Byrnes, K. Davidson, P. Hildebrand, H. Orozco, and R. Ortiz. Special thanks are due to the farmers in these rural areas who patiently and graciously taught me how they made their decisions, and to Hugh Gladwin who, through endless discussions and many readings of earlier drafts, led me to many of the hypotheses contained in this theory.

References

Anderson, Jock R.
    1974   Risk Efficiency in the Interpretation of Agricultural Production of Research. Review of Marketing and Agricultural Economics 42(3):131–184.
    1979   Perspective on Models of Uncertain Decisions. *In* Risk, Uncertainty, and Agricultural Development. J. A. Roumasset, J. Boussard, and I. Singh, eds. pp. 39–62. New York: Agricultural Development Council.
Anderson, Jock R., J. L. Dillon, and J. B. Hardaker
    1977   Agricultural Decision Analysis. Ames, Iowa: Iowa State University Press.
Arrow, Kenneth J.
    1951   Social Choice and Individual Values. New York: Wiley.
Barlett, Peggy F.
    1977   The Structure of Decision Making in Paso. American Ethnologist 4(2):285–307.

Benito, Carlos A.
1976 Peasants' Response to Modernization Projects in *Minifundia* Economies. American Journal of Agricultural Economics 58(2):143–51.
Boussard, J. M., and M. Petit
1967 Representation of Farmer's Behaviour under Uncertainty with a Focus–Loss Constraint. Journal of Farm Economics 49(4):869–880.
Boyd, John, and Kenneth Wexler
1973 Trees with Structure. Journal of Mathematical Psychology 10(2):115–147.
Byrnes, Kerry
1977 The Adoption of Innovations in Agricultural Production Technology in Developing Countries: The Case of Fertilizer Usage. Draft version. International Fertilizer Development Center, Muscle Shoals, Alabama.
Cancian, Frank
1972 Change and Uncertainty in a Peasant Economy. Stanford: Stanford University Press.
Chomsky, Noam, and George Miller
1963 Introduction to the Formal Analysis of Natural Languages. *In* Handbook of Mathematical Psychology (Vol. II). R. D. Luce, R. R. Bush, and E. Galanter, eds. pp. 269–322. New York: Wiley.
Coombs, Clydes, Robyn Dawes, and Amos Tversky
1970 Mathematical Psychology. Englewood Cliffs, NJ: Prentice–Hall.
Cyert, Richard, and James March
1963 A Behavioral Theory of the Firm. Englewood Cliffs, NJ: Prentice–Hall.
Debreu, Gerard
1954 Representation of a Preference Ordering by a Numerical Function. *In* Decision Processes. R. M. Thrall, C. H. Coombs, and R. L. Davis, eds. pp. 159–165. New York: Wiley.
1960 Review of R. D. Luce, Individual Choice Behavior: A Theoretical Analysis. American Economic Review 50:186–188.
Degerman, Richard
1972 The Geometric Representation of Some Simple Structures. *In* Multidimensional Scaling (Vol. I). R. N. Shepard, A. K. Romney, and S. B. Nerlove, eds. pp. 194–209. New York: Seminar Press.
Dillon, John L.
1971 An Expository Review of Bernoullian Decision Theory in Agriculture: Is Utility Futility? Review of Marketing and Agricultural Economics 39(1):3–80.
Dillon, John, and Pasquale L. Scandizzo
1978 Measurement of Altitudes Toward Risk. American Journal of Agricultural Economics 60(3):425–435.
Gladwin, Christina
1975 A Model of the Supply of Smoked Fish from Cape Coast to Kumasi. *In* Formal Methods in Economic Anthropology. S. Plattner, ed. pp. 77–127. A special publication of the American Anthropological Association, No. 4.
1976 A View of the Plan Puebla: An Application of Hierarchical Decision Models. American Journal of Agricultural Economics 58(5):881–887.
1977 A Model of Farmers' Decisions to Adopt the Recommendations of Plan Puebla. Doctoral dissertation, Stanford University.
1979a Cognitive Strategies and Adoption Decisions: A Case Study of Nonadoption of an Agronomic Recommendation. Economic Development and Cultural Change. 28(1):155–173.
1979b Production Functions and Decision Models: Complementary Models. American Ethnologist 6(4):653–674.

1980   The Future of Corn Production in the Altiplano of Western Guatemala: How Do the Farmers Decide? Guatemala: Informe del Instituto de Ciencia y Technologia Agricolas (ICTA).

Gladwin, Hugh
1971   Decision Making in the Cape Coast (Fante) Fishing and Fish Marketing System. Doctoral dissertation, Stanford University.
1975   Looking for an Aggregate Additive Model in Data from a Hierarchical Decision Process. *In* Formal Methods in Economic Anthropology. S. Plattner, ed. pp. 159–196. A special publication of the American Anthropological Association, No. 4.

Gladwin, Hugh, and Michael Murtaugh
1975   Understanding Understanding Decision Making. Paper given at Mathematics in the Social Sciences Board (MSSB) Conference on Standardization and Measurement in Anthropology, Davis, California.

Hall, R., and C. Hitch
1951   Price Theory and Business Behavior. *In* Oxford Studies in the Price Mechanism. T. Wilson, and P. Andrews, eds. pp. 107–138. Oxford: Oxford University Press.

Hammond, K. R.
1974   Human Judgment and Social Policy. Prog. Res. Hum. Judgment Soc. Interaction Report 170. Boulder: Institute of Behavioral Science, University of Colorado.

Henderson, James, and Richard Quandt
1971   Microeconomic Theory: A Mathematical Approach. New York: McGraw–Hill (2nd ed.).

Hildebrand, Peter E.
1976   Multiple Cropping Systems are Dollars and 'Sense' Agronomy. *In* Multiple Cropping. Special Publication #27. Madison: American Society of Agronomy: pp. 347–371.

Hildebrand, Peter, Sergio Ruano, Teodoro Lopez Yos, Esaú Samayoa, and Rolando Duarte
1977   Sistémas de Cultivos para los Agricultores Tradicionales del Occidente de Chimaltenango. Guatemala: Informe del Instituto de Ciencia y Tecnología Agrícolas.

Johnson, Allen
1971   Security and Risk-Taking among Poor Peasants. *In* Studies in Economic Anthropology, G. Dalton, ed. pp. 143–150. American Anthropological Association Monograph 7.

Kahneman, D., and A. Tversky
1972   Subjective Probability: A Judgement of Representativeness. Cognitive Psychology 3:430–454.

Lancaster, Kevin
1966   A New Approach to Consumer Theory. Journal of Political Economy 74:132–157.
1971   Consumer Demand: A New Approach. New York: Columbia University Press.

Luce, R. D.
1956   Semi-Orders and a Theory of Utility Discrimination. Econometrica 24:178–191.
1959   Individual Choice Behavior: A Theoretical Analysis. New York: Wiley.

Markowitz, Harry
1959   Portfolio Selection. New York: Wiley.

Miller, George
1956   The Magic Number Seven, Plus or Minus Two: Some Limits on Our Capacity for Processing Information. Psychological Review 63:81–97.

Miller, George, Eugene Galanter, and Karl Pribram
1960 Plans and the Structure of Behavior. Holt, Rinehart and Winston.
Moscardi, Edgardo R.
1979 Methodology to Study Attitudes Towards Risk: The Puebla Project. *In* Economics and the Design of Small-Farmer Technology. A. Valdes, G. Scobie, and J. Dillon, eds. pp. 30–42 Ames: Iowa State University Press.
Moscardi, Edgardo, and Alain de Janvry
1977 Attitudes Toward Risk Among Peasants: An Econometric Approach. American Journal of Agricultural Economics 59(4):710–716.
Murtaugh, Michael, and Hugh Gladwin
1979 A Hierarchical Decision-Process Model for Forecasting Auto-Type Choice. Manuscript, School of Social Sciences, University of California at Irvine.
Newell, Allen, J. C. Shaw, and Herbert A. Simon
1958 Elements of a Theory of Human Problem Solving. Psychological Review 65:151–166.
Ortiz, Sutti de
1979 The Effect of Risk Aversion Strategies on Subsistence and Cash Crop Decisions. *In* Risk, Uncertainty, and Agricultural Development. J. Roumasset, J. Boussard, and I. Singh, eds. New York: Agricultural Development Council.
Quinn, Naomi
1971 Simplifying Procedures in Natural Decision-Making. Paper presented at the Mathematical Social Science Board Seminar in Natural Decision Making Behavior, Palo Alto, California, November 22–25.
1975 A Natural System Used in Mfantse Litigation Settlement. American Ethnologist 3:331–351.
1978 Do Mfantse Fish Sellers Estimate Probabilities in Their Heads? American Ethnologist, 5(2):206–226.
Raiffa, Howard
1968 Decision Analysis. Reading, MA: Addison–Wesley.
Restle, F.
1961 Psychology of Judgement and Choice. New York: Wiley.
Roumasset, James
1976 Rice and Risk: Decision-Making among Low Income Farmers. Amsterdam: North Holland.
Roumasset, James, Jean–Marc Boussard, and Inderjit Singh
1979 Risk, Uncertainty, and Agricultural Development. New York: Agricultural Development Council.
Sellers, Stephen
1977 The Relationship between Land Tenure and Agricultural Production in Tucurrique, Costa Rica. Draft paper. CATIE, Turrialba, Costa Rica.
Simon, Herbert A.
1959 Models of Man, Social and Rational. New York: Wiley.
Slovic, Paul, Baruch Fischhoff, and Sarah Lichtenstein
1977 Behavioral Decision Theory. Annual Review of Psychology 28:1–39.
Tversky, Amos
1972 Elimination by Aspects: A Theory of Choice. Psychological Review 79(4):281–299.
Wagner, Harvey
1969 Principles of Operations Research. Englewood Cliffs, NJ: Prentice–Hall.

Chapter 4

The Statistical Behavior Approach: The Choice between Wage Labor and Cash Cropping in Rural Belize[1]

MICHAEL CHIBNIK

Over a decade ago, Salisbury (1968) reported that the major issue in economic anthropology at that time was to determine "to what extent different formal calculations of rationality or 'economizing' can be isolated in non-Western societies [p. 478]." In recent years, an increasing number of anthropologists have come to question the usefulness of this approach to the analysis of economic behavior. Convinced that real-life choices seldom approximate optimality, they have been studying how individuals actually made economic decisions. Sophisticated ethnoscientific methods have been used to isolate what people consider when deciding where to market fish (C. Gladwin 1975; H. Gladwin 1971; Gladwin and Gladwin 1971; Quinn 1978), where to reside (Geoghegan 1970) and where to plant different crops (Johnson 1974).

There is, however, a quite different way of studying economic decision making that requires neither assumptions about the complete rationality of economic actors nor extensive elicitation of rules governing choice. Statistical analyses can be made of the relationship between observed characteristics of economic actors and the choices they make. Analyses of this type often have been carried out by sociologists, geographers, and political scientists, but only occasionally by anthropologists.

---

[1] This chapter first appeared in *American Ethnologist* 7(1) February 1980 and is copyrighted by the American Anthropological Association.

AGRICULTURAL DECISION MAKING:
Anthropological Contributions to
Rural Development

ISBN 0-12-078880-2

In this chapter, I discuss and illustrate the advantages of such a "statistical behavior" approach to the investigation of economic decision making. After comparing various methods that have been used to study decisions, I describe how adult male residents of two Belizean villages allocate their labor between cash cropping and wage labor.[2] I argue that statistical analysis provides insights about this choice that cannot be discovered using a natural decision-making approach.

Studying Economic Choice

The principle of rationality, one of the cornerstones of economic theory, states that human beings will, given enough information, seek to maximize their gains by obtaining the highest possible return for any given resource or else will seek to economize (minimize) using the smallest quantity of a resource to obtain a given return (Cohen 1967:104). Anthropologists have been interested for many years in determining the degree to which tribal and peasant people are rational in their decision making (Cook 1973:842) and continue to be concerned with this question today (e.g., Ortiz 1973; Schneider 1974). The concepts of "rationality," "maximization," and "minimization" commonly are used in contemporary anthropological discussions of economic behavior, for example,

> The rational selection of garden sites in different environmental zones is linear because—assuming that households maximize labor efficiency—cultivators will travel the least distance between gardens in different environmental zones if households make garden site choices in a straight line perpendicular to the river [Rutz 1977:163–164]

> Such an attitude obviously serves as a social justification for choosing a strategy for maximizing income per unit of time rather than income per unit of land [Williams 1977:77].

> Predation is bound to have taken a heavy toll of hunters, perhaps selecting for a feeding strategy that minimizes the amount of time spent in hunting [Hardesty 1977:61–62].

Despite the frequent use of the concepts of rationality, maximizing, and minimizing, there are difficulties associated with their application to actual situations. As Burling (1962:817) pointed out more than 15

[2] This chapter is based on fieldwork carried out in the Stann Creek District of Belize in 1971–1972. This research was supported by a National Institute of General Medical Sciences grant administered by the Columbia University Training Program in Ecological Anthropology.

years ago, if we state that people act to maximize something very broad (e.g., "satisfactions") we say very little, but if we say people act to maximize one particular goal such as cash income, power, or prestige, usually we are wrong. Others (e.g., Plattner 1974; Simon 1955) have noted the limited information-processing abilities of human beings. They argue that even if people wish to maximize (or minimize) one particular goal and know all the relevant data, they will usually still be unable to do so. Furthermore, studies of Western industrial firms (Ansoff 1969:12; Green and D'Aiuto 1977; Simon 1976:3; Winter 1971:240) and contemporary North American commercial farmers (Dillon and Anderson 1971; Lin, Dean, and Moore 1974; Officer and Halter 1968) indicate that even in contexts where they might be expected to be most useful, profit maximization models often do not adequately describe actual behavior.

Much contemporary economic and psychological thought on choice emphasizes the multiplicity of goals of decision makers. For example, Simon points out:

> In the decision-making situations of real life, a course of action, to be acceptable must satisfy a whole set of requirements, or constraints. Sometimes one of the requirements is singled out and referred to as the goal of the action. But the choice of one of the constraints from many is to a large extent arbitrary. For many purposes it is more meaningful to refer to the whole set of requirements as the (complex) goal of the actor [1976:262].

In situations with multiple objectives, maximization approaches often seem inappropriate even though there exist methods such as linear programing and lexicographic theory (see Lin, Dean, and Moore 1974:503; Tversky 1969) that assume that decision makers try to maximize (or minimize) one goal subject to "satisfactory" levels of other goals. In such circumstances the "other" goals usually become so important that Simon (1976:262) remarks that "if you allow me to determine the constraints, I don't care who selects the optimization criteria." More sharply, White (1976:36) has expressed the similar thought that a "theory of how people make their economic choices is without interest and probably impossible until we have tackled the prior question of the factors determining what choices are available to them."

Dissatisfaction with the maximization approach to economic behavior has led several anthropologists to study natural decision making:

> This approach is inspired by Simon's early discussion of "bounded rationality" and subsequent work of cognitive psychologists who, with Simon, have rejected normative assumptions that real-life choices match or approach optimality. Instead, these scholars pursue the processes by which individuals actually make decisions [Quinn 1978:206].

The natural decision-making approach to the analysis of choice emphasizes determining, via ethnoscientific interviewing, the rules people use when making decisions. Students of natural decision making vary in the extent to which they examine the correspondence between the models they construct and observed behavior. The Gladwins (1971) construct a complex model predicting fish-sellers' behavior but devote only one paragraph (p. 137) to noting and explaining why their model does *not* work. In a later article, however, C. Gladwin (1975) gives considerably more space to a consideration of how well her model predicts actual behavior. Quinn, although stating that "actual decision outcomes provide crucial verification of a model based on verbal reports," devotes only one sentence (1978:222) to a comparison of her model of fish selling and behavior. Johnson (1974; 1978:179–183), in contrast, goes into considerable detail on the correspondence between rules and behavior about land use in northeastern Brazil.

Natural decision analyses do not seem particularly useful when choice makers have difficulty describing the factors influencing their behavior. As Pelto and Pelto have noted, day-to-day economic activity is one domain of decision making in which jural rules are often few and far between and people act "in terms of the varying efficacies of complex interrelated social, economic, physical, and psychological constraints [1975:11]." In such situations, individuals may have incomplete knowledge of the nature and effects of the relevant constraints. Since decision makers act from particular times and places, they may not know very well what they would do in different circumstances.

A Belizean farmer, for example, often will say that he would grow more rice if he could market the crop more easily. However, a farmer has difficulty responding when asked how much more rice he would grow if he lived in a neighboring village where there is better access to a market. Similarly, a Belizean farmer is likely to know that his age and the size of his household affect the amount of rice he plants, but can compare only impressionistically the relative importance of the two variables. He cannot sensibly reply to inquiries designed to determine how much rice he would plant if he were ten years older and had four more people in his household.

Statistical analyses of decisions are useful in situations where it is difficult to elicit "rules of choice." These analyses provide numerical indices of the extent to which different observed characteristics of decision makers influence particular choices. Statistical analyses do not ignore the verbal reports of informants since the variables examined in correlations are often chosen as a result of information gathered in interviews or casual conversations. Other variables, however, can be selected for theoretical reasons or as a result of nonverbal ethnographic observations.

The distinction between natural decision making and statistical behavior approaches can be seen by examining a hypothetical situation in which farmers have a choice of two towns where produce can be marketed. An anthropologist using the natural decision-making approach would devise an elaborate interview schedule intended to determine the conditions under which a farmer says he will market crops in town A rather than in town B, construct a model of economic choice based on verbal data, and only then compare the model with actual behavior. An anthropologist using the statistical behavior approach would record which farmers market in town A rather than in town B and then attempt to construct a statistical profile of people who market in town A (see Keesing 1967 for further elaboration).

Surprisingly few statistical analyses of economic behavior appear in anthropological reports. In the past several years there have been an increasing number of studies focusing on intracultural and intracommunity diversity, and a recent issue of the *American Ethnologist* (February, 1975) was devoted entirely to this topic. Of the ten articles in this volume, however, only one (DeWalt 1975 on differential adoption of new technology in rural Mexico) is concerned directly with economic choice; and despite exceptions (e.g., Barlett 1977; Cancian 1972; Cook 1970; Durrenberger 1976; Shapiro 1975), the great majority of economic anthropological studies continue to stress cultural homogeneity rather than diversity (see Johnson 1972; Pelto and Pelto 1975 for extended discussions).

In the remainder of this chapter, I attempt to demonstrate the advantages of a statistical behavior approach to economic decision making by analyzing choices adult male residents of two Belizean villages make between own-account cash cropping and engaging in wage labor for others. This analysis is of practical as well as theoretical interest because, despite the intensive efforts of the Belizean government to stimulate small-scale agriculture, rural residents have often preferred to work at wage labor in citrus estates, in sawmills, and on government-funded projects.

Economic Conditions in Rural Belize

Since the end of World War II, the expansion of agriculture has played an important role in development plans for Belize (formerly British Honduras). A decline in the lumber industry, traditionally the mainstay of the economy, has led to an unfavorable balance of trade, and until recently domestic shortages have forced the Belizean government to import rice and red kidney beans, the staples of much of the population (for details, see Ashcraft 1973). The extremely favorable human-cultivable land ratio (Romney *et al.* 1959; Tripartite Report 1966) in Belize, combined with the government's desire to stimulate exports and ensure domestic agricultural self-sufficiency, has led to a number of development projects aimed at increasing cash cropping. The government has built roads into remote villages, provided price supports for staple crops, sold land to villagers planting permanent crops, formed cooperatives, and attempted to encourage the use of fertilizer, weed killer, and new types of seeds.

Despite these incentives, small scale farmers often have not increased their cash cropping. Discussing the period 1955–1965, Ashcraft reports: "In contrast to expectations small scale producers have not responded to guaranteed price incentives, for their contribution to government purchases has been declining rather than increasing. It seems evident that most cereal farmers continue to sell only the excess over household needs and that excess is not increasing [1973:90–91]."

The failure of agricultural development schemes in parts of Belize is sometimes attributed to the economic history and "culture" of rural residents. Many projects intended to increase agricultural productivity are aimed at "Creoles," usually monolingual English-speakers of mixed African and European descent.[3] Creoles, the largest ethnic group in the country, comprising about half of the population, are descendants of slaves brought to Belize as woodcutters.[4] Although Creole forest workers sometimes grew rice, corn, and root crops using slash-and-burn methods, such production was primarily for home use rather than for sale. With the decline of the mahogany and chicle industries in the past 75 years, rural Creoles in the center and south of Belize have been forced

---

[3] To say that Creoles speak "English" is not quite accurate. The first-learned language of most speakers is an "English-based Creole" and there exists an English-Broad Creole continuum in Belize. Speakers have at their command a span along this continuum and may shift speaking style in response to the demands of the sociolinguistic situation (Kernan, Sodergren, and French 1977).

[4] There is great confusion concerning the population of the different Belizean ethnic groups. For conflicting interpretations of the 1970 Belizean census see the October and November, 1978, issues of *Brukdown*, a Belizean magazine. The 1960 census reported that 55% of the people in Belize were Creole. I do not know the methods census takers used to place people into the Creole category. Other major ethnic groups in Belize are "Spanish" (mesitzos), Garifuna (also known as Caribs or Black Caribs), and Mayas (Kekchi, Mopan, and Yucatecan). There are smaller populations of Arabs, Chinese, Mennonites, Europeans, and East Indians.

to emphasize agriculture more. Nonetheless, many development "experts" and other outside observers feel that Creoles "regard agriculture as an inferior and transitory occupation—a resting spell from the more vigorous forestry activities [Grant 1976:15]." Development personnel sometimes appear to regard implicity the supposed Creole attitude toward agriculture as an obstacle to rural projects:

> There is evidence . . . that the organization of forest work encouraged antagonism and even bred a contempt for agriculture; the relative freedom of the forester returning to the town for his considerable rest periods contrasted with the continuous attention which cultivation involved [Settlement Report 1948:253].

Other planners seem more ambiguous about the Creole's attitude toward farming. The authors of the Land Use Survey, for example, remark:

> It is often said that the Creole is a person without a farming tradition and this is advanced as one reason why agricultural development is so slow to take hold in British Honduras. This may not be a fair comment. Of the large number of Creoles who try to make a living from the land there are few indeed who have not learned their lessons from their early mistakes and most of them are careful to try and find out the causes when things go wrong [Romney *et al.* 1959:36].

But at another point in their report, the same authors write of Creoles: "They are, perhaps, not natural farmers and do not work the land continually and in harmony with it [p. 151]."

Two themes pervade discussions by development personnel of the supposed Creole preference for wage work over agriculture. First, when Creoles are described as not being natural farmers, they usually are being compared either explicitly or implicitly with the Mayas of Central America (about 10% of the population of Belize) who plant more extensive plots and work less frequently in wage labor. Second, and more directly relevant to the analysis of economic choice, whereas the Creoles' alleged preference for wage labor is thought to have been reasonable in the past when opportunities for forestry work were abundant and incentives for cash cropping virtually nonexistent, the continuing aversion for farming is thought to be an irrational obstacle to development.

Ashcraft (1973:91), disagreeing with most of the development experts, sees the lack of Creole response to incentives to increase agricultural production as quite sensible. He cites poor transportation, high costs of transport, and cumbersome means for obtaining fixed prices from the government as impediments to cash farming. Ashcraft also notes that increasing numbers of large sugar and citrus estates have been established in Belize in recent years. He suggests that people reasonably might prefer the ready cash from labor on these commercial farms to the uncertain returns from their own agriculture.

Research Sites

In 1971–1972, I conducted research on the choice between wage labor and cash cropping in Silk Grass and Sittee River, two predominantly Creole villages in east central Belize.[5] The villages are only 10 miles apart and many of the residents of Silk Grass emigrated there from Sittee River. People in the two villages farm similar land (Romney et al. 1959:57) and have access to the same wage labor opportunities, but they have had quite different exposure to government projects aimed at stimulating cash cropping.

The Belizean government established Silk Grass in 1962 specifically to encourage agriculture. The village is located on a major road, the Southern Highway, and trucks frequently make the round trip between Silk Grass and Dangriga (formerly Stann Creek), a town of 7000 people, 15 miles away. The approximately 400 people in Silk Grass thus can get easily to town to sell their crops and purchase consumer goods. As additional incentives to cash cropping, the Marketing Board of Belize visits Silk Grass at harvest time to purchase rice, and the government has sold land to village residents to encourage the planting of trees.

The approximately 350 Sittee River residents can neither market their crops nor purchase consumer goods as easily as their Silk Grass neighbors. Their village is located 6 miles from the Southern Highway, 26 miles from Dangriga. A truck makes the Sittee River–Dangriga round trip only once a week, and the road from the village to the Southern Highway is not always passable during the rainy season of June through September. The marketing board does not visit Sittee River to purchase crops, and, as most village residents do not own the land they farm, they have little incentive to plant trees.

Despite the greater incentives for cash cropping in Silk Grass, differences in agricultural production for sale between the two villages in 1971–1972 were not great. In a sample of randomly selected adult male farmers, the mean income in Silk Grass in 1971–1972 from rice, corn,

---

[5] Silk Grass is no longer predominantly Creole. When I did my research in 1971–1972, Mayas from the south of Belize were beginning to move into Silk Grass, but had not yet begun to farm extensively. In the last few years, Mayas have continued to migrate to Silk Grass and the village can now be more accurately described as ethnically mixed.

beans, plantains, and root crops was $253 (Belizean) compared to $232 in Sittee River.[6] The difference in mean income from oranges, grape-fruits, and coconuts was more marked ($93 in Silk Grass; $47 in Sittee River) but in both villages tree crops accounted for only about 10% of the total cash income of those surveyed. Intervillage differences in wage-labor income, however, were considerably greater. In Silk Grass, the mean wage-labor income of men included in my economic survey was $439; in Sittee River it was only $211.

Creoles in Silk Grass, as in other parts of Belize, have not responded much to incentives to increase cash cropping. Indeed, their major reac-tion to governmental agricultural development projects seems to have been to increase the amount of wage labor they do. The explanation for this apparently paradoxical response, I will argue, has little to do with the alleged Creole preference for wage labor. Instead, the relevant factors seem to be the increased cost of living associated with agricul-tural development in Silk Grass and certain characteristics of wage labor and cash cropping.

The Costs of Living in the Two Villages

In 1961, Hurricane Hattie devastated coastal Belize. The government established Silk Grass in 1962, with the intention of encouraging people living near the coast to move inland where the potential danger from hurricanes was less and the opportunities for cash cropping were greater. When Silk Grass was founded, villagers were given free land, houses, electricity, and transportation to town. In the late 1960s, how-ever, villagers were made to pay for these items. Some residents left the village, but most remained. In 1971–1972 almost all of the residents of Silk Grass were still paying for their houses and land. Houses, two-room wooden cottages with separate outside kitchens, cost $400 pay-able in monthly installments of $10. Land along a creek about 5 miles from the village cost, on the average, about $90 for a 10-acre block, payable over 5 years in semiannual installments. Since electricity cost about $1 a month, in 1971, most Silk Grass households annually paid the Belizean government at least $150.

Most Sittee River residents, in contrast, paid no money to the Beli-zean government in 1971–1972. The village has no electricity and few residents must meet house payments. Almost all Sittee River farmers

---

[6] In 1971–1972, $1 Belizean equalled about 60¢ US. Income figures are not precise and should be regarded as approximations.

use land owned by a citizen of the United States who has not attempted to collect rent in recent years.[7]

The greater overall cost of living in Silk Grass and the necessity for the residents of that village to make periodic cash payments affect economic decision making. Silk Grass residents, compared to their counterparts in Sittee River, have a greater need for a source of income that is reliable and offers fairly quick returns. A careful examination of the characteristics of wage labor and cash cropping, the major income sources, shows that wage work better meets these requirements.

The Choice between Wage Labor and Cash Cropping

The principal characteristics of wage labor and cash cropping that influence economic decision making are differences between the types of work in returns to labor, time commitment, physical requirements, and pleasantness of working conditions.

Returns to Labor

Adult men do wage work in privately owned sawmills, for the Forestry and Public Works Departments of the government, on citrus estates, and (infrequently) as agricultural laborers for their neighbors. Wage labor is ordinarily migratory, forcing men to leave their villages during the week. Calculating the monetary returns for this work is relatively straightforward. In 1971–1972 most wage laborers earned $21–24 for a 5–6-day work week. From these returns should be subtracted some of the approximately 90¢ per day workers living away from home must spend on food.[8]

Calculating returns per labor-day spent at agriculture is considerably more complex. Corn, rice, and plantains are the most important crops grown by slash-and-burn methods, but cassava (manioc), red kidney beans, and coco yams are also planted. In addition, orange, grapefruit, and coconut trees are raised. Returns to agricultural labor vary according to crops (or crop combinations) planted, types of land used, and market and weather conditions. Furthermore, farmers grow most crops for home consumption as well as for sale, and calculations of returns to

---

[7] The Belizean govermment has been negotiating to buy this land in recent years. When the land is purchased, it will be sold to Sittee River residents.

[8] The figure of 90¢ per day comes from interviews of a number of men who did wage work. Exactly how much of this should be subtracted is a complex question, depending on how much money is usually spent daily to feed a man when he lives at home.

agricultural labor should include "the value of subsistence production" as well as cash income (see Chibnik 1978 for details). Moreover, as citrus trees require substantial inputs of capital as well as labor, they cannot be compared easily to other crops or wage work.

Nonetheless, calculations (see Chibnik 1975:185–199) indicate that the average (expected over the long run) monetary return per labor day for fields of rice, corn, plantains, and beans range from $2.50–$4.00 (in these calculations, crops consumed at home are valued at retail price). Thus, the average returns per unit of labor input for agriculture and wage work are quite similar. These similarities, however, mask important differences between the two income sources in risk, speed of returns to labor, and composition of work force.

Returns to agricultural labor are less certain than those from wage work since there is some risk associated with farming in the two villages. The major dangers farmers must contend with, besides uncertain markets, are high winds, flooding, birds, animals, insects, and diseases. These dangers vary in relative importance for the different crops, and farmers assert that this is one of the reasons they prefer to plant several crops rather than just one or two. Rice and corn generally are safer than plantains and beans, but offer smaller returns per unit of labor input. A farmer could, in theory, if all went well, earn $5–6 per labor day by planting plantains and beans and buying rice and corn for subsistence, but because of the great risk of such a strategy, no one does this.

The returns to labor are also much less immediate for agriculture than for wage work. A wage laborer can expect to be paid every week or two while he works. In contrast, a farmer who begins clearing fields in March cannot expect monetary returns from rice and corn until at least September, and he must wait longer to receive returns from plantains and root crops. Trees are a special case, as monetary returns come no sooner than 5 years after planting.

Finally, most farmers do not do all the work involved in growing, harvesting, and selling crops because they can mobilize some labor from their wives, children, and occasionally other relatives and friends (for details on the division of labor, see Chibnik 1975:132–135). However, only adult males can engage in most types of wage labor. One consequence of this difference is that adult males have somewhat less autonomy in making choices about farming than they do for wage work. Because women and children help out at planting and harvest, they have some say in deciding which crops are grown, where they are grown, and how large fields are.

The agricultural division of labor has some advantages for adult men.

Since an adult male receives help on some agricultural tasks from other household members, from his point of view, the returns from labor in farming (as compared to wage work) are somewhat higher than input–output calculations would suggest.

Time Commitment

If a physically able, adult male villager wants wage work enough to be willing to travel considerable distances, he ordinarily can obtain a job any time he wants and work as long as he desires. Because of climatic conditions and crop characteristics, men participating in agriculture have less choice about when they work. Fields must be cleared and crops planted and harvested at particular times of the year. A wage laborer may choose which months of the year to work in; an agriculturalist planting rice and corn must work intensively in March, April, May, September, and October.

Farming offers less flexibility than wage labor from the standpoint of labor allocation over the course of a year, but in day-to-day time allocation, farming offers somewhat more flexibility. A farmer, unlike a wage laborer, can choose when during a day to work, when to pause to rest, and when to stop working. Since farmers do not have supervisors, they can choose to a certain extent which tasks to carry out on any particular day. Moreover, an adult male farmer can try to induce or coerce other members of his household to perform certain tasks or he can pay others to work on his fields.

Physical Requirements

Few men older than 55 can obtain wage work. They are either physically unable to perform the labor required by the job or potential employers regard them as too prone to injury or illness. Most agricultural tasks, however, can be undertaken by any adult male who is not seriously ill (the major exception to this generalization is cutting large trees). Although older men may not be able to work quite as long or as hard as younger men, it was common in the villages for men in their sixties or even in their seventies to grow several acres of crops, using shifting cultivation methods.

"Pleasantness" of Working Conditions

Although I did not interview systematically a carefully selected sample of residents of Sittee River and Silk Grass on their perceptions of the comparative pleasantness of wage work and agriculture, numerous informal conversations indicated that most village men do not share the stereotyped Creole aversion to farming (the stereotype may well be more accurate among urban Creoles). Most men in the villages do not feel that the returns they get from labor are satisfactory compensation for the work they do, considering the hardships they endure and the costs of consumer goods in Belize. Men complain about the costs of transportation and food when away from home, but the major source of discontent is that men miss village life and their women. Most middle-aged men seem to view wage work as a necessary evil that must be engaged in because of the inability of a man to support his household by agriculture alone. Young men, however, sometimes say they prefer wage work to agriculture because farming requires a commitment to stay in the village at certain times of the year. Young Creole men are quite mobile and prefer to be able to work where and when they want to, and wage labor is a convenient way to satisfy this preference.[9]

To summarize the most important differences between wage labor and cash cropping: Wage work offers quicker and more certain returns to labor and requires a shorter time commitment, but offers less chance of large profits, is regarded by most men (except those in their twenties) as more unpleasant, and is usually more demanding physically.

Statistical Analysis

The information presented to this point suggests that both the village a man lives in and his age will greatly affect his allocation of labor-time between wage work and cash cropping. Because of the higher cost of living, easier access to consumer goods, and greater incentives to cash cropping in Silk Grass, men from that village might be expected (all other things being equal) to spend more time than their counterparts in Sittee River at all income producing activity, especially agriculture. The physical limitations of older men (over 55) should lead them to participate in wage labor less than other men.

The size of a man's household also might be expected to affect his economic behavior. Men with large families to support (all other things being equal) should work more than men with small households.[10]

---

[9] Historically, Creole men rarely have begun farming extensive plots until reaching their thirties. A pattern in which young men move around doing wage labor has been a useful response to fluctuating economic conditions in different parts of Belize.

[10] Data were analyzed with reference to household size rather than to "consumer–worker ratio" (cf. Chayanov 1966, Sahlins 1972) because of difficulties in specifying the

Assuming that men with a higher cost of living will be less willing to engage in high-risk activities than others (because of the more serious consequences of failure), one might also expect that men with large households would devote a higher proportion of their labor time to wage work than men with small households. (Reasoning in the same way, one also would expect—all other things being equal—that Silk Grass men would devote a higher proportion of their labor time to wage work than Sittee River men; however, all other things obviously are not equal since there are greater incentives to cash cropping in Silk Grass.)

To examine variations in work patterns in the two villages, in 1971–1972, I conducted a small economic survey in Sittee River and in Silk Grass. Since my research focused on the effects of government plans to increase cash cropping, I restricted my sample to households in which there was present at least one adult man who did some farming. From a random selection of 31 (out of 108) such households (16 in Silk Grass; 15 in Sittee River), information was collected on income sources over the course of a year.[11]

Because obtaining direct data on labor inputs over a year for many people was difficult, I assume in the analysis that follows that a good measure of the relative amount of time different men devote to a particular economic activity is their comparative cash income from that activity.[12] For example, a man earning $300 from wage work is assumed to have spent more time at wage labor than one earning $200.

Tables 4.1 and 4.2 reveal considerable differences between the economic behavior of men in Sittee River and Silk Grass and between men who head households of different sizes (the *male head of a household* was defined as the man in the household who had the highest cash income over the course of the year). As was expected, Silk Grass men averaged higher total incomes, milpa crop incomes (agriculture other than trees),

number of "workers" in particular households. Households with the same age–sex composition in rural Belize can differ sharply in the amount of labor available for (say) agricultural tasks. For example, a boy of 15 in one household may help considerably in farming, whereas his counterpart next door does very little agricultural work.

[11] Technically, my selection of households was "systematic" (i.e., every *n*th household) rather than at random. Furthermore, a few households selected had to be discarded from my analysis since financial information was clearly inaccurate. Only three (out of 34) households were rejected from the sample because of the absence of an adult male (for details on sampling and data collection methods see Chibnik 1975:248–252).

[12] Note that I am *not* assuming that a man who earns $300 from wage labor and $200 from agriculture spends more time at wage labor than agriculture. Moreover, a man who earns $300 fron agriculture is assumed to have spent more time at *cash* cropping than a man who earns $200; he is *not* assumed to have spent more time farming (the second man might have been producing more for home consumption).

TABLE 4.1
Cash Income Sources of Men in the Two Villages

	Sittee River ($N = 15$)		Silk Grass ($N = 16$)	
Income source	Mean annual income ($)	Percentage of total	Mean annual income ($)	Percentage of total
Milpa (agriculture other than trees)	232	39	253	31
Tree crops	47	8	93	11
Wage labor	211	35	439	53
Other (fowl and eggs, livestock, fishing, hunting)	111	18	38	7
Totals	601	100	823	102[a]

[a] Percentages do not always sum to 100 because of rounding off.

and tree crop incomes than men from Sittee River. Men with large households had higher mean total incomes and earned a higher percentage of their income from wage labor than did men with smaller households. As noted earlier, an important and surprising finding was that the difference in mean total income between Silk Grass and Sittee River men was composed much more of differences in wage-labor income than of differences in the extent of cash cropping.

The differences in total and wage-labor incomes between men from different villages were significant at the .05 level (using the t test), as were differences in total income between men with large and small

TABLE 4.2
Cash Income Sources of Men with Households of Different Sizes

	Household size			
	1–6 ($N = 16$)		7–15 ($N = 15$)	
Income source	Mean annual income ($)	Percentage of total	Mean annual income ($)	Percentage of total
Milpa crops	237	38	244	30
Tree crops	67	11	72	9
Wage labor	251	40	410	50
Other	65	10	93	11
Totals	620	99[a]	819	100

[a] Percentages do not always sum to 100 because of rounding off.

households. Differences in wage-labor incomes between men with large and small households were significant at the .10 level.

The six men over 55 did, as had been expected, have a considerably lower mean income from wage labor ($143) than did the other 25 men in the sample (mean wage-labor income $373). This difference is significant at the .10 level.

As men over 55 do considerably less wage labor than others, it is important to examine the age structure of household heads in the two villages. If the proportion of men over 55 were higher in Sittee River than in Silk Grass, this would explain some of the intervillage differences in income sources noted in Table 4.1. However, the age structure of male household heads included in the survey was similar in the two villages. In Sittee River there were 3 household heads under 35, 9 between 35 and 55, and 3 over 55; in Silk Grass there was 1 household head under 35, 12 between 35 and 55, and 3 over 55. In Table 4.3, an intervillage comparison of income sources is made for only those household heads between 35 and 55. The considerable intervillage differences in wage labor and total income noted in Table 4.1 (for household heads of all ages) can also be observed in Table 4.3.

Because of the developmental cycle of domestic groups, adult men under 35 and over 55 head smaller households than their counterparts between 35 and 55 (see Table 4.4). Consequently, the effects of household size noted in Table 4.2 might be a result of the effects of the differential distribution of ages of men with large and small households. In Table 4.5, age is controlled for by restricting the sample to men

TABLE 4.3
Cash Income Sources of Men between 35 and 55 in the Two Villages

Income source	Sittee River (N = 9)		Silk Grass (N = 12)	
	Mean annual income ($)	Percentage of total	Mean annual income ($)	Percentage of total
Milpa (agriculture other than trees)	230	35	316	35
Tree crops	22	3	35	4
Wage labor	276	42	497	56
Other (fowl and eggs, livestock, fishing, hunting)	122	19	44	5
Totals	650	99[a]	892	100

[a] Percentages do not always sum to 100 because of rounding off.

TABLE 4.4
Relationship between Age of Male Household Head and Size of Household

	Household size			
Age	1–6	7–15	Totals	
Under 35 or over 55	9	1	10	$\chi^2 = 8.52$ significant at .01 level.
35–55	7	14	21	
Totals	16	15	31	

between 35 and 55. With a sample controlled for age, differences in work patterns of men with large and small households are quite small.

Tables 4.6 and 4.7 numerically express the relationship between the village a man lives in, the size of his household, and his choice of income sources. These tables use a measure of association, gamma, that can range between −1 and +1. In the tables, the closer a correlation coefficient is to +1, the greater the relationship between a particular income source and (depending on the row) residence in Silk Grass or having a large household (seven or more people). Table 4.6 more precisely expresses the results of Tables 4.1 and 4.2, whereas Table 4.7 does the same for Tables 4.3 and 4.5.

To summarize the most important findings of the statistical analysis:

1. The village a man lives in greatly influences the total amount of cash income he earns.
2. Despite the equal wage-labor opportunities in the two villages, and the greater incentives for cash cropping in Silk Grass, the

TABLE 4.5
Cash Income Sources of Men between 35 and 55 with Households of Different Sizes

	Household size			
	1–6 (N = 7)		7–15 (N = 14)	
Income source	Mean annual income ($)	Percentage of total	Mean annual income ($)	Percentage of total
Milpa crops	226	29	262	33
Tree crops	42	5	22	3
Wage labor	353	46	418	53
Other	154	20	93	12
Totals	775	100	795	101[a]

[a] Percentages do not always sum to 100 because of rounding off.

TABLE 4.6
Strength of Association (Gamma) between Village, Household Size, and Sources of Cash Income[a]

	Village (Silk Grass, Sittee River)	Household size (7 or more, 6 or less)
Milpa crops	+.19	+.19
Milpa crops plus tree crops	+.16	+.31
Wage labor	+.43*	+.33
Total income	+.63***	+.54**

[a] For an explanation of the statistic gamma and the methods used in its calculation here, see the appendix.
* Significant at .10 level.
** Significant at .05 level.
*** Significant at .01 level.

village a man lives in correlates more highly with wage-labor income than with cash-crop income.
3. If age is controlled for, the size of a man's household is virtually unrelated to the amount of wage work and cash cropping he does.
4. A man's age strongly affects the amount of wage labor he does.

Ethnographic Interpretation

For ethnographic purposes, neither village nor age are entirely satisfactory variables, since neither can be regarded as "single-dimensional." Villages vary with respect to access to market, cost of living, and land tenure, whereas men of different ages differ in their physical condition and willingness to stay in the village. For this reason, the statistical observation that, for example, Silk Grass men earn significantly more cash income than Sittee River men is incomplete. We

TABLE 4.7
Strength of Association (gamma) between Village, Household Size, and Sources of Cash Income for Sample Restricted to Men between 35 and 55

	Village (Silk Grass, Sittee River)	Household size (7 or more, 6 or less)
Milpa crops	+.25	.00
Milpa crops plus tree crops	+.09	+.06
Wage labor	+.33	+.13
Total income	+.53*	.00

* .1 > p > .05.

cannot know if the most relevant factor is access to market, cost of living, or some other, unmentioned difference. Unfortunately, it is very difficult (perhaps impossible) to quantify many of the separate dimensions of village and age (e.g., access to market, physical condition, willingness to commit oneself to the village). In contrast, it is not difficult, as has been demonstrated, to provide numerical indices of the strength of the relationship between village, age, and various economic activities. Thus, the interpretation of what the statistics mean (what are the "real" causes of economic variation in the villages) must in the final analysis be based on qualitative ethnographic observation as well as quantitative number crunching.

To be more specific, the principal question this chapter addresses is the effects of incentives to cash cropping in Silk Grass. A combination of statistical and ethnographic evidence strongly suggests that whatever small differences there may be in extent of cash cropping between Silk Grass and Sittee River should be attributed primarily to the higher cost of living in Silk Grass, rather than to the direct effects of agricultural development projects in that village. *Statistical* analysis reveals that intervillage differences in wage-labor incomes are considerably greater than intervillage differences in cash cropping. I explain these intervillage wage-labor differences with reference to the higher cost of living in Silk Grass because of my *ethnographic* observations that (a) wage-labor opportunities are the same in the two villages (see Chibnik 1975:99–104), and (b) no intervillage difference other than cost of living "seems" very relevant to allocations of labor time between wage work and other economic activities, including "leisure" (see Chibnik 1975:156–161). If their higher cost of living compels Silk Grass men to do more wage work than Sittee River men, it seems reasonable to assume that the same consideration would compel Silk Grass men to cash crop more.

Since villagers seldom express the reputed Creole distaste for agriculture, the question remains as to why men with high costs of living prefer wage labor to agriculture as a source of income. Interviews with Silk Grass men were not too informative on this point since villagers, for the most part, are not very aware that such a preference exists. Men said that the amount of time they spent on different economic activities varied considerably from year to year. Perhaps for this reason, villagers saw themselves as doing "a little of this and a little of that" to meet their cash needs and did not emphasize the relative proportion of time they spent on farming and wage labor.

I initially hypothesized that the labor demands of planting and harvesting precluded much expansion of cash cropping, forcing villagers needing money to engage in wage labor during slack agricultural periods. However, detailed analysis of inputs and outputs associated with agriculture (see Chibnik 1975:204–216) shows that villagers could increase greatly the amount of cash cropping they do, and even a cursory examination of the much larger plots planted by the Mayas of Belize supports this.[13]

What appears to be more relevant is the greater speed of returns to labor and more flexibility in disposal of labor time offered by wage work. Because of the seasonality of agriculture, a villager faced with specific cash needs such as house or land payments is more likely to be able to raise money quickly through wage work than by farming. The comparative risk of the two activities is also pertinent since, as has been mentioned, a villager with high cash needs might prefer the surer returns from wage labor to the less certain returns from agriculture. This last idea, interestingly, was mentioned only occasionally by villagers asked about the relative merits of wage work and farming.

Practical Implications

The findings presented here have implications for the agricultural policy of the Belizean government. Reports written by outside experts have varied greatly in their recommendations about the relative emphasis that should be placed on large-scale capital-intensive agriculture and small-scale peasant production. Some reports (Dumont 1963; Tripartite Report 1966) have argued that "the correct policy is taking advantage of British Honduras' extremely favorable man-cultivable land ratio and to specialize in the production of capital intensive crops which . . . depends in practice on private foreign investment in estate agriculture [Tripartite Report 1966:6]." Others have argued (Romney *et al.* 1959:51) that "peasant production should be the basis on which most of the agricultural exports are produced." The official response of the Belizean government has been to attempt to steer a middle ground by encouraging "the growth of small efficient farms side by side with large producing units [Comment on Tripartite Report 1966:3]."

[13] Since, historically, Mayas and Creoles have lived in different places in Belize, it is difficult to compare directly Maya and Creole economic patterns. However, the Mayas living in Silk Grass appear to plant larger plots than their Creole neighbors. Whether this pattern will continue, given Maya opportunities for wage work, is an interesting question. One relevant factor is that Maya farmers appear able to mobilize more labor from relatives than Creoles.

We have seen, however, that development projects aimed at small farmers in Silk Grass have stimulated wage labor more than cash cropping. Since some of this wage labor is on large capital-intensive citrus estates, it appears that government projects intended to stimulate peasant production have had the inadvertent consequence of providing cheap labor for foreign-owned, capital-intensive agricultural concerns (in the area where I did my research, the owners of citrus estates were Canadians and Jamaicans). Wage labor in all of Belize is performed increasingly on commercial sugar and citrus estates rather than in forestry. If other agricultural development projects in Belize are associated with higher costs of living, the government's developmental program may favor large farmers more than is indicated by official policy.

The preceding remarks should not be interpreted as stating that the agricultural-development projects of the Belizean government offer no incentives to Silk Grass farmers. Clearly, people from various parts of Belize have been willing to move into and stay in Silk Grass despite the cash investments in houses and land they must make. In the future, when most people own their houses and some land, Silk Grass residents may expand their cash cropping and reduce the amount of wage labor they do. In 1971–1972, however, the advantages of wage labor as a source of ready, reliable cash income seemed (as evidently they have in the past) to outweigh incentives to farming offered by the Belizean government.

The co-occurrence of development projects and a higher cost of living is not unique to either the Stann Creek District or to Belize. Agricultural-development programs commonly involve increasing the cash needs of farmers through loans that must be repaid and sale of fertilizer, seeds, weed killer, machinery, and land. Perhaps in many parts of the world increases in cash cropping attributed to developmental projects can be explained partially by the higher cash needs incurred by farmers induced or coerced to participate in projects.

Advantages and Limitations of Statistical Analysis

Although rural Belizeans are quite aware that a man's village, age, and size of household influence choices he makes between wage labor and cash cropping, they cannot make many statements about the relative importance and interrelationship of these variables. Men therefore find it impossible (or at least extremely difficult) to state rules specifying when they do wage work and when they remain at home and farm. For this reason a natural decision making analysis of the choice between wage work and cash cropping in rural Belize is of limited use.

Using a statistical behavior approach, however, I have shown that one observed characteristic of an economic actor, a man's village, predicts the amount of money he earns from wage labor and agriculture better than another observed characteristic, the size of his household, which a priori seemed equally relevant. Statistical analyses also strongly suggest that a man's age affects how he allocates his labor time between wage work and cash cropping more than does the size of his household.

Elsewhere (Chibnik 1975:150–184), I have used statistics to show that particular observed characteristics of economic actors are more relevant to some choices than to others. For example, in a comparison of Sittee River, Silk Grass, and a third village, Hopkins, I found that:

1. The size of a man's household predicts the amount of rice he plants as well as his village and better than his age. (A statement of this sort, of course, depends greatly on how household size and age were operationalized).
2. A man's age predicts the number of trees he plants as well as his village and better than the size of his household.
3. A man's village predicts the area he devotes to plantains better than either his age or the size of his household.

Patterns of this type are often not immediately obvious and can easily be missed if an ethnographer relies only on unsystematic observation and elicitation of rules of economic behavior.

Informants' unawareness of statistical regularities is not restricted to rural Belize. The average citizen of the United States, for example, would have great difficulty presenting a learned exposition on the distribution and causes of voting patterns. An economic anthropological analysis emphasizing actors' rules of decision making is, I would argue, as incomplete as an analysis of American elections based on individuals' rules of voting. In either case, a statistical analysis of behavior is also necessary.

Quantitative analyses in isolation from qualitative ethnographic data and anthropological theory are not particularly informative. Many anthropologists, noting this, have felt that statistical analyses carried out by sociologists and economists display insufficient concern for cultural context. To explain statistically significant differences between groups, social scientists must consider information other than quantitative data. Such examinations should be theoretically informed, and can involve interviewing people, participant observation, examining historical sources, and, in general, striving to achieve the holistic view anthropologists pride themselves on.

In analyses of choice, elicitations of rules of decision making often are particularly useful ethnographic aids in the interpretation of statistical data. In cases where firm rules exist and can be shown to be consistently followed, a reasonable argument can be made that the investigation of choice should stress rule elicitation. But for many economic decisions (certainly for the choice between wage labor and cash cropping in rural Belize), I have great doubts that clear cut, consistently followed rules exist. In such cases (and I suspect for peasant production decisions they are the majority), other types of ethnographic information must be used in interpreting statistical descriptions.

A final advantage of statistical behavior analysis concerns cultural change. Statistical analyses of sociocultural data usually have been, as in the present case, synchronic. Two (or more) groups of people are compared and statistical breakdowns are used to aid in the explanation of differences. However, statistical analysis, as was suggested some years ago (Firth 1961:80–86), also can be used to look at the economic behavior of one group at different points in time. Such data can be regarded as evidence of cultural change, and statistical methods can be applied to the data to infer some of the reasons for the change (Barlett [1976, 1977] is one of the few economic anthropologists who has attempted this even though "panel analysis," time series studies, etc. are common in other social sciences). It is difficult to see what inferences about the causes of economic and cultural change could be made from a diachronic natural decision-making analysis (i.e., ethnoscientific elicitation of rules of economic behavior from the same people at different times). Goodenough (1961) has argued that the rules can remain the same while conditions change, and the interaction of the rules and conditions produce variations in decision outcomes over time. I suspect, however, that changing economic conditions usually are accompanied by changing ideas about "appropriate" economic behavior (Keesing [1967:13] agrees). Elicitation of variations over time in rules of decision making may be quite informative about changes in ideology, but aid little in any search for the causes of such change.

Neither natural decision making nor statistical behavior analyses alone constitute a complete description of economic choice. The goals of the two analytic methods are complementary rather than antithetical. Natural decision analyses are attempts to describe the search procedures people use in making choices, whereas statistical behavior analyses are attempts to describe carefully intracultural variation. A combination of these analytic methods provides nearly complete information on choice in a particular place, given the opportunities that exist for decision making. Neither analytic method, however, can provide explanations of the world and national political and economic factors that affect the options available to choice makers.

Appendix: Statistical Methods

The goal of my statistical analyses was to compare the relative influence of various factors on the amount of income male household heads obtained from different sources. Ethnographic interviews and observations suggested that a man's village, the size of his household, and his age strongly influence his choice of income sources. If these variables were measurable at the interval or ordinal level, multiple regression techniques of analysis could be used (see Pelto and Pelto 1978:164–166). Village, however, can only be measured at the nominal level. Age also poses measurement problems as the relationship between age and various economic activities is sometimes curvilinear (for example, men between 35 and 55 do more farming than both men under 35 and over 55). Although age is theoretically measurable at the interval level, for practical purposes often it is better to treat the variable as being nominal (i.e., divide the population into three age groups such as under 35, 35–55, and over 55).

The level of measurement of certain key variables thus necessitated the use of nonparametric analytic tools suitable for data in which at least one of the variables being examined is nominal. Tables 4.6 and 4.7 use one such statistic, gamma, as a measure of association. Many textbooks (e.g., Harshbarger 1977:467, Thomas 1976:414) either assert directly or imply that gamma can be used only when measurement of both variables is ordinal. This is not quite accurate. Gamma also can be used for 2 × N tables in which the level of measurement for the dichotomized variable is nominal, as long as the level of measurement for the other variable is ordinal (in the special case of a 2 × 2 table, gamma can be used where there is nominal level measurement on both variables; in such situations, gamma reduces to Yule's Q [Blalock 1972:298]).

In calculations of gamma in Table 4.6, incomes from various sources were divided into three groups, the highest 10, the middle 10, and the bottom 11 (for how to calculate gamma and its level of significance, see Harshbarger 1977:467–476). Villages (obviously) were divided into two groups (Sittee River and Silk Grass) as were households (7–15 members and 1–6 members). Thus income and household size were measured at the ordinal level, but village could be measured only at the nominal level.

In Table 4.6, gammas were calculated for 2 × 3 tables in which one variable was village and the other variable was income in dollars from a particular source. For these tables, the dichotomized variable was measured on the nominal level; the trichotomized variable on the ordinal level. Gammas also were calculated for 2 × 3 tables in which one variable was household size and the other variable income in dollars from a particular source. For these tables, measurement of both variables was on the ordinal level.

Table 4.7 uses the same divisions of household size, income, and village as Table 4.6. In this table, the distribution of cases in each income source category (high, medium, low) is not as even as in Table 4.6. The distribution of cases in each household size category (7–15, 1–6) is also less even.

Besides problems posed by level of measurement, there were two other major statistical difficulties. First, the small sample size created problems in examining the effect of age on economic activities. For analytic purposes, men surveyed were divided into three age groups, under 35, 35–55, and over 55. These divisions were chosen since informal ethnographic observation suggested that the economic behavior of men in these three groups differs markedly. Few men are willing to farm extensively until they reach their midthirties, and men over 55 usually are unable physically to participate in most forms of wage labor. Of the household heads surveyed, only four were under 35, and six were over 55. These small numbers often precluded direct analysis of the effect of age on economic activity. Instead, at several points in the text, a comparison is made of the behavior of all men in the sample with the behavior of men between 35 and 55. By this indirect method, some of the effects of age on economic activity can be seen.

Second, it was necessary to determine if the three variables, village residence, household size, and age, were independent of one another. Village and age and village and household size were found to be unrelated, but there was a substantial relationship between age and household size (see Table 4.4). This necessitated the inclusion of tables controlling for age by restricting the sample to men between 35 and 55.

Acknowledgments

I wish to thank Peggy Barlett, Gerald Britan, Carole Browner, and Paul Durrenberger for their helpful comments on earlier drafts of this chapter.

References

Ansoff, H. Igor
   1969   Toward a Strategic Theory of the Firm. *In* Business Strategy. H. Ansoff, ed. pp.
           11–30. New York: Penguin.
Ashcraft, Norman
   1973   Colonialism and Underdevelopment: Processes of Political Economic Change in
           British Honduras. New York: Teachers College Press.
Barlett, Peggy
   1976   Labor Efficiency and the Mechanism of Agricultural Evolution. Journal of An-
           thropological Research 32(2):124–140.
   1977   The Structure of Decision Making in Paso. American Ethnologist 4(2):285–307.
Blalock, Hubert
   1972   Social Statistics. New York: McGraw–Hill.
Burling, Robbins
   1962   Maximization Theories amd the Study of Economic Anthropology. American
           Anthropologist 64:802–821.
Cancian, Frank
   1972   Change and Uncertainty in a Peasant Economy. Stanford: Stanford University
           Press.
Chayanov, A. V.
   1966   The Theory of Peasant Economy. (Translated from the Russian and German
           editions published in the 1920s.) Homewood, IL.: R. Irwin.
Chibnik, Michael
   1975   Economic Strategies of Small Farmers in Stann Creek District, British Honduras.
           Doctoral dissertation, Anthropology. Columbia University.
   1978   The Value of Subsistence Production. Journal of Anthropological Research
           34(4):561–576.
Cohen, Percy
   1967   Economic Analysis and Economic Man. *In* Themes in Economic Anthropology.
           R. Firth, ed. pp. 91–118. London: Tavistock.
Comment on Tripartite Report
   1966   Comment on Report of the Tripartite Economic Survey of British Honduras.
           Unpublished.
Cook, Scott
   1970   Price and Output Variability in a Peasant–Artisan Stoneworking Industry in
           Oaxaca, Mexico. American Anthropologist 72:776–801.
   1973   Economic Anthropology: Problems in Theory, Method, and Analysis. *In* Hand-
           book of Social and Cultural Anthropology. J. Honigmann, ed. pp. 795–860.
           Chicago: Rand McNally.
DeWalt, Billie
   1975   Inequalities in Wealth, Adoption of Technology, and Production in a Mexican
           Ejido. American Ethnologist 2(1):149–168.
Dillon, John, and J. Anderson
   1971   Allocative Efficiency, Traditional Agriculture, and Risk Aversion. American
           Journal of Agricultural Economics 53:26–32.
Dumont, Rene
   1963   Report on the Economic and Natural Features of British Honduras in Relation to
           Agriculture, with Prospectives for Development. Belize City: unpublished.
Durrenberger, E. Paul
   1976   The Economy of a Lisu Village. American Ethnologist 3(4):633–644.

Firth, Raymond
1961 Elements of Social Organization. Boston: Beacon.
Geoghegan, William
1970 Residential Decision-Making among the Eastern Samal. University of California, Berkeley: Manuscript.
Gladwin, Christina
1975 A Model of the Supply of Smoke Fish from Cape Coast to Kumasi. In Formal Methods in Economic Anthropology. S. Plattner, ed. pp. 77–127. Washington, D.C.: American Anthropological Association.
Gladwin, Hugh
1971 Decision Making in the Cape Coast (Fante) Fishing and Fish Marketing Systems. Doctoral dissertation, Stanford University.
Gladwin, Hugh, and Christina Gladwin
1971 Estimating Market Conditions and Profit Expectations of Fish Sellers at Cape Coast. In Studies in Economic Anthropology. G. Dalton, ed. pp. 122–142. Washington, D.C.: American Anthropological Association.
Goodenough, Ward
1961 Comment on Cultural Evolution. Daedalus 90:521–528.
Grant, C.
1976 The Making of Modern Belize. New York: Cambridge University Press.
Green, Justin, and Joan D'Aiuto
1977 A Case Study of Economic Distribution Via Social Networks. Human Organization 36(3):309–315.
Hardesty, Donald
1977 Ecological Anthropology. New York: Wiley.
Harshbarger, Thad
1977 Introductory Statistics: A Decision Map. New York: Macmillan.
Johnson, Allen
1972 Individuality and Experimentation in Traditional Agriculture. Human Ecology 1:145–159.
1974 Ethnoecology and Planting Practices in a Swidden Agricultural System. American Ethnologist 1:87–101.
1978 Quantification in Cultural Anthropology. Stanford: Stanford University Press.
Keesing, Roger
1967 Statistical Models and Decision Models of Social Structure: A Kwaio Case. Ethnology 6(1):1–16.
Kernan, Keith, John Sodergren, and Robert French
1977 Speech and Social Prestige in the Belizean Speech Community. In Sociocultural Dimensions of Language Change. B. Blount and M. Sanches, eds. pp. 35–50. New York: Academic Press.
Lin, William, G. Dean, and C. Moore
1974 An Empirical Test of Utility vs. Profit Maximization in Agricultural Production. American Journal of Agricultural Economics 56:497–507.
Officer, R. R., and A. N. Halter
1968 Utility Analysis in a Practical Setting. American Journal of Agricultural Economics 50:257–277.
Ortiz, Sutti de
1973 Uncertainties in Peasant Farming: A Colombian Case. New York: Humanities Press.
Pelto, Pertti, and Gretel Pelto
1975 Intracultural Diversity: Some Theoretical Issues. American Ethnologist 2(1):1–18.
1978 Anthropological Research (2nd ed.). Cambridge: Cambridge University Press.

Plattner, Stuart
   1974   Formal Models and Formalist Economic Anthropology: The Problem of Maximization. Reviews in Anthropology 1(4):572–582.

Quinn, Naomi
   1978   Do Mfantse Fish Sellers Estimate Probabilities in their Heads? American Anthropologist 5(2):206–226.

Romney, D., A. Wright, R. Arbuckle, and V. Vial
   1959   Land in British Honduras, Report of the British Honduras Land Use Survey Team. London: Her Majesty's Stationery Office.

Rutz, Henry
   1977   Individual Decisions and Functional Systems: Economic Rationality and Environmental Adaptation. American Ethnologist 4(1):156–174.

Sahlins, Marshall
   1972   Stone Age Economics. Chicago: Aldine.

Salisbury, Richard
   1968   Anthropology and Economics. In Economic Anthropology: Readings in Theory and Analysis. E. LeClair and H. Schneider, eds. pp. 477–485. New York: Holt, Rinehart and Winston.

Schneider, Harold
   1974   Economic Man. New York: The Free Press.

Settlement Report
   1948   Report of the British Guiana and British Honduras Settlement Commission. London: His Majesty's Stationery Office.

Shapiro, Kenneth
   1975   Measuring Modernization among Tanzanian Farmers. In Formal Methods in Economic Anthropology. S. Plattner, ed. pp. 128–148. Washington, D.C.: American Anthropological Association.

Simon, Herbert
   1955   A Behavioral Model of Rational Choice. Quarterly Journal of Economics 69:99–118.
   1976   Administrative Behavior. New York: The Free Press.

Thomas, David
   1976   Figuring Anthropology. New York: Holt, Rinehart and Winston.

Tripartite Report
   1966   Report of the Tripartite Economic Survey of British Honduras. Unpublished.

Tversky, Amos
   1969   Intransitivity of Preferences. Psychological Review 76(1):31–48.

White, Benjamin
   1976   Production and Reproduction in a Javanese Village. Doctoral dissertation, Columbia University (Printed in Bogor, Indonesia: Agricultural Development Council).

Williams, Glyn
   1977   Differential Risk Strategies as Cultural Style among Farmers in the Lower Chubut Valley, Patagonia. American Ethnologist 4(1):65–83.

Winter, S.
   1971   Satisficing, Selection, and the Innovating Remnant. Quarterly Journal of Economics, May.

Chapter 5

The Attentive–Preattentive Distinction in Agricultural Decision Making

HUGH GLADWIN
MICHAEL MURTAUGH

Introduction

The study of natural information processing in agricultural decision making is important because it focuses attention on the farmer actually making the decision and avoids externally imposed normative assumptions. By focusing on the individual farmer, it provides a research alternative that complements the statistical analysis of aggregate behavioral outcomes (an approach presented in several chapters of this volume). By initially eliminating normative assumptions, it begins with an analysis of what the farmer does, thus avoiding the confusion of descriptive and normative models rightly condemned by Cancian (Chapter 7, this volume). However, descriptive decision models and normative models also can be complementary if used together. Natural decision models have proven to be accurate predictors of individual choice in a number of agricultural settings, and thus have the empirical power of useful tools in agricultural development research.[1]

---

[1] It should be noted that natural decision-making researchers have various goals: some, like Quinn, are interested in finding out the general heuristics and procedures widely used in a culture and across cultures to make decisions. This is a crucially important task for cognitive anthropology, but it is somewhat separate from the issues discussed in this book. Agricultural decision-making studies must be able to predict farmers' behavior if they are to be useful for policymaking.

AGRICULTURAL DECISION MAKING:
Anthropological Contributions to
Rural Development

Christina Gladwin's Chapter 3 in this book presents a theoretical basis for, and a number of models from, natural decision research in agricultural contexts. This chapter continues the discussion by placing the study of decision making in a wider psychological perspective. By taking a broader view, it is possible to analyze the functions of information, experience, and different presuppositions affecting farmers' decisions. It is then possible to show how agronomists and other development agents, in communicating with farmers and in evaluating the results of changes in agricultural technology, can affect farmers' decision making in unexpected ways.

This chapter takes the point of view advocated by Herbert Simon over the past 30 years. Simon has argued that the place to observe decision making is at the interface between human rationality (with its limitations) and the complex environments in which decision makers find themselves. Essential to Simon's view is the idea that problem solvers (including decision makers) do not pay attention to the full complexity of the environment in solving problems or in making decisions. "Choice is always exercised with respect to a limited, approximate, simplified 'model' of the real situation [March and Simon 1958:139]."

The focus of this chapter is on the *processes* by which decision makers arrive at the simplified model of the real situation. These are referred to as *preattentive processes*, in contrast to *attentive processes* that include the decision maker's calculations, heuristics, and decision rules for consciously manipulating information within the simplified model.

The importance of understanding preattentive processes will be demonstrated from examples of farmers' decision making, agronomists' research and extension work, and studies by social scientists. The chapter also presents a series of methods that have been developed to elicit preattentive process information and incorporate it into models of agricultural decision making. The presentation of these ideas will follow this scheme:

Another research area important in social and behavioral science is the unconscious in its various forms. In this chapter the words conscious and unconscious are frequently used, and it is true that attention can be considered to be conscious, and preattention unconscious, in some way. But it must be emphasized that the concern of the chapter is with the immediate making of a particular decision, rather than the long-term contents of the unconscious mind. A past lesson learned such as "the soils in this area hold too little moisture for early planting" may be unconsciously applied outside the attention of a particular and immediate choice problem, and yet be easily retrieved to conscious attention when a question is asked. Other preattentive processes conform more closely to the classical definition of unconscious in that they are more difficult to verbalize, even when probed for by a specific question.

1. Definition of preattentive processes and some agricultural examples.
2. How the decision situation is set up; farmers' assumptions (presuppositions).
3. Preattentive assumptions made by farmers and agronomists; resulting difficulties in communication and innovation.
4. Evaluation of new technology.

Definition of Preattentive Processes and Some Agricultural Examples

> It is a profoundly erroneous truism, repeated by all copy books and by eminent people when they are making speeches, that we should cultivate the habit of thinking of what we are doing. Civilization advances by extending the number of important operations we can perform without thinking about them.
> —ALFRED NORTH WHITEHEAD [1911:61]

In this chapter, *preattentive process* refers to any information processing that is outside of a decision maker's ordinary attention and awareness. The terminology is borrowed from the psychologist Ulric Neisser (1967), but no attempt will be made in this chapter to deal with issues primarily of concern to cognitive psychologists. Rather, this section will provide examples of how preattentive, *unconscious* processing underlies the routine decisions of everyone, including farmers.

There is a great deal of evidence from everyday life that people are continually engaged in preattentive, unconscious processing of their environment. For example, a person involved in a conversation at a cocktail party will suddenly hear his or her name spoken in another conversation. A person driving a car will immediately notice when a child runs onto the street (but might not be aware of an airplane flying overhead). A mother more likely will hear the cries of her baby than outside traffic noises. Village farmers in Mexico's Puebla valley notice from a distance the presence of a small worm that eats the roots of the maize plant, and the small white marks on corn leaves that point to recent hail damage. This ability to suddenly shift attention suggests that humans are continually monitoring their environment for matters of immediate importance, processing to some degree a variety of information unconsciously.

Such unconscious processing serves not only to call attention to events of importance; it also can control some apparently complex activities. Many of us are able to ride a bicycle or drive a car while thinking about something other than our numerous hand, foot, and eye movements, not to mention the complex coordination among them. The same preattentive, unconscious processing underlies the performance of more specialized skills when performed by experts. These would include ballet dancing, playing a musical instrument, typing a letter, or harvesting a field of corn by hand. In each of these activities, experts use conscious attention judiciously, at the most difficult moments of their performance, while preattentive processes handle routine behaviors and prepare the actor to anticipate what happens next.

In all of the above examples, preattentive processes serve to create and maintain an actor's *sense of situation*. The parent who notices a baby's faint cries at home may not notice the different cries of a baby when away from home. Individuals are prepared to attend to only the limited number of important events that are likely to occur at any given point in the sequence of an activity. Over time, actors become increasingly skilled at behaving in familiar situations. As their routines are continually fine tuned, the more basic skills are proportionately removed even further from conscious attention.

A good example of how a sense of situation underlies skilled behavior is found in Chase and Simon's study of chess players. The researchers attempted to characterize how master players process information about chess differently than do other players. They found that chess masters do not differ from average players in the number of alternative moves seriously considered. Rather, "Masters invariably explore strong moves, whereas weak players spend considerable time analyzing the consequencs of bad moves. The best move, or at least a very good one, just seems to come to the top of a master's list of plausible moves for analysis [1973:216]." In our terms, the master is unconsciously preattending to a vast amount of information about chess, including the criteria for determining a good or bad move.

Farmers proceed through the agricultural cycle as master players proceed through a chess game, using an extensive body of knowledge to define potential problems and alternative solutions at each point in the cycle. Farmers are not necessarily conscious of the criteria that determine possible courses of action and may appear to communicate very imprecisely about them. When village farmers in Puebla are asked how they decide which ears of corn to save for seed, most reply that they save "the best." The same answer is given when they are asked which of the leaves that envelop the ear will be saved and sold as tamale wrappings. A farmer often will communicate to his workers the amount of fertilizer to be used on his field by showing them the palm of his hand and placing his thumb at some point along his fingers. These comments and gestures are not sufficient to enable a novice to select corn seed, harvest a field, or apply fertilizer, but they serve as sufficient cues for farmers who are already experts in these matters.

A more extended example suggests how ordinary interaction between two farmers requires preattending to each other's conversation. On one occasion, a farmer was working in the field when a second farmer approached him and said simply, "How ugly." The other farmer replied, "I don't care how it looks, this is just forage." The visiting farmer had noticed that the corn leaves in this field were extremely yellow—an "ugly" color for a corn plant. Since the owner of the parcel shared the basic knowledge and assumptions of the other, he understood the remark and was able to respond appropriately. His own remark implied that he had overplanted the field, had used very little fertilizer, and would harvest the crop before the ears would begin to develop. The visiting farmer could also have inferred that the other owned several large animals which he relied on for a major part of his income, and that the farmer also had other fields where he could raise corn for his family.

It should be clear from these examples that preattentive processes underly all of a skilled person's communication and performance. What this chapter and Chapter 3 by C. Gladwin, this volume, refer to as "Stage 1" is a major (but not the only) area where the preattentive processes of skilled individuals are involved in decision making.

Stage 1 and "Ethnoagronomy"

A major function of farmers' preattentive processing involves setting up what they consider to be the real decision, that is, specifying the set of feasible options that they attentively consider in making the decision. For example, a farmer may be asked how he or she decided to plant a particular crop. The response might begin with (in the case of some areas of highland Guatemala): "I considered the alternatives: carrots, lettuce, and cabbage. . . ." These three crops represent the feasible set of alternatives as the farmer perceives them. But in reality, a much larger set of crop alternatives are available to the farmer (e.g., maize, wheat, and other vegetables). A researcher interested in predicting the farmer's decisions over the whole set of crops cannot model only the choice between carrots, lettuce, and cabbage. The other alternatives somehow have been eliminated and the researcher must find a way of understanding this preliminary decision.

One formal decision model that explains the elimination of inappro-

priate alternatives has been proposed by the psychologist Amos
Tversky (1972) and is called "elimination by aspects." This model essentially claims that alternatives are presented to the decision maker and
each alternative is considered against various criteria or aspects. An
alternative must be acceptable on all aspects or it will be rejected. Thus,
for example, one could imagine a choice between maize, wheat, and
tomatoes, where for each of these three alternatives, the following
questions (aspects) were applied:

1. Is there sufficient water to grow this crop here?
2. Will growing this crop require no more credit than I can obtain?
3. Will this crop, if it yields well, provide enough money to pay for a
   significant amount of the annual cash needs of my family?

Elimination by aspects assumes that if the answer of any of these
questions (aspects) is no for a given crop, the crop will be eliminated
from consideration. Economists will be surprised to note that this
model does not provide for any form of trade-offs among aspects, in the
way that an indifference curve in a standard microeconomics text assumes that a low score on one aspect can be traded-off against a high
score on another aspect. Tversky's model applies each aspect independently, and makes no allowance for trade-offs.[2]

[2] One of the long-standing debates in social science concerns the extent to which people
are able to estimate accurately utility magnitudes and probabilities. The microeconomic
models illustrated with indifference curves assume that people can estimate the utility or
desirability of alternatives and use these estimates to calculate trade-offs. Accurate calculations of probability are likewise assumed by some models, such as those generated by
Bayesian decision theory (Knight 1965; Raiffa 1968).

   If people do not in fact make utility and probability calculations, one might argue that
they unconsciously follow some analogous process, which leads to the same result. This
"as if" hypothesis has a long tradition in neoclassical economics (Friedman 1953). Unfortunately, the weight of the psychological evidence in recent years has been turning
against this view. Experiments conducted during the 1960s on probability judgments
consistently demonstrated that people underestimate the probability of very likely events
and overestimate the probability of very unlikely events. These data have led some
researchers to modify decision theory models and conclude that humans are "conservative Bayesians" (Edwards 1968).

   In recent years, experimental psychologists have begun to search for new models to
explain the divergence between subjects' judgments and the predictions of decision
theory. Kahneman and Tversky have proposed that humans employ a number of "heuristics" which they use in making probability judgments (cf., Kahneman and Tversky 1972;
Tversky and Kahneman 1974).

   Some researchers concerned with modeling the behavior of individual agricultural
decision makers, and who have been influenced by the recent work in cognitive psychology, have constructed decision models that do not carry all of the assumptions of traditional decision theory. C. Gladwin (1977, this volume) has found that in the case of

Christina Gladwin (1975, 1977) rejects Tversky's model as too simplistic to be a complete account of situations where decision makers are carefully attending to choices made difficult by competing aspects.[3] But

---

utility judgments, different "utility dimensions" or aspects can be ordered in importance for any given set of alternatives. This simple ordering of aspects is usually all that is needed to make a model predict correctly decision outcomes. Occasionally it is necessary to introduce a magnitude judgment like "X is more than twice as profitable as Y," but precise calculations of utility magnitudes are not needed. Moreover, it is not necessary to compare the "utility magnitude" of one aspect (e.g., economy) with that of another (e.g., quality).

With respect to probability judgments, Quinn (1978) goes even further than Gladwin in minimizing the assumptions about actual calculations made by decision makers. She criticizes models of Fante (Ghanaian) fish sellers presented by H. Gladwin (1971) and C. Gladwin (1975) which assume some level of probability estimation on the part of the sellers. Quinn argues that sellers have heuristics ("rules of thumb") which enable them to make decisions without the necessity of calculating probabilities. For example, instead of estimating the probability of a given supply of fish at the principal market (Kumasi), sellers need only answer specific questions, such as "Are there six or more lorries leaving Biriwa?" If not, the seller goes to Kumasi.

For routine agricultural decisions, heuristics are clearly more compatible with preattentive processing than are assumptions of elaborate unconscious calculations. As argued earlier, preattentive processes allow actors to continually fine tune their skills and to use their attention more and more efficiently. Farmers with several years experience have developed heuristics for handling the problems that regularly occur throughout the agricultural cycle. For example, one Puebla farmer explained that he plows his fields in late fall, "When the volcano (Popocatepetl) is free of clouds." The farmer has learned that this heuristic works well for conserving soil moisture, but there is no reason to assume that the farmer ever makes a probability calculation about the likelihood of one more rainfall.

The routine decisions of the agricultural cycle always involve the possibility of unfavorable outcomes (i.e., the risk of sustaining an economic loss). But farmers need not treat risk as an outside constraint acting on normal profit-making activities. The farmer's sense of situation must be based on adaptation to the long-run consequences of actions, whether this adaptation has been developed by each individual through experience or has been acquired nearly intact from the experience of others.

A very nice example of the way risk aversion becomes incorporated into routinized specific cropping decisions is shown in data on choice of potatoes grown in Peru (Faust 1976, Bolton and Faust 1979). Data were gathered on the planting choices of potato varieties from Quechua–Qolla Indians in Peru. These choices were analyzed with descriptions of potato characteristics. The results show that a cropping mix is preferred in which some potatoes are grown that are frost resistant but less desirable on other dimensions. With these are grown potatoes that have desirable qualities (sweet, floury) but are not frost resistant. Here, as in Stage 1 aspects discussed earlier, farmers do not have to recalculate the probability of loss every time they consider the crop mix to plant.

[3] Tversky's (1972) results can possibly be explained by the fact that he was modeling data obtained from experiments in which subjects presumably did not have much at stake in the choices. In such a situation, "elimination by aspects" is the fastest decision procedure for subjects to use.

she has shown that the model is adequate for dealing with Stage 1 of decision making discussed in Chapter 3 in this volume, where alternatives are eliminated in the process of setting up the "real decision" over the set of feasible alternatives (C. Gladwin 1977:26–33, 1978). In the previous example, each of the three reasons could be seen as aspects of the alternative that are sufficiently constraining to eliminate the alternative. With some alternatives eliminated, the decision maker goes on to Stage 2, where there are usually a number of feasible alternatives that have different (and often competing) aspects affecting their desirability or undesirability. Stage 2 is discussed in Chapter 3 in this volume; it is of less concern to the topic of this chapter since it covers that part of the decision process people attend to and most readily talk about. Stage 1 in agricultural decision making provides the best situation in which to explore the effect of preattentive processes.

Stage 1 of decision making is more like setting up the feasible ways for solving the decision problem, rather than problem solving itself. In Stage 1 the personal preferences of individual actors are likely to be of less importance than the environmental constraints that determine admissible solutions. The product of Stage 1 decision making, referred to earlier as the sense of situation, is analogous to the model of the choice situation described in the introduction (March and Simon 1958:139). Elsewhere, Newell and Simon have called this model a "problem space [1972:789]," which includes the goals, feasible operations, and specific information available to the problem solver at the start of the solution process. Stage 1 thus produces in the actor a sense of situation, based on preattentive considerations of constraints such as basic needs, budget, and physical laws of the environment, which determine the small set of possible options that a decision maker will consider.

The decisions of farmers in one of the highland Guatemala towns studied by C. Gladwin suggests an example of how a decision situation might be represented in Stage 1. In this town, vegetables are grown on the valley floor around the town, where there is ample irrigation. Rainfed maize is grown on the hillsides. The Stage 2 cropping choices here involve vegetables. But it is unlikely that the farmer deliberately ponders whether or not it is appropriate to grow vegetables on the hillsides. It is rather more likely that the cropping decision problem is set up after Stage 1 as a situation in which the farmer already perceives the valley floor and hillsides, with maize and vegetables growing in their appropriate places.

If Stage 1 involves preattentive processes generating a usually nondeliberate and nonverbalized sense of the decision situation, can such knowledge later be verbalized by the decision maker (and accessed by outside investigators)? It is conceivable that potential alternatives go unmentioned because farmers simply are unable to talk or reason about them. Fortunately for social scientists and agronomists, much current research that requires the elicitation of reasons for making decisions suggests that this is not the case (Quinn 1975, 1978; Randall 1976; Young 1980). Farmers can be asked why an alternative was eliminated in the process of identifying a feasible set of alternatives. Thus a question like: "You are considering carrots, lettuce, and cabbage. Why aren't you also considering maize?" usually elicits a quick and quite consistent answer from farmer to farmer. A typical answer in this case would be "I can make twice as much money selling these vegetables as the maize is worth, and I have other nonirrigated land where I can grow my maize." This research indicates that the reasons for eliminating alternatives at the beginning of decision making can be elicited if the farmer is asked specifically about them.

An example of Stage 1 in agricultural decision making is the model of Stage 1 in cropping decisions across diverse ecological zones in western highland Guatemala presented by C. Gladwin in Chapter 3 in this volume.

Similar findings apply to a recent study of decision making in U.S.A. automobile purchases. Research found that the buyers' preattentive Stage 1 decisions included three criteria or aspects, narrowing the list of potential cars to the small number of options that each buyer carefully considered (Murtaugh and Gladwin 1979). Size was one of the aspects that eliminated cars; to determine this, buyers had to be asked questions like "Why didn't you consider a Honda?" (for people who felt the choice was between two larger sedans). Only then did buyers volunteer information on the reason why the smaller alternatives were eliminated. Once they were asked, however, a long discussion of the space and seating capacity needs of their families usually ensued. Like the Guatemalan farmers described earlier, they had considered the constraints that eliminated cars in Stage 1 so obvious that they had not mentioned them. An interview procedure asking about all possible car choices ("contrastive eliciting") provided a rich source of information from which a model of decisions made across all car types was constructed.

The view of preattentive processing in Stage 1 decision making presented here has three important implications for the methodology of studying agricultural decisions. First, it implies that farmers often *can* talk about the reasons for eliminating alternatives. They often do not talk about the reasons simply because interviews or questionnaires do not ask about them. A major element of good interviewing should be contrastive eliciting (Gladwin 1971; Young 1978b) in which decision makers are asked to recount why and how decisions are made across the entire range of choices open to them. Second, the detailed information on Stage 1 obtained by careful interviewing about earlier experiences can be put in verbal statement (propositional) form in the decision model. This is true even though in the actual decision situation the information might be the result of earlier experience (Ortiz, Chapter 8 this volume) and remembered now primarily as a visual image of where things ought to be growing (e.g., vegetables only on the valley floor). Farmers often can recall the episodes or categories of learned information (Abelson 1976) that led to their current beliefs about the correct agricultural practice. Thus the current practice may be represented simply by the statement "Do not grow vegetables on the hillside." But farmers may respond to contrastive questions (e.g., "Why would you grow corn and not vegetables on the hillside?") with the reasons why they follow the practice (e.g., "Hillside land has no irrigation and is not moist enough for the two crops per year I need to grow to make enough money from selling vegetables."). These verbal statements must be cross-checked as carefully as any post hoc verbal data, but they provide a rich source of information on farmers' own perceptions of agronomic facts (ethnoagronomy). In this way, Stage 1 information can be converted into ethnoagronomic propositions so the model can be used to make predictions, given differing soil, altitude, moisture conditions, and economic factors (markets). Third, if farmers can talk to social scientists about Stage 1 considerations, they also clearly can talk to each other and to agronomists and extension agents. This reasoning agrees with much current cognitive research indicating that the view of the tradition-bound native unable to talk (or even think) about new options that have not been tried recently can be a figment of poor research methodology (Lave 1977; Reed and Lave 1979).

Preattentive Assumptions Made by Farmers and Agronomists; Resulting Difficulties in Communication and Innovation

An Example from a Plan Puebla Demonstration Tour

The same preattentive processes that underlie Stage 1 decision making also underlie farmers' communications with outside agricultural agents. This section will use examples from a tour of agricultural demonstration plots in the state of Puebla, Mexico, in 1975. The agronomists conducting the tour represented the Plan Puebla, an innovative extension program which has had considerable success bringing credit and fertilizer to small farms in the Puebla Valley (see CIMMYT 1974 for a favorable review of the project). Conflicts that occurred during the tour can be traced to differing implicit assumptions and differing perceptions about the purpose of the tour itself held by farmers and agronomists.

One disagreement that arose between farmers and agronomists concerned the association of maize and beans (*frijol*). Agronomists had recommended that farmers plant larger quantities of beans alongside their normal amounts of maize. This recommendation was a change from the traditional practice of intercropping a few seeds of beans in every other hill of corn. But when an agronomist pointed to one field as a successful example of *maíz–frijol* association, a farmer called his attention to a lodged (toppled) corn plant entangled in a *frijol* vine. The farmer noted that this was precisely the danger in planting *frijol alrededor,* 'using the cornplant as a stalk for the vine'. The agronomist acknowledged the farmer's argument but he emphasized the need to consider the whole field. He argued that 10 or 20 damaged plants over an entire hectare (ha) were not important. What really mattered was that expected yields would be higher following the recommended practices.

The agronomist then went on to address what he believed to be the major problem underlying the farmer's objections. He pointed out how the plants toppled by the *frijol* vines in this field had thin and poorly developed stalks, which often resulted from an insufficient dosage of fertilizer. The problem, therefore, could be eliminated by following the plan's fertilizer recommendations. Nonetheless, several other farmers subsequently voiced similar objections to the practice at a demonstration at another site.

Several days later, the farmer who challenged most vocally the *maíz– frijol* recommendation volunteered further comments on the matter. He argued that the experiment generally worked quite well for the *ingenieros agrónomos*, since they planted in better soils—much deeper soils—than those of his own village. When shallow soils are heavily soaked with rain, strong winds tend to topple the corn. If every plant is wrapped in a *frijol* vine, the likelihood of toppling increases. Consequently, he persisted in defending his own practice of planting only a small quantity of *frijol* scattered throughout his field (similar to the practice of most farmers of the region).

In this case, farmers and agronomists confronted the same physical evidence but were drawing different conclusions. These conclusions were based less on the evidence at hand than on the differing experiences that molded their respective assumptions about proper farming practices.

The demonstration tour has the potential to bring the farmer's pre-attentive knowledge to the conscious awareness of both sides. The tour has similarities with the contrastive-eliciting technique discussed earlier in that both present the farmer with concrete examples that serve as cues for calling to mind relevant previous choices and events. In this case, the tour was able to generate a disagreement about one agricultural practice, but why was that disagreement not resolved? Here we must turn to the presuppositions that each party holds concerning the nature of the tour itself.

From the farmer's point of view, the purpose of the tour is to test the adequacy of the agronomist's recommendations. The agronomist, on the other hand, tends to view the tour as a concrete demonstration of the superiority of a new technology. The agronomists are justified in not viewing the demonstration as a true test since not every test plot will be statistically representative. Nevertheless, to win converts to the new technology, they foster the illusion that a visual glance at each experimental parcel will reveal the superiority of the plan's recommendations.

At the end of the tour, one plan recommendation met with a very favorable response from farmers, though there was no attempt to present it in a concrete visual demonstration. This recommendation emerged from a discussion in which the agronomists elicited comments from farmers about how they choose their seed for the following year. Several answered simply that they choose the best ears from their previous harvest and save them for the next year's planting. They select these ears after the harvest is completed and all the ears have been brought back to the house.

At this point, the agronomist interjected that the farmers were right in saving the best ears for seed, but that they should select these ears before the harvest, not after. Many large ears do not come from better plants but are especially large because they come from plants which grew alone (farmers usually drop three or four seeds in each hole at planting but not all germinate). These plants thus had no competition for nutrients. The agronomist recommended that the farmer select a smaller ear from a very bad part of the field. This plant probably had characteristics that resisted the unfavorable conditions afflicting the others nearby.

Also, choosing ears in the field would enable the farmer to select from plants having more than one ear on the stalk. Finally, selecting ears in the field would allow the farmer to choose from small stalks. Selecting for smaller plants will save time and effort during harvesting as well as increase the amounts of nutrients that develop the ear itself.

The farmers responded well to these suggestions and agreed with the underlying principle. In this case, there was no need to construct a concrete situation for farmers because both groups shared the same background knowledge. In contrast, the *maíz–frijol* recommendations were rejected as unsuitable to the environment in which the farmers were operating. The concrete, visual demonstration was therefore readily dismissed, and since neither side pursued the preattentive criteria of the other, the conflict in conclusions was neither explored nor resolved.

An Example from Crop Insurance

Failure to share preattentive presuppositions is one of the many reasons for problems with programs requiring contact between farmers and outside agencies (e.g., banks and distant government programs). An example is provided by crop insurance, administered in the Puebla area by an agency that most often has little contact with the agronomists who regularly work with the farmers and that the farmers see as providing *seguro* 'insurance' for the bank in case their crops fail (resulting in default) but not for food and income lost to their families.

This conflict in presuppositions was evidenced in a dispute between the local crop insurance agency and farmers in Puebla. During a meeting between agronomists from Plan Puebla and a group of farmers, the insurance agency's refusal to provide coverage for some of the farmers became a heated issue. As of 1976, Puebla farmers long had complained of unfair treatment by the insurance agency and had resisted government regulations that made crop insurance compulsory for any farmer seeking credit (cf. Díaz 1974). On one occasion, agronomists from the Plan Puebla called a meeting with a group of angry farmers for the purpose of familiarizing the farmers with the insurance agency's regulations.

During the meeting, farmers listened with forced attention as the Plan Puebla agronomist explained the insurance agency's policies, such as the list of damages officially insured against and the maximum indemnity covered by the agency. The discussion became far more animated when the topic shifted from official policy to the practical difficulties of being accepted for insurance—although insurance was compulsory, farmers still could be rejected by insurance agency inspectors if their parcels failed to meet agency guidelines.

Several farmers presented problems that had occurred in their fields that year, and explained the innovative solutions they had attempted. The Plan Puebla agronomist examined these practices with some reservations. Those who did "unusual things," such as one farmer who threw on an extra dosage of fertilizer when he noticed the poor development of his crop, were warned that they would have problems during the insurance inspection. The agronomist said the insurance agent's basic reaction would be "Last year you followed the recommendation for this area and you had no problems; this year you're doing things differently and you have problems." The agronomist explained that the farmer's best judgment considering the unique and adverse conditions facing his crop would not be highly valued by the insurance agents.

The discussion then turned to other problems that contradicted the insurance agency's strict interpretation of the plan's recommendations. One farmer explained that he had chosen to alternate rows of corn and beans. Another farmer quickly pointed out that the insurance agency does not accept that practice. Another farmer's field had been damaged very badly by excess moisture and he had abandoned it, as it would be useless to perform the first and second cultivations on the parcel. The inspectors also probably would find the farmer's evaluation unacceptable because, as the agronomist informed him, insurance agency regulations required farmers to work all of their fields. Consequently this farmer too would be rejected.

The farmers who took part in this meeting eventually took the bold step of refusing as a group to accept crop insurance, despite their statements to the bank earlier in the year when they had received credit. They stated that this action was justified because the insurance agency failed to inspect their fields on time, so, when the inspection actually occurred, failing crops already would be evident to the inspector.

Two areas of conflict in presuppositions are relevant to understanding the interaction at this meeting. First, the environmental conditions assumed by the farmers to justify their courses of action were not assumed by the insurance personnel. Farmers were expected to go through the motions of proper agricultural procedures regardless of how pointless these may have appeared to the farmers under given conditions.

This exchange between farmers and agronomists hints at a second set of presuppositions—the consequences of failing to adopt plan technology. In order to obtain credit, farmers were required to buy insurance that year. In order to obtain insurance, farmers had to follow recommended agricultural practices. Failure to obtain insurance would leave farmers vulnerable to increased indebtedness toward the bank (as explained earlier), and the risk of not obtaining credit in the future. As most village farmers need credit each year, their adoption of the agronomists' planting and fertilizer recommendations is not purely voluntary. Thus the farmers themselves ultimately cannot evaluate and decide whether or not to adopt the recommended changes in agricultural technology.

The crop insurance example shows farmers innovating in response to environmental conditions and Plan Puebla agronomists officially committed to innovation in agricultural technology. But because of conflicting presuppositions, the crop insurance program introduced conflict over the value of the two notions of innovation (not to mention over the meaning of 'insurance'—*seguro*).

Evaluation of New Agricultural Technology

The discussion and examples presented in this chapter have clear implications for the way in which the recommendations of new agricultural technologies should be evaluated. *Evaluation* is defined here as the determination of the costs and benefits of changing an agricultural practice. Evaluation, particularly cost–benefit evaluation focused on economic factors (Gittinger 1972), typically has been the task of the outside expert.

Despite much current discussion about "participant evaluation," most evaluations available to the participant, the farmer, are still ones in which the actual judgments about the value of new recommendations are made by experts other than farmers. Even in the case of the Plan Puebla demonstration tour already discussed, the final judge of the efficacy of the recommendation to increase the amount of beans, *frijol*, grown with corn was the agronomist, not the farmers on whose fields the demonstration plants were grown. Such procedures, it is argued in this chapter, often leave farmers and agronomists talking about the same recommendation and the same observed effects, but with such different presuppositions at the preattentive level that the farmers may reach an entirely different conclusion than that intended by the agronomist. And the farmers may be correct in their conclusions, as their presuppositions may include knowledge of facts not known to the agronomist that determine the value (to *them*) of the recommendation and its effects.

Even more problematic is the case in which the farmers only hear the opinions given by agronomists about a recommendation and do not see a demonstration of its effects. In these cases, the problem is that only the farmer fully knows his or her own situation, and the agronomist is only successful to the extent that the farmer's situation is known.

It follows from the argument of this chapter that there are two ways of approaching the problem of evaluation. One approach assumes agronomists need to develop better procedures than they now have for determining the preattentive presuppositions underlying the farmer's perception of the recommendation's effects. The other approach relies on the farmer to do the ultimate evaluating, based on the farmer's own full (preattentive, for the most part) knowledge of the agricultural situation.

The first solution presupposes that the agronomist either be closer to the experience of the farmers on an intuitive basis or be better trained in the task of determining the preattended factors affecting the farmer's evaluation. For example, some of the agronomists for the Plan Puebla were known by the authors to have grown up in rural families, and they were indeed close and sympathetic to the experience of the farmers. They were thus in a position to share many of the preattentive presuppositions of the farmers. But even those agronomists often tended to rely heavily on the evaluation procedures learned in their postgraduate training in agronomy and to discount farmers' opinions. Part of the reason for this is the highly verbal mode of school learning and of agricultural extension efforts. As was discussed earlier, well learned and highly skilled activity such as farming often operates without much deliberate attentive thought or explicit talk. Without realizing it, the agronomists often created a "schoollike" setting in which farmers found it difficult to express their ideas or objections (see Lave 1977 for a discussion of this effect).

It also is hard to ask agronomists to be better social scientists when the task of determining preattended factors in decision making and evaluation is not easy. One conclusion of this chapter is that agronomists can make considerably better evaluations both by being closer and more sympathetic to the experience of the farmers, and by using more precise techniques in determining information considered by farmers at a preattentive level. But, given the small number of agronomists working in Third World countries, the gulf that often exists between their experience and that of poor farmers, and the time constraints under which they work, it seems very unlikely that the evaluation problem can be solved by agronomists alone.

This leaves the second solution to the evaluation problem: that the farmers themselves do more of the evaluating. The reason for this approach, in terms of the argument presented in this chapter, is that only the farmers can fully preattend to the effects of a recommendation on their own crops and for their own goals. Whereas agronomists can consider long-term effects, which farmers might not know, only the farmers can fully evaluate the effect of a recommendation. What this suggests, concretely, is that farmers be given the facilities to try newly recommended practices (and not be hampered by differing institutional presuppositions about what is considered innovation, as in the crop insurance example). After the recommendation has been tried and the results obtained (e.g., by harvesting the crop), the farmer is the one who should decide whether or not the recommendation has been a success. This means that the agronomist only can discuss with the farmer why a rejected recommendation was rejected, and must as a matter of policy accept the farmer's judgment. Note the difference between this procedure and that of "test plots" where only the agronomist judges the success of a recommendation.

One example shows that application of the second solution to the evaluation problem by governmental agencies in developing countries can be successful. The Instituto de Ciencia y Tecnología Agrícolas (ICTA) in Guatemala has a program of farmers' tests (*parcelas de prueba*), in which the farmers are the final judges of whether the recommendations should be accepted (Fumagalli and Waugh 1977; R. Ortiz 1979). This program of farmers' tests on their own fields was started after agronomists and social scientists in ICTA became disenchanted with demonstration plots, for many of the same reasons explained in the earlier example. They thus developed a procedure for having a sample of farmers test, on their own fields, a new technical practice or set of technical practices that previously had been generated by agronomists' experiments (*ensayos de finca*) also on farmers' fields. The acceptance of the new practice then was measured by seeing how many farmers continued to use the new practice in the year *following* the test, on their own. The new technical practice thus became an ICTA recommendation only if a significant number of farmers continued to use it. The benefits of such a procedure particularly would appear to outweigh the costs in an environment such as highland Guatemala, where the wide ecological variation means that experienced farmers have widely varied preattentive presuppositions about local conditions.

Timmer (1976) discusses an analogous situation for the development of small-scale agricultural technology in China in the 1968–1978 decade. Evaluation and development of new technology is left largely to local groups. He reports that Chinese officials feel the costs of loss of centralized control are worth the benefit of having technology development and evaluation be fully responsive to local conditions. In the view of this article, Timmer's point could be restated as saying that the full knowledge farmers have on the preattentive level should support rather than conflict with the development of agricultural technology and its evaluation in a given setting. Such a policy also implements the learning of scientific procedures and attitudes within the framework of preattentive knowledge already held by peasant farmers.

Impact evaluation of technological change (e.g., strip mining) that affects land use and thus agriculture, must also include participant evaluation in the ways described. One study did this by using a decision model that incorporated Navajo perceptions of the impact of proposed mining operations (Schoepfle *et al.* 1978).

Agronomists are not the only outside experts who have trouble with unshared presuppositions between themselves and farmers. The same difficulties can face social scientists as they make observations, give out questionnaires, and even apply statistical models. D'Andrade (1974) has given abundant evidence that, even in seemingly direct observations of behavior (e.g., of children's interactions), the observer's presuppositions about the relationships between categories of behavior will strongly bias the frequency with which different behaviors are observed. Werner *et al.* (1978) have pointed out that any mode of data gathering that involves the asking of questions, such as a questionnaire, is in fact a conversation between interviewer and interviewed. Preattentive presuppositions operate in any conversation; there is no reason to suppose they will not operate with questionnaires. An example is a (hopefully apocryphal) story concerning a questionnaire given to Mexican peasant farmers that began with the question "How many agricultural periodicals do you receive in your home?" One only can imagine the presuppositions with which the farmer began the long task of figuring out what the interviewer wanted with this questionnaire. These examples indicate that it is unlikely that most forms of questionnaire data can be free of bias induced by unanalyzed preattentive presuppositions. This is true of presuppositions of farmers as well as of observers and social scientists.

Statistical methods used by social scientists for analyzing data on decisions are another area in which the social scientist's lack of awareness of preattentive assumptions about the statistical model can lead to serious error. Social scientists often do not realize that statistical procedures assume an underlying model. Thus erroneous presuppositions might be introduced by the model itself. As an example, one might want to analyze the aspects that eliminate unfeasible crops in the Stage 1 crop-mix choice in Guatemala presented by C. Gladwin in Chapter 3 in this volume with an additive (and thus erroneous) multiple regression model as follows:

$$\text{crop } Y \text{ choice} = a + b_1X_1 + b_2X_2 + b_3X_3 + b_5X_5 + b_6X_6 + b_7X_7$$

where each Xi is one of the aspects (such as sufficient market demand). But this model assumes that a low value on one aspect can be balanced-off (traded-off against) higher values on other aspects. In this particular example, such a model would be completely wrong, since a low value on *any* aspect Xi will eliminate crop Y without consideration of any other aspects. The correct model would actually have the form:

$$\text{crop } Y \text{ choice (1:accept,0:reject)} = X_1 * X_2 * X_3 * X_4 * X_5 * X_6 * X_7$$

where each Xi has the value zero if the crop is unacceptable on that aspect and otherwise has the value of one. H. Gladwin 1975 has used a simulation program to show that erroneous multiple regression models such as the additive example described can appear to be highly significant predictors of decision outcome data even though they do not resemble at all the decision process that generated the data. Thus statistical significance is no protection against assuming an erroneous model of how people make decisions.[4]

In this example, a researcher's ignorance of Stage 1 preattentive processes could lead to the development and seeming verification of a highly erroneous model. And the policy results could be disastrous since an additive trade-off model, such as the illustration, implies that a recommendation having one undesirable aspect may be accepted if it is balanced by its other better aspects. Stage 1 decision making, on the other hand, implies rejection of a recommendation if it fails any aspect. Thus an agronomist working from results of an erroneous analysis such as this may try to push a recommendation which inevitably will be rejected.

Conclusion

Skilled farming is an activity requiring great amounts of learning and experience. The argument of this chapter is that the skilled activity of

[4] Of course, one may use a statistical model only to look at the relationship between variables such as a correlation between farm size and labor efficiency. Here, an additive model may be valid even if it has nothing to do with the way farmers make decisions. The problem comes in when the model is used to impute motives or adaptive mechanisms to farmers. When this happens, the social scientist is trying to make the model make a prediction given some assumptions about the form of the "black box"—the farmer's individual or collective decision process.

those who practice all civilized pursuits, including agriculture, has been advanced, as noted by Whitehead (1911:61), "by extending the number of important operations which we can perform without thinking about them." Agronomists and social scientists must develop better techniques for becoming aware of the operations performed in agricultural decision making that involve preattentive knowledge rather than attentive thought. Only then can the full range of factors affecting the decisions of farmers be understood.

Acknowledgements

The authors wish to thank the staff of the Plan Puebla and many individual farmers and their families in the Puebla valley for the great amount of cooperation and information provided during 1974–1976.

Some of the ideas in this chapter were developed by the authors in 1975 and presented at a Mathematics in the Social Sciences Board conference organized in that year by Jerry Moles. These ideas have been developed in the course of many conversations, particularly with Christina Gladwin and Naomi Quinn.

Many of the notions of preattentive processing in routine decision making and problem solving have been sharpened in the work of the Adult Math Skills Project at the University of California, Irvine, supported by a grant from the National Institute of Education. In addition to the authors, project members are Jean Lave, Olivia de la Rocha, and Katherine Faust.

The perceptive and thorough editing of earlier drafts of this chapter by Peggy Barlett is greatly appreciated, as is the careful reading by Christina Gladwin and Gloria Sanchez.

References

Abelson, Robert P.
    1976   Script Processing and Attitude Formation in Decision Making. *In* Cognition and Social Behavior. John S. Carrol and John W. Payne, eds. pp. 33–45. New York: Wiley.
Bolton, Ralph, and Katherine Faust
    1979   Cognitive Variation in the Domain of Potatoes. Pomona College, Pomona, California, Ms.
Chase, William, and Herbert Simon
    1973   The Mind's Eye in Chess. *In* Visual Information Processing. William G. Chase, ed. pp. 215–281. New York: Academic Press. Proceedings of the Eighth Annual Carnegie–Mellon Symposium on Cognition, Carnegie–Mellon University, Pittsburgh, PA, 1972.
CIMMYT—Centro Internacional de Mejoramiento de Maíz y Trigo
    1974   The Puebla Project: Seven Years of Experience: 1967–1973. El Batan, Mex.: Mexico. [Apdo. Postal 6-641, Mexico 6, D.F.].
D'Andrade, Roy G.
    1974   Memory and the Assessment of Behavior. *In* Measurement in the Social Sciences. Hubert M. Blalock, ed. pp. 159–186. Chicago: Aldine–Atherton.

Díaz Cisneros, Heliodoro
   1974   An Institutional Analysis of a Rural Development Project: the Case of the Puebla
          Project in Mexico. Doctoral dissertation, University of Wisconsin, Madison,
          Wisconsin.
Edwards, Ward
   1968   Conservatism in Human Information Processing. *In* Formal Representation of
          Human Judgment. B. Kleinmuntz, ed. pp. 17–51. New York: Wiley.
Faust, Katherine
   1976   Papas: a Semantic Domain. B.A. Honors Thesis, Pomona College, Pomona,
          California.
Friedman, Milton
   1953   Essays in Positive Economics. Chicago: University of Chicago Press.
Fumagalli, Astolfo, and Robert K. Waugh
   1977   Agricultural Research in Guatemala. Presentation at the Bellagio Conference,
          Bellagio, Italy. Available from Instituto de Ciencia y Tecnología Agrícolas,
          Guatemala (Ciudad), C.A.
Gittinger, J. Price
   1972   Economic Analysis of Agricultural Development Projects. Baltimore: Johns Hop-
          kins University Press.
Gladwin, Christina
   1975   A Model of the Supply of Smoked Fish from Cape Coast to Kumasi. *In* Formal
          Methods in Economic Anthropology. Stuart Plattner, ed. pp. 77–127. Special
          Publication of the American Anthropological Association, No. 4.
   1976   A View of Plan Puebla: An Application of Hierarchical Decision Models. Ameri-
          can Journal of Agricultural Economics 58(5):881–887.
   1977   A Hierarchical Model of Limited Risk-taking. Paper given at American Agricul-
          tural Economics Association symposium, August 1977.
Gladwin, Hugh
   1971   Decision Making in the Cape Coast (Fante) Fishing and Fish Marketing Systems.
          Doctoral dissertation, Anthropology, Stanford University.
   1975   Looking for an Aggregate Additive Model in Data from a Hierarchical Decision
          Process. *In* Formal Methods in Economic Anthropology. Stuart Plattner, ed. pp.
          159–196. Special Publication of the American Anthropological Association, No.
          4. Washington, D.C.: American Anthropological Association.
   1979   Arithmetic Problem Solving, Situational Memory, and the Mayan Calendric
          System. Adult Math Skills Project, School of Social Sciences, University of
          California, Irvine.
Gladwin, Hugh, and Christina Gladwin
   1971   Estimating Market Conditions and Profit Expectations of Fish Sellers at Cape
          Coast, Ghana. *In* Studies in Economic Anthropology. George Dalton, ed. pp.
          122–142. Anthropological Studies 7. Washington, D.C.: American Anthropologi-
          cal Association.
Kahneman, Daniel, and Amos Tversky
   1972   Subjective Probability: A Judgment of Representativeness. Cognitive Psychol-
          ogy 3:430–454.
Knight, Frank H.
   1965   Risk, Uncertainty, and Profit (2nd ed.). New York: Harper and Row.
Lave, Jean
   1977   Cognitive Consequences of Traditional Apprenticeship Training in West Africa.
          Anthropology and Education Quarterly 8(3):177–180.
March, James, and Herbert Simon
   1958   Organizations. New York: Wiley.

Murtaugh, Michael, and Hugh Gladwin
  1979   A Hierarchical Decision-Process Model for Forecasting Automobile Type–
         Choice. School of Social Sciences, University of California, Irvine.
Neisser, Ulric
  1967   Cognitive Psychology. New York: Appleton–Century–Crofts.
Newell, Allan, and Herbert A. Simon
  1972   Human Problem Solving. Englewood Cliffs, NJ: Prentice–Hall.
Ortiz Dardón, Ramiro
  1979   La Generación y Validación de Tecnología y su Relación con un Proceso Efec-
         tivo de Transferencia. Agronomía Publicación de Colegio de Ingenieros Ag-
         rónomos, Avenida Elema 14–45, Zona 1, Guatemala (Ciudad), C. A.
Quinn, Naomi
  1975   A Natural System Used in Mfantse Litigation Settlement. American Ethnologist
         3(2):331–351.
  1978   Do Mfantse Fish Sellers Estimate Probabilities in Their Heads? American
         Ethnologist 5:206–226.
Raiffa, Howard
  1968   Decision Analysis. Reading, MA: Addison–Wesley.
Randall, Robert A.
  1976   How Tall Is A Taxonomic Tree? Some Evidence for Dwarfism. American
         Ethnologist 3(3):543–553 (Special Issue on Folk Biology).
Reed, H. J., and Jean Lave
  1979   Arithmetic as a Tool for Investigating Relations between Culture and Cognition.
         American Ethnologist 6:568–582.
Schoepfle, G. Mark, K. Y. Begishe, R. T. Morgan, J. John, H. Thomas, and P. Reno
  1978   A Study of Navajo Perception of the Impact of Environmental Changes Relating
         to Energy Resource Development. Navajo Community College, Shiprock, New
         Mexico 87420. Final Report on U.S. Environmental Protection Agency funded
         project.
Simon, Herbert A.
  1956   Models of Man: Social and Rational. New York: Wiley.
Timmer, C. Peter
  1976   Food Policy in China. Food Research Institute Studies (Stanford University)
         15(1):53–69.
Tversky, Amos
  1972   Elimination by Aspects: A Theory of Choice. Psychological Review 79(4):281–
         291.
Tversky, Amos, and Daniel Kahneman
  1974   Judgement under Uncertainty: Heuristics and Biases. Science 185:1124–1131.
Werner, Oswald, Judith Remington, and Kathleen Gregory
  1978   How Language Makes Us Know: Problems of Knowing in Ethnoscience. North-
         western University, Evanston, Illinois. Presentation at American Anthropologi-
         cal Association Annual Meeting, Los Angeles.
Whitehead, Alfred North
  1911   An Introduction to Mathematics. London: Williams and Norgate.
Young, James C.
  1978   Illness Categories and Action Strategies in a Tarascan Town. American
         Ethnologist 5:81–97
  1980   A Model of Illness Treatment Decisions in a Tarascan Town. American Ethnologist
         7(1):106–131.

Chapter 6

Cost–Benefit Analysis:
A Test of Alternative Methodologies

PEGGY F. BARLETT

One can view cost–benefit analysis as anything from an infallible means of reaching the new Utopia to a waste of resources in attempting to measure the unmeasurable.

—PREST AND TURVEY (1966:200)

Decision making involves the evaluation of different options, usually followed by an assessment that one option is preferable. To understand agricultural decisions, the approach used to measure how farmers balance the costs and benefits of alternative choices is crucial. The allocation of resources to alternative ends is seen in all aspects of agricultural production: choices about land use, allocation of labor and capital, and decisions about markets, prices, and technology. Agricultural decisions also are weighed by governments and international agencies, who evaluate agricultural development plans and projects and use similar information about the costs and benefits of each agricultural option. These resource allocations and attendant balancing of costs and benefits are the subject of this chapter.

The appropriate methodology with which to measure agricultural decisions depends on the goal of the research. Given that the goal is to describe and predict the behavior of farmers, three methodologies have been presented in the literature (*a*) qualitative assessments that do not attempt to measure costs and benefits (Beals 1974; Bennett 1969; Clayton

AGRICULTURAL DECISION MAKING:
Anthropological Contributions to
Rural Development

1968; Halperin and Dow 1977); (b) traditional economic calculations[1] that quantify returns to all factors of production (Edwards 1961; Epstein 1962; Greenwood 1976); and (c) Chayanovian calculations that compute profitability subtracting cash costs only (Barlett 1977; Cancian 1972; Moerman 1968; Ortiz 1973). Most researchers have chosen their methodology without substantiating its validity and such dogmatism is characteristic of anthropologists as well as of economists.

The purpose of this chapter is to explore the value of these three methodologies in light of data from Paso, Costa Rica.[2] Chayanovian theory will be explored in depth and several criticisms presented. The data show, however, that traditional economic calculations may distort some aspects of agricultural decisions that are illuminated by Chayanov's theory. In other situations, the results of these two methodologies are similar. The importance of a qualitative approach also will be explored. Data from Costa Rican farmers suggest that Chayanovian calculations, combined with qualitative assessments of agricultural options, will provide the most accurate tool for understanding agricultural decisions.

Research Goals

The appropriate methodology to use when studying agricultural decisions depends on the goal of the research. Economists and anthropologists tend to differ in their goals; anthropologists generally seek *to describe* agricultural choices, and economists, often implicitly, seek *to evaluate* them. "Unlike anthropologists, economists have not ordinarily been interested in finding out *whether* people economize intelligently, but only in figuring out *how* they can economize more intelligently [Burling 1962:819]."

Such a normative element in decision making research may distort the extent to which cost–benefit methodology accurately describes real behavior. Yang discusses traditional economic analyses of farmer's options, as providing "the essential information to enable producers to make the *right* management decisions [Yang 1965:87, emphasis added]."

Textbooks in agricultural economics tend to assume that if farmers do

---

[1] I have chosen to use the label "traditional economic calculations" to describe the procedure of quantifying all costs and benefits. This usage does not mean to imply that all economists use this methodology nor that there are no economists who use and prefer Chayanovian calculations. These "traditional calculations" stem from the analysis of decision making in capitalist firms, and often have been uncritically applied to small farm decisions by anthropologists as well as by economists.

[2] "Paso" and all personal names used here are pseudonyms.

not follow the recommendations of such cost–benefit analyses, they should be persuaded to do so (Bishop and Toussaint 1958; Cochrane 1974; Doll, Rhodes, and West 1968; Heady 1952). Doll, Rhodes, and West clarify that, if a farmer does not make a profit above their approved level of returns to the farmer's own labor and capital, "he could better move to the city [Doll, Rhodes, and West 1968:28]."

Economists who use cost–benefit analysis to evaluate proposed agricultural projects distinguish between financial analysis, which is supposed to determine the desirability of the proposed changes to farmers, and economic analysis, which weighs proposals in light of the costs and benefits to the larger (usually national) economy. Financial analysis depends on farm plans that "usually represent a careful judgment of agriculturalists about the 'optimum' or most profitable farming activities and cropping patterns. . . . These considerations reflect themselves in the rate at which farmers can be expected to adopt new practices [Gittinger 1972:131]." The financial analysis of farm plans is not only supposed to provide a way to compare farming activities (a normative goal), but also is assumed to reflect farmers' decision making processes. In general, procedures for financial analysis do not allow for the possibility that the recommended calculations are based on criteria of evaluations that are not shared by farmers, and therefore may lead to inaccurate understandings of farmers' choices (Gittinger 1972; Scandizzo 1978; Squire and Van der Tak 1975; UNIDO 1978; Yang 1965).

The goal of this discussion of alternative cost–benefit methodologies is to explore their value in understanding real decisions among farmers in the Third World today. Such farmers are the focus of massive development efforts, but the criteria for their ongoing decisions are not always established clearly before attempts are made to alter those decisions. This chapter assumes that all the relevant variables of the institutional, local, and national context already are understood adequately and focuses instead on the farmer's choice process itself. The methodologies contrasted here may or may not reflect accurately the cognitive process of each farmer. More important is the extent to which each methodology describes accurately and predicts the way in which decision criteria are taken into account and the way they result in certain patterns of choices.

Traditional Economic Analysis

Traditional agricultural economics and many development specialists adapt the cost–benefit analysis used for capitalist firms to evaluate the agricultural options of farms (Anderson, Dillon, and Hardaker 1977; Gittinger 1972; Little and Mirrlees 1974; Scandizzo 1978; Squire and Van der Tak 1975; UNIDO 1978). The general procedure for such calculations begins by converting benefits to a cash value of agricultural production, either by using the sale price of goods produced, the consumer purchase price (sometimes used to value subsistence production; see Chibnik 1978), or some other calculation of the value of farm production. Other benefits such as increases in farm value or permanent investments also are translated into monetary values and added to the gross income of the option being assessed.

From this gross income are subtracted all costs. Cash costs such as fertilizer are deducted at their purchase price. Tools and other costs whose life continues beyond the time period under assessment usually are depreciated over the expected life of the investment. Any resources not actually paid for must be given an *opportunity cost*, since the farmer is seen to be foregoing the opportunity to use those resources elsewhere. The opportunity cost is defined as the return from any resource in its next best use. Thus, capital invested on the farm foregoes interest in a bank account. The opportunity cost of land usually is calculated by using the rent that could be obtained from it, if it were not being used by the farmer. Likewise, the opportunity cost of unpaid family labor usually is calculated at the going wage, though some researchers realize that its true opportunity cost may be closer to zero. All of these paid and imputed costs are subtracted from the gross proceeds.

Other factors such as risk or the unpleasantness of the work involved also can be incorporated into traditional cost–benefit analysis. Scandizzo (1978) outlines a procedure to add a "risk premium" to discount overall proceeds according to climatic factors or other causes of risk. Unpleasant work can be expressed as a higher wage imputed to labor. Once these other factors have been incorporated into the analysis, the remaining profit is the return to the farmer's entrepreneurship. Any profit over zero, by this traditional economic calculation, means that the farmer is able to make something over and above a "fair return" to his resources. In using this methodology, it is assumed that farmers compare these profits and choose options with the highest return.

Chayanov's Theory of Agricultural Decisions

Chayanov rejects the traditional economic cost–benefit calculations for the study of family farms (Chayanov 1966). His evidence from massive research on Russian farmers suggests that family farms do not behave according to the calculations appropriate to capitalist firms.

Chayanov says that imputing monetary values to nonmonetary costs distorts the decision-making process of these households. He proposes the labor–consumer balance as a tool to understand family farm decisions, and specifies the criterion of *returns to labor* as the appropriate methodology with which to understand agricultural decisions. Before looking at this methodology, the subject matter of Chayanov's theory on decisions must be made clear.

Chayanov agrees that the family farm does make certain kinds of allocation decisions, especially labor, but he rejects the idea that some farms have variable allocations of land as well. Chayanov distinguishes between "nonmonetary farms," that produce the vast majority of their own needs and exchange relatively little of their produce for cash, and the "commodity farms" that sell the majority of their produce. For the more self-sufficient farms, each crop grown is needed to meet a specific consumer need and therefore is not compared to other crop options (Chayanov 1966). "The question of whether it is more advantageous to sow rye or mow hay, for example, could not arise, since they could not replace each other [p. 124]." The farmer does not quantify precisely how much is needed, but rather uses a more qualitative estimate: "there's enough" or "there's not enough [Chayanov 1966:124]." In a commodity farm, where production is weighed in cash terms, the farmer can make a more "quantitative" decision, since money can be used to meet family needs.

Chayanov's assertion that crops are not substitutable in the non-monetary farm is not completely accurate, and makes an unnecessary distinction between qualitative and quantitative decision making. He reifies farmers' common choice or preferred choice as the *only* choice, but farm needs are more flexible than he suggests. Whereas it may be true that a certain amount of hay is desirable to feed cattle, there are other crops that can be used, even to the point of feeding cattle food-grains such as wheat. The farmer may prefer to raise cattle for meat, milk, manure, and traction, but there are always the alternatives of raising pigs for meat and manure, foregoing the milk, and borrowing a neighbor's oxen. Given that the Russian farmer has determined that rye, hay, and cattle are his optimal land uses, Chayanov may still be incorrect in saying that the amounts of rye and hay are fixed on the nonmonetary farm. Ortiz's (1973) work with the Paez Indians of Colombia shows that complex calculations are behind a qualitative statement of having "enough" or "not enough," even for subsistence crops. For our purposes here, we will reject Chayanov's distinction between allocating land and labor, and assume that farmers make quantitative decisions about how best to allocate all the factors of production.

The key point in Chayanov's divergence from a traditional economic calculation of household decisions is the issue of whether farmers impute a value to unpaid household labor when weighing the costs of different choices. Chayanov cites the fact that the capitalist firm must pay for each of the factors of production (rent to land, wages to labor, interest to capital) and operates within a market system that establishes prices for each of these factors. In a household economy, however, the category of wages is missing (by definition, Chayanov is studying the 90% of farms in Russia that employ little if any outside labor), and to attribute a wage to unpaid family labor is to distort their decision process. Chayanov asserts that peasants make decisions based on the gross product of the whole farm enterprise minus cash costs (1966:5). This is the family "labor product" and it only can be divided by the amount of labor invested to obtain it—the "degree of self-exploitation [1966:6]." No calculation of an imputed wage is valid because farmers do not conceive of their labor product as divisible or as earning a separate wage. Chayanov, therefore, rejects any calculation of the opportunity cost of family farm labor, on the grounds that such a calculation does not conform to the observed reality of how decisions are made. His evidence will be discussed shortly.

Chayanov's "labor–consumer balance" is his alternative to explain how families actually make labor allocation decisions. The household weighs its unmet consumption needs together with the drudgery of labor required to meet those needs, until an equilibrium is reached. The family's needs are determined by the size and composition of the family "and the urgency of its demands [Chayanov 1966:6]." The drudgery required is determined by market forces and the specific farm conditions (such as soil quality, or climate). Household labor is invested according to this "subjective" equilibrium and not necessarily according to the highest net profit (1966:7). The highest returns to labor are the goal of allocation decisions in the family farm (Chayanov 1966:86). These returns to labor are usually the same as the highest net profit by a traditional economic calculation, but not always.

To test the labor–consumer balance, data must illustrate that when family needs rise, increasing amounts of labor are invested to meet them. Conversely, when needs decline, labor investment should fall (1966:41). If forces external to the household affect the efficiency or productivity of labor, then the labor invested by the family also should rise or fall accordingly. This pattern only can be observed if market characteristics and population density remain constant, and Chayanov discusses the impact of variations in these "social factors [Chayanov 1966:110–115]." Less attention, however, is given to the level of consumption. Chayanov assumes that consumption demands per person remain constant: "Another less important, yet essential social factor, is the traditional standard of living, laid down by custom and habit, which determines the extent of consumption [Chayanov 1966:12]."

The assumption of constant consumption standards is essential for the labor–consumer balance to be testable. If family "needs" vary according to advertising campaigns or personal taste, then labor investment could go up while household size remains constant or even drops. Unless consumption levels are fixed, the labor–consumer balance becomes tautological, since all allocations of labor (level of drudgery) can be seen as a response to changing family "needs." For our purposes of understanding agricultural decisions in the Third World today, this requirement poses a serious problem. Although in many areas, rural standards of living are dropping in real terms, consumption goals nevertheless are changing rapidly. Desires for radios, bicycles, purchased medicines, manufactured cloth, Coca–Cola, or similar products are seen in even remote areas of "subsistence" farming. The labor–consumer balance is subsequently hard to operationalize for most agricultural decision makers today.

It is also questionable, however, that standards of living were as fixed as Chayanov asserts for Russia in the early 1900s. Although he found that the "annual personal budget is strongly correlated with family size [1966:106]," there were always differences of wealth among rural households. Even in areas without a developed kulak class, some farmers surely were aware of differences in diet, dress, houses, and farm investments. It is hard to accept the idea that, as a couple's children reached adulthood and could spread among themselves the required family labor, labor investment per person would necessarily drop so that consumption levels would remain constant (Chayanov 1966:218), while other households, nearby, enjoyed meat more frequently, wore nicer clothes, or were able to give their daughters bigger dowries. Although it is not necessary to assume that all agriculturalists throughout history behave from a desire to endlessly increase material possessions, it is doubtful that in areas of class differences among farmers, where consumption standards in the community vary, poorer households are always content with their lot and do not seek to increase their standard of living. This last point must, of course, be seen within the context of stable market forces and population density specified by Chayanov. The closed corporate community in Mesoamerica may be an example in which consumption standards were fixed from external and internal pressures (Wolf 1957). Failing to find such pressures on household consumption, however, this aspect of Chayanov's theory raises some doubts.

Distinctive Characteristics of the Family Farm

Chayanov's theory was developed in a situation very different from current conditions in most of the world today. Although stable market prices, relatively abundant land, and few nonagricultural alternatives for rural labor no longer characterize the environment in which most farmers make decisions, his fundamental points for understanding household economy still may be accurate. His methodology gives primacy to the returns to household labor and was derived from observed differences between the behavior of family farms and that of capitalist farms. Chayanov lists six examples of how a family farm acts differently from a capitalist firm that must evaluate profits based on paid wages (1966:39–48).

Without going into detail on each of the six examples cited by Chayanov, three of them represent evidence that family farms sometimes choose options that give returns to labor that are lower than the going wage. Such a method for evaluating family decisions is not incompatible, however, with traditional economic calculations. Chayanov gives good evidence that the opportunity cost of labor may be lower than the going wage, and therefore households will make decisions differently from a firm that must pay that wage. Although such decisions are different, the evaluation methodology used to reach them need not be. Costs of labor clearly are lower for the family farm, but both enterprises may be weighing a cost to labor.

One of Chayanov's examples will suffice: Peasants in one area rejected new threshing machines because displaced family labor had no alternative use at that time of year. The cost of the machine was therefore seen as unattractive, though a capitalist firm would have found it very attractive, since its purchase price was considerably lower than the wages of the laborers it replaced (Chayanov 1966:39). Chayanov cites this rejection of the threshing machine as an illustration that capitalist accounting methods cannot explain the "peculiarities in peasant farm organization." This case is not incompatible, however, with capitalist accounting, so long as the opportunity costs of labor can be seen as lower than the going wage. Since the labor displaced from threshing has "no alternative," its value can be placed at or near zero. This value is clearly below the opportunity cost of the capital needed for the machine, thereby making an allocation of labor in threshing cheaper than an allocation of capital. Two others of the six examples can be similarly reconciled.

Another two of Chayanov's examples deal with the family farm's capacity to move labor out of agriculture into crafts and trades, should returns there exceed those in farming. Whereas most capitalist firms do not have quite the flexibility of the household organization, diversification is always possible, and capitalist firms may have a certain advantage in that labor can be fired and the firm's "needs" reduced quite quickly in that manner. Neither of these two examples requires Chayanov's alternative cost–benefit method. Chayanov's sixth example involves his observation that consumption levels of the family farm will remain constant, and that labor investment will drop if prices (income) rise. Regardless of its accuracy for the Russian peasants described, there is insufficient evidence to conclude that constant standards of living are the goal of Third World agricultural decision makers today.

Although Chayanov's six examples of the "peculiarities" of the peasant farm are not compelling evidence in favor of his proposed methodology, his points about agricultural decisions may still be useful. Data from research in Paso, Costa Rica, a farming community of 77 households, will be used to test the following six aspects of Chayanov's theory:

1. The family cycle, as expressed in the size and composition of the family, plays an important role in agricultural decisions.
2. Profits for family farms are calculated by subtracting cash costs from the gross proceeds of the farm (opportunity costs for noncash items are not included).
3. Family farm resources are allocated according to which option gives the highest returns to labor.
4. The attribution of a wage to family labor distorts the decisions being studied.
5. Smaller farms will accept lower returns to labor than households with more abundant land.
6. Bigger families will accept lower returns to labor than smaller families.

The Family Cycle and Agricultural Decisions

The influence of the family cycle on agricultural decisions can be measured most clearly in Paso by looking at the decision to plant tobacco. Tobacco is a very labor-intensive as well as capital-intensive crop, requiring (per land unit) three and one-half times the labor and two and one-half times the capital cost of coffee, the next most intensive option. A decision to adopt tobacco production represents a decision to attempt a difficult, risky, and costly crop, but one that produces much higher profits and better grain yields (in rotation with the tobacco harvest). Tobacco production also represents greater technological sophistication, since the soil is ridged into contour terraces and the farmer must apply large amounts of insecticide and chemical fertilizer.

For the purposes of testing Chayanov's theory, agricultural decision data are used only from small (.20–4.80 ha) and medium (5.30–19.30 ha) farms. Landless farmers and large farmers have such different resource bases from the farms studied by Chayanov that their situations may be said to distort this test of Chayanov's method (see Barlett 1977). The small and medium sized Pasano farms generally are owned and operated by nuclear families. Separate households in Paso are formed at marriage. Some couples receive gifts of land from their parents at that time, but most have purchased land prior to marriage or do so as soon as possible thereafter. Land accumulation continues for some households up to the death of the husband. Unmarried sons usually stay and work with their fathers, though some may be given personal plots from which they can accumulate savings toward their marriage. The family organization of these Costa Rican farms is thus quite similar to the Russian households studied by Chayanov.

The small and medium households in Paso can be divided into five life-cycle categories, according to the number and age of the males in the household. Since agriculture traditionally is considered the male domain, male workers are of greater importance in determining agricultural decisions than are female workers. The effect of the total number of consumers (to test the labor–consumer balance) will be discussed. Three households in the sample can be categorized as having no men; all are widows with no children living at home, or with only young children. A fourth household is made up of a widow and her adult sons. Of households consisting of a couple with their children, there are three significant groups: those in which the husband is still under age 50; those in which he is over 50 but has adult sons to help him; and those in which the husband is over 50 but has no adult sons at home. Since tobacco requires strenuous labor to build the terraces, and long hours even in seasons of lighter work, having a strong man in the house who is capable of doing tobacco work makes an important distinction between these households. Table 6.1 shows the percentage of each of the family-cycle categories that have chosen to produce tobacco.

Table 6.1 shows that 81% of households with a man under 50 have chosen to intensify by producing tobacco. None of the nine households without a man under 50 has chosen to do so, and many have stated "Tobacco is too much work." Discussions with Pasanos clearly reveal that household composition affects the tobacco decision (see Barlett

TABLE 6.1
Tobacco Production by Family Category

	Total N	Percentage producing tobacco		Total N	Percentage producing tobacco
Widow with adult sons	1	100	Widows, no adult sons	4	—
Husband under 50	22	77			
Husband over 50, with adult sons	8	88	Husband over 50, no adult sons	5	—
Total with man under 50	31	81	Total, no man under 50	9	0

1977). In comparing the household of a man 45 and that of one 55, where all other conditions are the same, neither the traditional economic cost–benefit analysis nor Chayanov's analysis of profit would predict that one would choose the more strenuous work whereas one would not. Thus, the data on agricultural intensification among the small and medium farmers in Paso suggest that the degree of drudgery involved is important in considering a crop, and that decisions are not made solely on profit calculations.

Chayanov would predict, however, that households choosing not to produce tobacco would be smaller and have lower consumption needs. Table 6.2 compares the total adult equivalents of each of the household categories. Adult equivalents are calculated by including children under 10 as .5 of an adult and children 10–15 as .75, following Schultz 1945:114. Table 6.2 supports Chayanov in that larger households tend to intensify their agricultural production ($p = .008$ by two-sample t test). The average number of adult equivalents for households with men un-

TABLE 6.2
Adult Equivalents by Family Category

	Mean AE		Mean AE
Widows, with adult sons	2.0	Widows, no adult sons	3.6
Husbands under 50	5.8		
Husbands over 50, with adult sons	7.2	Husband over 50, no adult sons	2.6
Total, with man under 50	6.1	Total, no man under 50	3.0

der 50 is 6.1. In contrast, households with no man under 50, average only three adults. Table 6.2 demonstrates the importance of household labor resources in predicting which groups are more likely to increase their labor and capital investments.

The relationship between adult equivalents and the tobacco decision itself is slightly less strong. The average adult equivalent for households who grow tobacco is 6.2. The average for those who do not grow tobacco is 4.0. Within these averages, there is considerable variation, however. Households who choose not to plant tobacco vary from one adult equivalent to 9.25. Tobacco growers vary from 2.0 to 11.75 adult equivalents, and four households have fewer than four adult equivalents. Thus, there is no fixed cutoff at which intensification becomes necessary, though the general tendency is clear; neither would it be possible to assert that all small and medium farmers have chosen to intensify with tobacco to maintain a constant standard of living, as Chayanov would assert. Many of these households are increasing their standards of living through the high profits of tobacco. The data do support, however, the general importance of both family cycle and household size in the decision to undertake this one key agricultural option. Traditional economic calculations would be unlikely to elucidate these relationships, since family "needs" are not included in the profit calculation. The utility of this methodology depends on the opportunity cost that is charged for farm labor. This question will be explored in the next section.

Profit Calculations and Returns to Labor

To compare the value of the Chayanovian methodology with the traditional economic cost–benefit calculation, profits for two land uses were computed by both methods. Tobacco is rotated with corn and beans in Paso and competes directly with traditional corn and bean production. These two crops are not only the most important crops for small and medium farmers, but are also the most directly comparable of all the land uses. These family farms allocate a total of 39.5 *manzanas* to tobacco and 15.25 *manzanas* to traditional grains (one *manzana* equals .69 ha and 1.7 acres).

To obtain a profit figure according to traditional economic conventions, a "rent" was charged to all land used, based on an average of current rents for both crops. Tools, tobacco sheds, sprayers, and other equipment needs were seen as capital investments, and their cost was depreciated over their estimated life. Unpaid family labor was valued at the going wage, and interest was calculated on all capital investments and also for inputs purchased on credit. Transportation costs, paid labor, and purchased inputs also were deducted. The gross income of the crop was derived by valuing produce that was sold at its selling price and produce consumed at home at its consumer purchase price (Chibnik 1978). All costs and benefits were prorated per *manzana*. Thus, the profit figure left, after returns to each of these factors of production were subtracted, represents the traditional economic calculation of returns to entrepreneurship.

Chayanovian profit figures were obtained by calculating the gross value of the harvest in the same way, but by subtracting only paid labor, cash inputs, transportation, and other costs that actually were sustained in the year under study, such as for a tool shed or sprayer. Table 6.3 presents the results.

Capitalist calculations severely lower the profitability of each crop, but the farmers' preference for tobacco is clear from both calculations. Tobacco, according to Chayanov's profits, is 4.3 times as profitable as traditional grains, whereas, with a capitalist calculation, it is 5.1 times as profitable. The low return per *manzana* for traditional corn and beans by a capitalist calculation, approaches the rental value of the land (average ¢285 if used for tobacco, corn, and beans). The farm family, however, benefits much more than just ¢296 from a *manzana* of traditional grains. The imputed rent, wages, and interest join this profit figure as a combined family "income," a figure that is obviously much higher than the possible rental value. By comparing these two options in Table 6.3, we can conclude that neither method is better to predict the desirability of tobacco over traditional grains. Since the traditional economic calculation, however, involves much more work on the part of the researcher, and, if both methods are equally valid, Chayanovian profit figures are preferable in the interests of researcher efficiency.

TABLE 6.3
Profitability by Two Methods of Calculation (per Manzana, in Colones)

	Chayanovian calculation	Traditional economic calculation
Tobacco, corn, and beans	¢3887	¢1509
Traditional corn and beans	¢901	¢296

Chayanov's profit calculation can be transposed to compare the over-all returns per day of labor,[3] once cash costs are paid. Comparing the returns to labor of tobacco, corn, and beans with the traditional corn and bean production shows that tobacco gives only a slightly higher return—¢16.02 versus ¢15.45. For both crop options, however, there are two farmers who make extraordinary profits. Dropping the top two cases from each category, the return per labor unit in tobacco is some-what more profitable in comparison to traditional grains—¢11.35 versus ¢9.59. Both of these returns are obviously far above the going wage at the time of this research—¢6.00.

This second use of Chayanov's calculations presents an important contrast to the profitability figures presented. Table 6.3 leads the re-searcher to assume that tobacco is so vastly preferable an option that no one would continue to plant traditional corn and beans. Yet the farmers studied here planted almost half as much land in traditional grains as in tobacco. The returns per labor unit show that tobacco is preferable, but that family labor can be profitably invested in the traditional crops as well. Whereas a capitalist farm that uses hired workers would be more interested in the overall returns to the farm, rather than the returns to labor, in the family farm where labor resources are more fixed, the profitability of traditional grains per labor unit becomes important. Since family farms cannot hire large numbers of workers to put all their land in tobacco (nor can they lay off household members), the returns to year-round family laborers must be considered. The use of opportunity cost calculations in a traditional economic analysis can accommodate some of these decision criteria, but they would involve complex esti-mates of labor value for each crop, throughout the year.

Possible Distortions in the Traditional Economic Calculations

Turning now from these aggregate figures to individual cases, the capitalist accounting method may tend to distort the decisions of some households. Of the total 23 small and medium households, four obtain negative profits from their yearly activities, when calculated by the traditional methods. One case provides an example. Paco Gutierrez

---

[3] "Days" of labor as used here refer to the traditional *jornal* of 6 hr. At most times of the year, farmers work from 6:00 A.M. until the rains come around noon. Only rarely is agricultural labor possible in the afternoon, and in such cases, the day's work is counted as more than one "day."

owns three and one-fourth *manzanas* of land and the total profit from his farm using Chayanovian calculations in 1972 was ¢4181, although traditional calculations showed him ¢−246.

This negative profit figure can be interpreted two ways. First, Paco can be seen as a "poor" farmer or a "bad" entrepreneur; he does not manage to obtain even the opportunity cost of the resources invested in his farm. The traditional calculation makes it easy to forget that household income includes the imputed wages, rents, and interest as well as "profit." Paco is not destitute; his family is well clothed and fed. Paco travels by bus to shop every week, and is generous in donating time to community projects.

The second interpretation is that the calculation itself is incorrect. This case can be seen to fit Chayanov's assertion that some households will accept lower returns to labor. The key to the negative profit figure is the assumption that the opportunity cost of labor is equal to the going wage. By assuming the opportunity cost of Paco's family labor to be zero, the farm's profits jump to ¢3137. Using the Chayanovian method of calculating returns to labor (which does not subtract opportunity costs for any noncash use of resources), Paco gets overall ¢8.01 per day of labor on his own farm. He gets ¢7.16 per day of labor even in the labor-intensive tobacco, corn, and beans rotation. Although these figures are clearly below the community average, they are still higher than wage labor.

The preceding figures show that if the calculation of profit is changed to reward family labor at slightly below the going wage, Paco can get full returns to land, capital, and entrepreneurship. Likewise, the figures can be shuffled to show lower returns to capital or land, and a regular return to labor. In any case, it is true that this household conforms to Chayanov's expectation that some families will accept lower returns to labor than would be possible for a capitalist firm. All four of the cases showing negative profits can be made to show positive profits by lowering the opportunity cost of labor.

To understand the meaning of these negative profit calculations, we must turn from a consideration of land use choices to a consideration of labor allocations. Some farmers "lose money" because they invest "too much" family labor, given their overall production. This issue again clarifies the difficulty of determining the appropriate opportunity cost for unpaid family labor. Some opportunity costs will show many farmers making "bad decisions" in keeping labor on the farm, when it could make more money elsewhere. Other levels of imputed wages will show that all farm activities are very profitable, even activities that farmers see as marginal. The fact that opportunities vary from family to family and

from month to month further compound the difficulties of determining
an accurate value. Whereas a researcher using the traditional calculation
may know that wage work is available nearby and uses that wage as the
opportunity cost of family labor, such a figure does not take into account the family's assessment of the desirability of that opportunity.
Working conditions may be such that the going wage is too low, in their
estimation. Perhaps farmers feel that farming provides more security
than job opportunities. Qualitative factors such as working conditions
also must be incorporated into an assessment of opportunity cost, as
will be discussed in the last section.

Returns to Labor—Smaller Farms

Another way of looking at this return to labor is to ask how much
time the household invests in each crop. Chayanov would expect that
poorer households, with their subsistence needs unmet, would invest
more labor, even though the returns to that labor may be relatively
marginal. Table 6.4 confirms Chayanov's expectations, comparing the
four households with negative profits to those with positive profits.

Table 6.4 shows that families with negative profits invest considerably more labor than the average of cases with positive profits.[4] For
tobacco and grains, they use 34% more labor, and with traditional
grains, 41% more. These higher labor investments also correlate with a
larger household size; cases with negative profits have higher average
adult equivalents than those with positive profits. This table suggests

TABLE 6.4
Labor Investment and Family Size

	Tobacco and grains		Traditional grains		Total N
	Mean days invested per manzana	Mean AE	Mean days invested per manzana	Mean AE	
Cases with negative profits	421	8.5	77	6.9	4
Cases with positive profits	314	6.4	55	5.4	19

[4] Because there are only four cases of negative profits, and because the Kruskal–Wallis
H test does not have much power, the differences between these cases and the 19 cases of
positive profits are not statistically significant. The total N in Table 6.4 is lower than in
Tables 6.1 and 6.2 because comparative data on tobacco production and traditional grains
were not available for all small and medium farms. Although the means in Table 6.5 are
interesting, they are likewise not statistically significant.

the possibility that some households do not obtain sufficient income to repay themselves a standard wage because they have excess labor in the family and invest it in agricultural pursuits without expectation that it will be repaid fully.

Given that the market for wage labor in Paso is limited, Chayanov would expect families with smaller farms to invest more labor per land unit than families with abundant land. This hypothesis tests the extent to which the four cases just discussed are part of a larger group. The data from Paso, however, contradict this idea. The average labor investment of small farmers in the tobacco, corn and beans rotation is 322 days, whereas medium farmers invest an average of 330 days. Although the labor investments of these two groups are similar, family size is not; tobacco-growing small households average 5.9 AE, whereas medium households average 7.2. The return per labor unit is lower for the smaller farms—¢10.01 per day—than for the medium farms—¢13.19 per day. Thus, the data show that households with less land do invest more labor in tobacco, corn, and beans per family member (but not per family or per land unit), and they get less profit back from this work than do medium landholders. It cannot be concluded, however, that small farmers in Paso show a higher level of "self-exploitation." When the whole farm and all its activities are taken into account, farmers with medium-size farms cannot be said to work less hard than those with small farms. Second, medium farmers have been in tobacco production longer and have mastered the technology better than many of the small farmers who are new to this difficult crop. Their higher profits and lower self-exploitation are thus partly attributable to experience. Nevertheless, it is clear that the small farmers are willing to accept a lower average profit and a lower return per labor unit.

Returns to Labor—Bigger Families

Chayanov has asserted that family size is the most important determinant of labor investments for family farms. For the small and medium farmers studied here, family size correlates not only with the amount of labor invested in tobacco, but also with the profit obtained per *manzana* (by a Chayanovian calculation). Returns to labor, however, are remarkably constant for all three of the family size groups in Table 6.5.

Table 6.5 shows that larger families spend more days at work per *manzana*, despite the fact that they tend to grow more *manzanas* of tobacco than smaller families. These larger households have produced tobacco longer than the smaller households, and this experience ac-

TABLE 6.5
Aspects of Tobacco Production by Family Size

	AE under 5.75 (N=5)	AE 5.75– 6.60 (N=6)	AE over 6.60 (N=6)
Days per *manzana*	282	330	360
Average profit per *manzana*	¢3432	¢3732	¢4422
Average profit per day	¢ 12.17	¢ 11.31	¢ 12.28

counts for their higher profits. The return per hour does not increase with family size, however. Overall, those households with larger families do not get a better labor return in tobacco, but neither are they forced to accept lower returns, as Chayanov would predict. Larger families invest considerably more labor than small families, but their returns are about the same.

Qualitative Costs and Benefits

So far, several problems have been noted with assessing the opportunity cost of labor. The difficulty of tobacco work is a factor that inhibits households with no man under 50 from adopting tobacco as a land use. The desirability of employing family labor year-round is a factor in assessing land uses and their profitability. A low return on tobacco production also can be seen as an investment in the future mastery of that technology. These and other factors make the calculation of specific figures for all costs and benefits very difficult. Part of this difficulty stems from the values inherent in the traditional economic methodology: Either implicit or explicit, priority is given to returns to capital (Gittinger 1972:6). All aspects of costs and benefits are assumed to be reducible to market prices and are expressed in a quantitative calculation. Such a quantification procedure, however, may tend to ignore values attached to the different options.

Belshaw and Hall (1972:45) note the multiplicity of goals that characterize decisions in "small-scale peasant agriculture": a range of material comforts, stable food supply, reasonable social status, satisfactory family life, participation in customary social activities, and some security against misfortune. Households also may forego certain profitable agricultural options to invest resources in social institutions, which often act as buffers against risk (Berry 1977:26; also see Berry, Chapter 13 in this volume). These "social constraints" are often seen as "noneconomic" factors affecting peasant farmers, but in fact they may have important economic roles. Since farmers must maximize many goals simultaneously, any attempt to force the complexity of their decisions into a profit orientation will necessarily leave out some considerations. In addition, the diverse resources, knowledge, and goals of each household affect their evaluations of each option, thereby adding another problem to the standardization of opportunity costs.

A number of writers have expressed doubt about the utility of traditional cost–benefit calculations for Third World farmers. Yang noted that there may be no alternative uses for labor, farm buildings, or by-products and "in consequence, the concept of opportunity cost becomes unrealistic, particularly in underdeveloped areas. . . . Secondly, the determination of opportunity costs can be arbitrary and, as a result, the production cost derived from opportunity costs can be erratic [Yang 1965:88–89]." Epstein's research on Indian farmers showed that their own evaluations of labor, manure, and fodder did not match her opportunity cost estimates. Epstein concluded that an accurate description of farmers' own decision processes is not compatible with opportunity cost calculations: "To the farmer . . . the difference between *total output* and cash input are the important criteria in his choice of crops cultivated [Epstein 1962:222]." Edwards tried to apply traditional cost–benefit analysis to Jamaican small farmers and concluded: "The great practical difficulty of determining the opportunity cost of such resources and the differing quality and layout of the land combined to make the comparison of managerial ability extremely difficult for many of the farmers [Edwards 1961:182]."

Scandizzo responded to similar problems with financial analysis of agricultural projects by proposing that costs and benefits should be assessed from the farmers' point of view, which may "then be a useful complement to the largely fictional budget required by the traditional financial analysis [Scandizzo 1978:6]." On the basis of the research just cited and the Costa Rican data presented here, the value of this dual methodology is not clear. If traditional economic methodology that quantifies all opportunity costs does not predict behavior as well as the simpler Chayanovian calculations, then perhaps the "traditional financial analysis" is a waste of time. Those development experts who rely on financial analyses will have to decide if its benefits are outweighed by the oversimplifications and distortions inherent in the calculation of opportunity costs.

Thus far, we have been discussing alternative methodologies for financial analysis of small farms, but many of these same issues are inherent in economic analysis as well. There is also a wide range of variables to take into account when assessing costs and benefits of development projects and many of them are hard to quantify. Dickinson's example of the opportunity costs incurred in a hydroelectric dam project is a good illustration. The dam's value must be weighed against the costs of "the displacement and marginalization of a productive agricultural population; the permanent loss of fertile irrigated land and its production; the intensification of pastoral activity on the watershed area, leading to erosion and a shorter life for the reservoir; and the downstream loss of riverborne nutrients and consequent reduction in fishing and flood-plain agriculture [Dickinson 1978:90]."

In light of the difficulty of including all these qualitative costs and benefits in a quantitative calculation, it is preferable for either financial or economic analysis to keep quantified estimates of profitability or the desirability of any one option in a relatively simple form. The cost–benefit analysis then can be supplemented by qualitative discussions of the more complex opportunity costs and assessment criteria that must be taken into account.

It might be assumed that these limitations in the traditional economic methodology are due more to the imperfect factor markets of a developing country than to any problems inherent in the methodology. Yet the problems with quantifying opportunity costs for the Costa Rican situation described here exist because each farm not only has different land, labor, and other resources, but also has different and sometimes conflicting goals and needs. This same complexity is expectable in the analysis of virtually any farm or firm, suggesting that rigid use of traditional economic methodology in any country often may distort the decision process under study and also, as shown here, may be less useful for predicting some kinds of behavior. Chayanov's methodology of cost–benefit analysis, since it is based on actual prices paid and costs incurred, leaves less room for researcher bias than do traditional economic calculations that must impute values to unpaid costs. Since an outsider's assessment of opportunities foregone may miss some of the qualitative factors noted, the possibilities for distortion are greater, and the chances of accurately portraying farmers' choices are thereby decreased. The data from Paso have confirmed the utility of Chayanov's formulations and suggest that his approach is preferable to more elaborate economic calculations.

Conclusion

Like farmers, researchers must weigh alternatives in the use of resources. To understand agricultural decisions, there are three common choices: qualitative analysis, traditional economic profit calculations, and Chayanovian profit calculations. This chapter has weighed the costs and benefits of each methodology, to determine the utility of each in accurately describing and predicting farmers' decisions in developing countries today.

To evaluate traditional economic cost–benefit analysis, it first must be noted that its purposes are often normative, though they may also be descriptive, and these goals may sometimes be in conflict. The required quantification of all the costs and benefits that farmers weigh is shown to be difficult, for a variety of reasons. It also has been shown that the process of estimating opportunity costs can distort farmers' decisions by leaving out important criteria in their assessment of different opportunities.

In the assessment of Chayanov's theory, several criticisms are put forward. The distinction between the nonmonetary farm and the commodity farm introduces a false dichotomy in land use decisions. Second, the labor–consumer balance as proposed by Chayanov is inherently untestable for research today, since it requires that family consumption needs remain constant. Third, none of the six examples presented by Chayanov to illustrate the "peculiarities in peasant farm organization" are found to require a cost–benefit methodology different from the traditional economic calculations.

These criticisms do not automatically invalidate the cost–benefit methodology that Chayanov proposes and the unique characteristics of the family farm that he feels make it behave differently from a capitalist firm. Six aspects of Chayanov's theory are tested with data from a Costa Rican village. First, the important role of the family cycle in agricultural decisions is seen in the willingness of households with a man under 50 to intensify their agricultural production by growing tobacco. Chayanov's expectation that intensification would begin at a clear point in the growth of family size is not substantiated, but a general relationship between large families and labor-intensive land use choices is demonstrated. Neither pattern would be predicted with traditional economic cost–benefit calculations.

In comparing the Chayanovian profit methodology with traditional economic calculations, both show that tobacco is more profitable than traditional grains, and farmers in Paso do, in fact, plant considerably more tobacco. This comparison of the profitability of two alternative land uses shows that neither methodology is more accurate.

When profit calculations are transposed into returns to labor, however, the greater profitability of tobacco is reduced, and Pasanos' continuing interest in the traditional grains option is consistent with its relatively high returns to labor. Traditional economic calculations would not predict the observed degree of interest in traditional grains. This finding suggests that family farms in Paso do make decisions on the overall needs and resources of the family, and these decisions may not be appropriate for a farm based on wage labor alone.

Looking at specific cases in which the traditional economic calculation shows negative profits, the opportunity cost of labor is seen to be the key factor. The estimation of the opportunity cost of family labor is easily distorted by the researchers' assumptions or values. Profits are closely linked to the amount of labor invested, which presumably reflects the farmers' assessment of its opportunity cost.

Chayanov predicts that smaller farms will accept lower returns to labor, and the Paso data partially support this expectation. Returns per labor unit in tobacco are lower for small farmers than for medium farmers, though the labor investment per land unit is the same. Taking the whole farm into consideration, however, there is no evidence that small farmers experience a higher degree of self-exploitation.

Chayanov also predicts that larger families will accept lower returns to labor. Although larger families do invest more labor per land unit in tobacco production than do smaller families, their higher profits (from longer experience with tobacco) result in returns to labor remaining roughly the same for all sizes of families.

Turning finally to consideration of the qualitative methodology of assessing agricultural decisions, many aspects of farmers' decisions are seen to be very difficult to quantify. With multiple goals for each household, complex resource mixes, and the diverse characteristics of the choices being considered, estimations of opportunity cost often may be more an expression of researchers' values than an estimation of the behavior of the people under study. Since qualitative factors are crucial in many agricultural decisions, they are best combined with a simple Chayanovian profit calculation. Such a combination takes advantage of the strengths of each methodology while avoiding the costs of researcher bias or distortion. The result, at least as far as the Paso data are able to illustrate, will be a more profitable understanding of agricultural decisions in the Third World today.

Acknowledgments

I am especially indebted to Donna Brogan, Leonard Carlson, Michael Chibnik, Christopher Curran, Kaja Finkler, Lowell Jarvis, Allen Johnson, Bonnie Anna Nardi, and William Shropshire for their helpful comments and criticisms on an earlier draft of this chapter.

References

Anderson, J. R., J. L. Dillon, and J. B. Hardaker
1977 Agricultural Decision Analysis. Ames, Iowa: Iowa State University Press.
Barlett, Peggy F.
1977 The Structure of Decision Making in Paso. American Ethnologist 4(2):285–308.
Beals, Alan R.
1974 Village Life in South India: Cultural Design and Environmental Variation. Arlington Heights, Illinois: AHM.
Belshaw, D. G. R., and Malcolm Hall
1972 The Analysis and Use of Agricultural Experimental Data in Tropical Africa. East Africa Journal of Rural Development 5(1–2):39–71.
Bennett, John W.
1969 Northern Plainsmen. Chicago: Aldine.
Berry, Sara S.
1977 Risk and the Poor Farmer. Economic and Sector Planning Division, Technical Assistance Bureau, AID, Washington, D.C.
Bishop, C. E., and W. D. Toussaint
1958 Agricultural Economic Analysis. New York: Wiley.
Burling, Robbins
1962 Maximization Theories and the Study of Economic Anthropology. American Anthropologist 64(4):802–821.
Cancian, Frank
1972 Change and Uncertainty in a Peasant Economy. Stanford: Stanford University Press.
Chayanov, A. V.
1966 (1925) The Theory of Peasant Economy. D. Thorner, R. E. F. Smith, and B. Kerblay, eds. Homewoood, Illinois: Richard D. Irwin.
Chibnik, Michael
1978 The Value of Subsistence Production. Journal of Anthropological Research 34(4):551–576.
Clayton, E. S.
1968 Opportunity Costs and Decision Making in Peasant Agriculture. Netherlands Journal of Agricultural Sciences 16(4):243–252.
Cochrane, Willard W.
1974 Agricultural Development Planning: Economic Concepts, Administrative Procedures, and Political Process. New York: Praeger.
Dickinson, Joshua C., III
1978 A Geographer's Perspective on Economic Development. In The Role of Geographical Research in Latin America. William M. Denevan, ed. pp. 83–100. Muncie: Conference of Latin Americanist Geographers. Pub. No. 7.
Doll, John P., V. James Rhodes, and Jerry G. West
1968 Economics of Agricultural Production, Markets and Policy. Homewood, Illinois: Richard D. Irwin.
Edwards, David
1961 An Economic Study of Small Farming in Jamaica. Glasgow: MacLehose. The University Press.
Epstein, T. Scarlett
1962 Economic Development and Social Change in South India. Manchester: Manchester University Press.

Gittinger, J. Price
    1972   Economic Analysis of Agricultural Projects. Baltimore: Johns Hopkins University
           Press.
Greenwood, Davydd J.
    1976   Unrewarding Wealth: The Commercialization and Collapse of Agriculture in a
           Spanish Basque Town. Cambridge: Cambridge University Press.
Halperin, Rhoda, and James Dow, Eds.
    1977   Peasant Livelihood: Studies in Economic Anthropology and Cultural Ecology.
           New York: St. Martin's Press.
Heady, Earl O.
    1952   Economics of Agricultural Production and Resource Use. New York: Prentice
           Hall.
Little, I. M. D., and J. A. Mirrlees
    1974   Project Appraisal and Planning for Developing Countries. New York: Basic
           Books.
Moerman, M.
    1968   Agricultural Change and Peasant Choice in a Thai Village. Berkeley: University
           of California Press.
Ortiz, Sutti R. de
    1973   Uncertainties in Peasant Farming: A Colombian Case. New York: Humanities
           Press.
Prest, A. R., and R. Turvey
    1966   Cost–Benefit Analysis: A Survey. In Surveys of Economic Theory (Vol. 3). Re-
           source Allocation. American Economics Association and the Royal Economic
           Society, eds. pp. 155–207. New York: St. Martin's Press.
Scandizzo, Pasquale L.
    1978   Financial Analysis of Rural Development Projects: An Economic Approach.
           World Bank Staff Working Paper, October.
Schultz, Theodore W.
    1945   Food for the World. Chicago: University of Chicago Press.
Squire, Lyn, and Herman G. Van der Tak
    1975   Economic Analysis of Projects. Baltimore: Johns Hopkins University Press.
United Nations Industrial Development Organization
    1978   Guide to Practical Project Appraisal: Social Benefit–Cost Analysis in Developing
           Countries. New York: United Nations.
Wolf, Eric
    1957   Closed Corporate Peasant Communities in Mesoamerica and Central Java.
           Southwestern Journal of Anthropology 13:1–8.
Yang, W. Y.
    1965   Methods of Farm Management. Investigations for Improving Farm Productivity.
           Revised ed. FAO Agricultural Development Paper No. 80. Rome: FAO.

Chapter 7

Risk and Uncertainty in Agricultural Decision Making

FRANK CANCIAN

There are close parallels between microeconomic analysis and the anthropological study of agricultural decision making. Anthropologists can learn about their own prospects for success by reviewing what economists have done. In particular, students of agricultural decision making should attend to the distinction between normative and descriptive economics. Much analysis of economic behavior, and much economic analysis of all behavior, suffers from its association with normative economics, for the goals of normative analysis are different from the goals of descriptive analysis. In the end, anthropologists usually are interested in the descriptive end of things, that is, in understanding what people do and why they do it. Economists, on the other hand, are often interested in determining what people ought to do, given specified goals and constraints that are very complex. That is, they are interested in normative problems, not description. In this chapter, I will argue that we (anthropologists) should distinguish our goals from those of economists; and that we should avoid confusing substance with logic, description with prescription, and behavior in practice with rules for making optimal decisions.

Peasants and the analysis of peasant economic behavior have both repeatedly suffered from failure to distinguish the prescriptions of normative economics from generalizations about peasant behavior in everyday situations. Normative economics applied as management sci-

AGRICULTURAL DECISION MAKING:
Anthropological Contributions to
Rural Development

ence to peasant decision-making situations yields prescriptions for rational, maximizing behavior. At the same time, it makes it possible to identify as irrational those peasants who do not follow its prescriptions. Most students of peasant agriculture are familiar with this exercise. In it, peasants' understandings of their situations are sacrificed to the pseudoincisiveness of a simple model constructed by outsiders to help them decide what they ought to do (Cancian 1966, 1972, 1974).

As I will argue in what follows, I believe that the attempt to harness the impressive techniques of normative economics in the service of descriptive analysis is misguided. It seems sensible to begin behavioral analysis with the prescriptive model, especially if we believe people are motivated by "profit" in some form or another—but there are crucial differences between calculating and behaving. They make basic normative principles misleading first approximations to behavioral generalizations. The false sense of security induced by the path through prescription to description is particularly great with regard to the concepts of risk and uncertainty that are the focus of this chapter.

My argument involves three steps. First, I will give a brief and somewhat oversimplified description of normative decision analysis that reveals why and how it obviates the risk–uncertainty distinction. Then, I will review an empirical study of agricultural decision making that uses the risk–uncertainty distinction to considerable advantage. This second step involves data on more than 3000 farmers in 16 communities in eight countries. Finally, I will treat the difference between the first step and the second step as a paradigm switch involving the risk–uncertainty distinction and other characteristics of normative analysis that are more widely recognized to be problematic in descriptive analysis. Together these characteristics make the normative decision-making model a particularly poor guide to the study of agricultural decision making.[1]

The distinction between *risk* and *uncertainty* that I want to use was made famous by the University of Chicago economist Frank Knight. In 1921, Knight distinguished measurable uncertainty or risk from true, unmeasurable, uncertainty (1971:20 and Chapter VII). In simple terms, it is risk in situations in which one knows the probabilities of various possible outcomes of an action; uncertainty in situations in which one

[1] Since I believe that most people are like farmers in that much of the time they act under substantial uncertainty, my arguments are meant to apply to the analysis of most behavior. The importance of uncertainty in decision-making behavior is not just some specialized concern that will become rarer and rarer as science and education dispel our remaining bits of ignorance about the future. It is a feature of everyday life that will continue to make it difficult to unify normative and descriptive analysis.

cannot specify the probabilities.[2] Many contemporary economists do not follow Knight. Instead, they commonly use the terms to refer to different aspects of the same situation, and sometimes they argue that the concepts cannot be distinguished in rational practice.[3] As I will try to show in the following discussion, the characteristics of microeconomic theory that lead economists to these conclusions stem from the demands of normative analysis.

Normative Economics: Decision Making without Distinguishing Risk and Uncertainty

Suppose you are a rational individual faced with alternatives that invite choice. Normative decision analysis can help you get the most out of what you know. It is intended to help you determine the choice that will maximize attainment of your goals. It provides a powerful calculus that is not limited to money profit.[4]

In its simplest version, the normative approach starts with perfect knowledge about the alternatives. From this and your goals or values, the appropriate choice can be calculated.

Where risk is involved, that is, where you lack perfect knowledge and certainty, but do know probabilities of various alternatives, normative analysis may be complicated to produce an expected utility solution. Probabilities may be used to specify the decision that will maximize your goal attainment over the long run. For example, even though you may lose at first, you are advised to toss a fair coin with a person who will give 51¢ for heads and take 49¢ for tails. Risk is involved, but by assuming that outcomes can be treated as aggregatable independent events, it is possible to make probabilities equivalent to the perfect

[2] Although I use Knight's distinction, our purposes are radically different. Knight sought to explicate a quasi-philosophical framework for understanding the origin of profit. I want to predict behavior from social position.

[3] Hirshleifer and Shapiro (1977:181n) refer to a tradition that "attempts to formulate a distinction between risk and uncertainty based on ability to express the possible variability of outcomes in terms of probability distribution" but reject it: "This distinction has proved to be sterile. Indeed we cannot in practice act rationally without summarizing our information (or its converse, our uncertainty) in the form of a probability distribution [Savage 1965]."

[4] See, for example, Kassouf's text *Normative Decision Making* (1970). Herbert Simon's foreword begins: "This volume is aimed at providing a clear and concise introduction to modern ways of conceptualizing the decision-making process—ways that have provided powerful analytic tools for the complex decision-making tasks of today's managers and organizations [p. iii]."

information employed in the simple decision-making model. This transformation is the first step down the road that leads away from simple utility maximization under certainty.

The next step is the big one. The real problem for normative analysis comes when the decision maker does not know the probabilities, for the apparatus of calculation depends on being able to fix the odds (or their functional equivalents). What is to be done when the odds are not known? That is, what is to be done when there is uncertainty?

Savage's (1954) work on subjective probability solved the problem for normative economics. Simply stated, Savage showed that, under uncertainty, the best way to achieve the maximum is to use whatever you do know to make your best guess about the probabilities. That is, he showed that uncertainty mitigated by even the little bit of "information" in the unspecifiable intuition of an expert (or client) is distinctly better than no information at all in reaching the goal of maximization (given the aggregation assumption). To use this approach to advantage, the decision maker must summarize his or her knowledge in a single probability estimate. In practice, this probability estimate must be based on an act of will that transforms uncertain reality into calculable probability—uncertainty into risk. With this step, Savage effectively expelled the noncalculable part of uncertainty from normative economic analysis:[5] This made it sensible for economists to talk about "risk and uncertainty" as a single concept in economic analysis, for, from the subjective probability point of view, they cannot be distinguished.

At the same time, it is this step that makes it necessary for normative and descriptive analysis to part ways. Though it may be phrased in many ways, the problem is really at the paradigmatic level, and it is best explored by starting afresh as I will in the next section.

First it may be useful to glance at one result of applying a subjective probability approach in descriptive analysis. This example reveals, I think, why *best guesses* that are extremely useful in normative analysis may be extremely misleading in descriptive analysis.

In his study of Mexican corn farmers, O'Mara (in press) asked adopters and nonadopters of a new program to state their beliefs about likely yields from the old and new practices. Each farmer estimated (in the form of best guesses about subjective probability distributions) the

---

[5] The expulsion of uncertainty may be traced in texts like Kassouf 1970. Kassouf discusses decision making "in the absence of probabilities" but introduces the subject by saying that "there is no completely satisfactory way to handle these situations [p. 66]." Though I am not technically competent to evaluate the approaches to decision making in the absence of probabilities that he covers, and the more sophisticated ones he does not cover, none seems to reverse the reduction of uncertainty inherent in Savage. Like Savage, they require the decision maker to crystallize complexities by an act of will.

probability of various yield levels under each set of farming practices. Users and nonusers of the new practices differed: Among farmers estimating yields for both old and new practices, adopters of the new practices gave much lower estimated yields for the old practices than did users of the old practices. There was no corresponding difference in estimates for the new practices; users and nonusers produced similar estimates.

When each farmer's behavior was compared with best practice estimated by using his own subjective probability estimates of yields, 55 of the 66 farmers were found to be following practices that were unambiguously more profitable to them than the alternative. In large part, of course, the high proportion of "economic" decisions flows from the difference in beliefs (subjective probability estimates). As students of decision-making behavior, we are faced with the problem of determining whether the decisions followed the beliefs, or the beliefs followed the decisions.[6] Have we learned that farmers are rational decision makers who use a subjective probability logic, or simply that they can think through decision problems in an expected value rhetoric when confronted with the need to "rationalize" their behavior?

Such a confusion is never a concern for normative analysis, for, by definition, the belief comes before the decision. But, treating the farmer's subjective probability estimates (beliefs about yields) as a full summary of his concerns is not as desirable when it comes to descriptive analysis—for two reasons. First, as just noted, if taken after the fact, the estimates may, understandably, be rationalizations more than anything else. Second, insofar as they are pure best guesses about yields, they may fail to incorporate other factors that may be important to the farmer. I want to explore two such factors, uncertainty and social position, or rank, as they interact. Neither is customarily included in decision-making analysis based on a parallel to normative economic analysis. Both, however, are important to descriptive analysis.[7]

[6] O'Mara points out this problem in an earlier paper (1972).

[7] This quick characterization in this section of the chapter fails to reflect the sensitivity and power of efforts by contemporary economists interested in analysis of agricultural behavior under risk and uncertainty. Roumasset (1976), especially Chapter 2, "Decision Theory and Appropriate Models of Choice under Uncertainty for Peasant Farmers," is an important contribution. Roumasset (1979) is an excellent collection of papers from the Agricultural Development Council Conference on Uncertainty and Agricultural Development, Mexico, 1976. Berry (1976) is particularly sensitive to distortions in descriptive work that are introduced by decision-making analysis oriented to normative theory.

It is also worth calling the reader's attention to Barlett (1977) and Quinn (1978) and the work in anthropology they review, and to the work in psychology reviewed by Slovic *et al.* 1977.

Decision Making under Uncertainty: Predicting Behavior

Standard decision-making analysis like that just described does not permit a meaningful distinction between risk and uncertainty. It always reduces uncertainty to something that approximates risk. Thus, in order to even discuss a useful distinction, something of a paradigm shift is required. The shift is made in the following section and discussed later.

The classic diffusion of agricultural innovations studies done by rural sociologists (Rogers and Shoemaker 1971; Rogers 1976) provide a good opportunity to specify a distinction between risk and uncertainty. Suppose a new, high-yielding seed is being introduced (it could be hybrid corn in Iowa in 1935 or dwarf wheat in Pakistan in 1968). Individual farmers must decide between buying seed and sticking with local varieties for at least another year. In this situation, it is likely that they face both risk and uncertainty.

The risk they face in the production process has principally to do with climate. Rainfall, for example, can severely affect the outcome of any decision a farmer makes. Since the experienced farmer knows the approximate likelihood of different rainfall patterns in his local area, the variability in production due to rainfall represents risk. The frequency of different rainfall patterns over the long run is known and can be used to inform decision making each year.

The uncertainty farmers face is quite different; it stems from the new practice being considered. The adoption of new seed involves uncertainty, for nobody really knows what the seed will yield under local conditions; there is no local experience with which to estimate its response to locally available inputs. The probable outcomes cannot be calculated very well compared to the probable outcomes for the traditional seed. Farmers must work out the implications of this ignorance for their farming decisions.

This specification of the distinction between risk and uncertainty may be incorporated into a theory relating adoption behavior to economic rank in such a way that confirmation of the theory yields an illustration of the importance of making the distinction. As will be apparent, the distinction is a crucial part of the theory that predicts changes in the adoption behavior of richer and poorer farmers over stages in the spread of an innovation. Since the full theory has been stated elsewhere (Cancian 1972), I will give only a sketch of the principal features that are relevant to the present discussion. In earlier statements of the theory (Cancian 1967, 1972, 1979a) the conceptual and terminological distinction between risk and uncertainty is not as fully developed as it is here.

The theory is based on three fundamental notions: First, that all farmers prefer higher economic rank in their community to lower rank; second, that, in general, rich farmers are more likely to adopt new practices than are poor farmers, because their resources make them more able to afford fixed (indivisible) costs involved in any innovation, and more likely to be able to survive anticipatable fluctuations in output; and third, that under uncertainty, all farmers face the possibility that they may suffer heavy losses, and each must compare what he has to gain against what he has to lose in what is essentially a random draw. Given these principles, the poor farmer may be more willing than the rich farmer to adopt when there is uncertainty because, whatever the potential loss, he cannot sink much lower in the local socioeconomic structure. Since starvation is unlikely for social reasons, the uncertainty is less of a threat to him than it is to the rich farmer. Thus, under uncertainty, as distinguished from risk, we should expect the poor farmer to adopt more than the rich farmer.[8]

This conceptualization can be somewhat confusing at first because two opposite tendencies are seen to have a single empirical manifestation in adoption behavior. How is the rich farmer's tendency to adopt more under risk to be distinguished from the poor farmer's tendency to adopt more under uncertainty? The answer to this question is provided by a natural situation where uncertainty varies while risk stays fairly constant.

When an innovation is introduced to a community of farmers from outside, some farmers adopt it immediately, and some adopt it in later years. Later adopters usually use the experience of early adopters to inform their decision. Thus, uncertainty is greater for the earlier adopters than it is for the later adopters. Risk remains fairly constant.

This gives us a critical test in the form of two predictions. First, if uncertainty is meaningfully distinguished from risk, poor farmers should adopt more, relative to rich farmers, in the early stages of the

[8] Many people find it hard to accept this conceptualization of the implications of rank for behavior under uncertainty. Some tend to psychologize the logic, attributing different personality characteristics to people at different ranks (which just pushes effective explanation another level away). My argument employs a "universal person" whose behavior is rank-seeking in every rank position. It characterizes ranks, not the people who occupy them. It may be useful to think of the ranks as a ladder and the actors under uncertainty as people stepping off it in darkness hoping that their feet will land on a platform from which there will be a better view come morning. Under these conditions, the lower-ranking person has an equal chance of improving his or her position and a lesser risk of injury from falling. In any case, of course, the rationale for the prediction should be separated from the prediction itself. If the prediction is confirmed, alternative explanations (rationales) are possible. If it is not confirmed, the problem disappears.

spread of an innovation. Second, in later stages, the rich should be relatively faster adopters. These points are illustrated in Figure 7.1.

In the real world, a variety of special considerations apply to the very rich and to the very poor (Cancian 1967, 1972, 1979a). Thus, I have confined my predictions to the behavior of the middle of the wealth continuum in agricultural communities. These predictions are illustrated by the solid lines in Figure 7.2. In the language of concrete research, it is predicted: First, that the low middle rank will have a higher adoption rate than the high middle rank in Stage 1 of the adoption process, and second, that the relationship will reverse in Stage 2 of the process.

Data and Tests

These hypotheses can be tested with data from diffusion of innovation studies (Rogers and Shoemaker 1971). Most such studies include all the farmers in a community, or a representative sample of them. For each farmer, the date he adopts an innovation is usually recorded along with information on a variety of other variables including size of farm or amount of farm income. Thus, we can ask how date of adoption (or the number of innovations adopted by a given date) relates to a farmer's economic rank in his community, and how that relationship changes as the adoption process progresses.

Figure 7.3 displays a concrete example that tests the predictions made above. Using a histogram that corresponds to a crosstabular specification of the theory, it represents the behavior of 173 Wisconsin dairy farmers studied by Fliegel (1957). Stage 1 shows the way the first approximate 25% of the farmers to adopt are spread across the economic ranks. (The ranks themselves are specified as the best approximations to quartiles given the discontinuities in the data.) In Stage 2, the number in each bar is the cumulative percentage adopting in that rank after about 50% of all farmers in the community have adopted. These data are displayed here because they provide an ideal confirmation of both predictions.

In Figure 7.4, data on 105 Pakistani wheat farmers studied by Rochin (1971) provide a second clear confirmation of both predictions. The scattergram is based on a moving average calculated on a 20% band of percentile rank. The approximately 81 points in the scattergram result from calculating the percentage of farmers adopting (approximately) 81 times as the midpoint of the band moved from 10 to 90 on the percentile rank scale. This moving average calculation is a better test of the theory

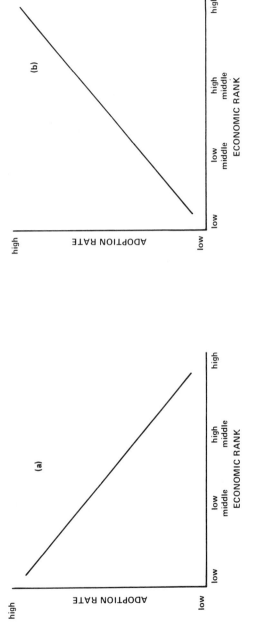

FIGURE 7.1. Theory. (a) Early stages: uncertainty high relative to risk. (b) Later stages: uncertainty low relative to risk.

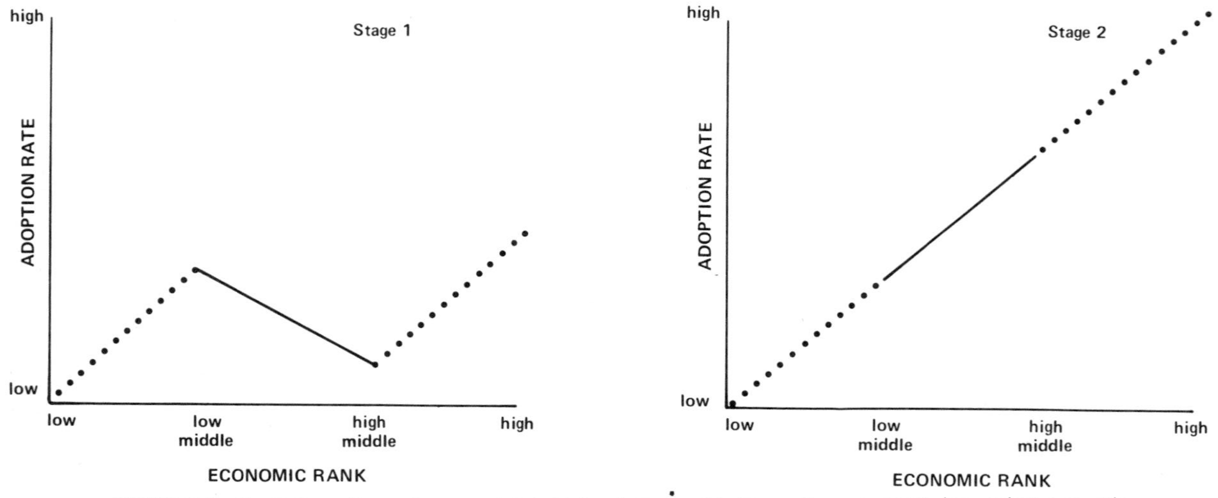

FIGURE 7.2. Predictions. Stage 1—uncertainty high relative to risk. Stage 2—uncertainty low relative to risk.

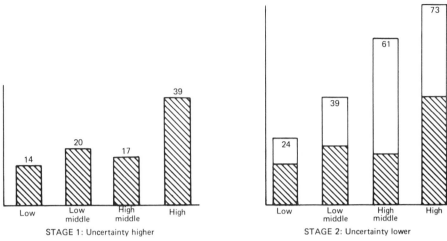

FIGURE 7.3. Histogram test (173 Wisconsin dairy farmers). Percentage adopting.

because it avoids the arbitrary division into rank quartiles that is used in the histogram test. Unfortunately, it is difficult to find data that permit the moving average calculation.

In a tabulation of studies of 16 communities in eight countries done by various rural sociologists, anthropologists, and agricultural economists, 14 cases confirm the first prediction and 12 cases confirm the second prediction.[9] Though compromises made in the process of selecting comparable cases weaken the results somewhat, the confirmation of both predictions is clear and strong (Cancian 1979a). *The relation of adoption behavior to rank clearly changes as uncertainty changes.* When uncertainty is relatively high, farmers of low middle rank adopt more than farmers of high middle rank, and this pattern *reverses* when uncertainty is reduced. During this change in degree of uncertainty, risk remains fairly constant.

[9] Of the 16 communities, one each came from India, Japan, and Kenya, two each from Mexico, the Philippines, Pakistan, and Taiwan, and five from the United States. Full documentation on the data sources and manipulation of the data is given in Cancian (1979a). Here, I want to thank the original investigators who gave permission to use their data and others who helped me collect and analyze the data sets, especially Joseph Alao, Susan Almy, Randolph Barker, Frederick Fliegel, John Gartrell, Peter Gore, Barbara Grandin, Herbert Lionberger, Max Lowdermilk, G. Parthasarathy, Robert Polson, Refugio Rochin, Everett Rogers, Alice Saltzman and Eugene Wilkening.

Each of the cases includes a reasonable approximation to a random sample or a complete census of a community of farmers. Sample size ranges from 91 to 540. In the communities outside the United States, the median size of holding for the full-time farmers included in the analysis was in no case more than two and one-half ha.

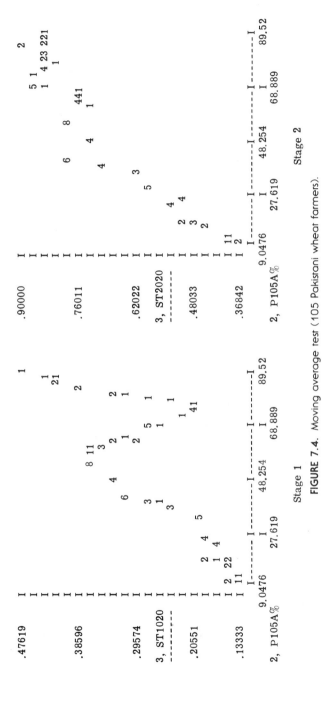

FIGURE 7.4. Moving average rest (105 Pakistani wheat farmers).

Discussion

The patterns just described have important policy implications. They suggest that poorer farmers would take a greater role in technological change than they have often been accorded, and that past hesitancy on the part of farmers who are well off in local terms may be more rank protection than peasant intransigence. More specific implications for both technology development and information diffusion programs are discussed elsewhere (Cancian 1979a, 1979b).

The important conclusions for the present chapter have to do with the differences between the traditional microeconomic approach and the approach used here to set off the risk–uncertainty distinction. I have chosen to call this approach an alternative "paradigm" because it is different from the traditional microeconomic paradigm in three fundamental ways.

First, the alternate paradigm abandons the effort to approximate perfect information and explores the meaning of uncertainty. It seems clear that normative economics will continue to find an important role for a best guess about probabilities of alternative outcomes. This fact about the appropriate needs of normative analysis should cease to constrain descriptive analysis.

Second, the alternative paradigm abandons the assumption that actors will behave as if they are making one of a large number of independent choices that can be aggregated over the long run. Normative decision-making approaches and microeconomic analysis cannot abandon this assumption; they would be too severely and fundamentally altered if they did so. Descriptive decision-making approaches have accommodated to the unreality of the aggregation assumption through the use of "safety-first" rules of thumb and like modifications (see Roumasset 1976), but the assumption nevertheless continues to impede descriptive analysis of behavior, and it should probably be rejected explicitly.

Third, the alternative paradigm abandons the assumption that individual choices are independent in the social sense. The paradigm used in this chapter sees the decision maker as a social person who expresses social position and relationships (his or her interdependence) in each decision.[10] The idea that choices a person makes are not fundamentally

---

[10] "Aggregative interdependence" (the fact that, for example, in a city where every individual produces just a "little" pollution, every individual suffers from a "lot" of pollution) should be distinguished from "relational interdependence" (the fact that not everybody can be the boss). The two types of interdependence are discussed, under other labels, by Hirsch (1976), and in my review of Hirsch's book (Cancian 1979c). Here I am focusing on relational interdependence.

dependent on choices other people make, or on relations to other people, is useful for normative analysis where the decision maker sums up social and other concerns in probability estimates, and both actor and analyst take for granted enormously complex institutional constraints—but it is simply misleading when trying to understand actual behavior. By using rank rather than absolute wealth as the independent variable, the analysis in this chapter makes the actor interdependent with people in the community: What people do is seen as influenced importantly by community position.[11]

Any experienced manipulator of the microeconomic approach knows that all of the difficulties just noted can be comfortably handled within the approach. For example, in using rank to predict behavior, I could be described as using rank as an estimator of risk preferences that can then be used to make predictions with the traditional model. In this sense, the microeconomic model is tautologically true, and can handle description of any behavior pattern derived from data gathered on individuals. Insofar as it is tautologically true, it is trivial, and contributes nothing substantive to descriptive analysis (Burling 1962, Cancian 1966). Understanding of the substantive world cannot proceed by means of tautologically true propositions.

In sum, a number of characteristics of normative economic analysis contribute to making it a misleading beginning point for analysis of human behavior. This chapter focused on the benefits of replacing the perfect information assumption and related ideas about subjective probability used in normative analysis with a distinction between risk and uncertainty that is productive in analysis of agricultural decision making.

Further attention to uncertainty and its implications for behavior may seem superfluous since Herbert Simon received a Nobel Prize for pioneering work on the limits of calculability, but it seems that descriptive analysis continues to be hampered by an element in our culture that might be called the conceptual dominance of normative economic analysis. We seem to believe that people generally act on knowledge—that

[11] The decision-making approach embedded in the microeconomic model also inherits related problems characteristic of marginal analysis or exchange theory. These approaches ask: Given everything else the same, what difference will this little transaction make? This phrasing of the question tends to obscure the importance of existing relationships (including institutional forms), and existing desires to change these relationships. Whether conceived of as relations of production or as conspicuous consumption, these are enduring aspects of social life that provide the options for many decisions for which they are (sometimes only temporarily) part of the taken-for-granted context. Approaches to the analysis of behavior that encourage attention to the isolated characteristics of isolated individuals in isolated transactions hamper descriptive analysis.

they use this knowledge to calculate, and having calculated, act. The fact of the matter is that they very often are called on to act before they can know.

Acknowledgments

An earlier draft of this chapter was delivered in the session on "Agricultural Decision Making in Developing Countries" at the annual meetings of the American Anthropological Association, Los Angeles, November 1978. I am indebted to the Center for Research on International Studies of Stanford University, the Ford Foundation, and the Rockefeller Foundation for financial support and to Francesca Cancian, Kenneth Chomitz and Stuart Plattner for comments on early drafts.

References

Barlett, Peggy F.
  1977 The Structure of Decision Making in Paso. American Ethnologist 4:285–307.
Berry, Sara
  1976 Risk and the Poor Farmer. Paper prepared for Economic and Sector Planning, Technical Assistance Bureau, USAID.
Burling, Robbins
  1962 Maximization Theories and the Study of Economic Anthropology. American Anthropologist 64:802–821.
Cancian, Frank
  1966 Maximization as Norm, Strategy, and Theory: A Comment on Programmatic Statements in Economic Anthropology. American Anthropologist 68:465–470.
  1967 Stratification and Risk-taking: A Theory Tested on Agricultural Innovation. American Sociological Review 32:912–927.
  1972 Change and Uncertainty in a Peasant Economy: The Maya Corn Farmers of Zinacantan. Stanford: Stanford University Press.
  1974 Economic Man and Economic Development. In Rethinking Modernization. John Poggie, Jr. and Robert N. Lynch, eds. pp. 141–156. Westport, Connecticut: Greenwood Press.
  1979a The Innovator's Situation: Upper Middle Class Conservatism in Agricultural Communities. Stanford: Stanford University Press.
  1979b A Useful Distinction between Risk and Uncertainty. Proceedings of the Workshop on Socio-Economic Constraints to Development of Semi-Arid Tropical Agriculture. Hyderabad, India: International Crops Research Institute for the Semi-Arid Tropics.
  1979c Consuming Relationships. Reviews in Anthropology 6:301–311.
Fliegel, Frederick G.
  1957 Farm Income and the Adoption of Farm Practices. Rural Sociology 22:159–162.
Hirsch, Fred
  1976 Social Limits to Growth. Cambridge, Mass.: Harvard University Press.

Hirshleifer, Jack, and David L. Shapiro
  1977 The treatment of Risk and Uncertainty. *In* Public Expenditure and Policy Analysis. (2nd ed.) Robert H. Haveman and Julius Margolis, eds. pp. 180–203. Chicago: Rand McNally.
Kassouf, Sheen
  1970 Normative Decision Making. Englewood Cliffs, N. J.: Prentice–Hall.
Knight, Frank
  1971 Risk, Uncertainty and Profit. Chicago: University of Chicago Press.
O'Mara, G.
  1972 A Decision-Theoretic View of an Agricultural Diffusion Process. West Lafayette, Indiana: Purdue Workshop on Empirical Studies of Small-farm Agriculture in Developing Nations, November 13–15. (mimeo).
  in press The Microeconomics of Technique Adoption by Smallholding Mexican Farmers. *In* Programming Studies for Mexican Agricultural Policy. Roger D. Norton and Leopoldo Solis, eds. Washington: World Bank.
Quinn, Naomi
  1978 Do Mfantse Fish Sellers Estimate Probabilities in Their Heads. American Ethnologist 5:206–226.
Rochin, Refugio
  1971 A Micro-economic Analysis of Small Holder Response to High-Yielding Varieties of Wheat in West Pakistan. Doctoral dissertation, Michigan State University. University Microfilms No. 72–16502. Ann Arbor: University Microfilms.
Rogers, Everett M.
  1976 New Product Adoption and Diffusion. Journal of Consumer Research 2:290–301.
Rogers, Everett M. (with F. Floyd Shoemaker)
  1971 Communication of Innovations: A Cross-Cultural Approach (2nd ed.). New York: The Free Press.
Roumasset, James A.
  1976 Rice and Risk: Decision Making Among Low-Income Farmers. Amsterdam: North–Holland.
  1979 Risk, Uncertainty and Agricultural Development. James A. Roumasset, Gene-Marc Boussard, and Inderjit Singh (eds.). New York: Agricultural Development Council.
Savage, L. J.
  1954 The Foundations of Statistics. New York: Wiley.
  1965 The Foundations of Statistics. New York: Wiley. (cited by Hirshleifer and Shapiro).
Slovic, Paul, Baruch Fishoff, and Sarah Lichtenstein
  1977 Behavioral Decision Theory. Annual Review of Psychology 28:1–39.

Chapter 8

Forecasts, Decisions, and the Farmer's Response to Uncertain Environments[1]

SUTTI ORTIZ

Introduction

Access to power and resources certainly will improve the position of small farmers and peasants. But political solutions are not the end of the story. Farmers, whether powerful or powerless, still must face the difficult task of planning an enterprise in the face of uncertain weather, demand and market conditions. How they behave under adverse circumstances is of great importance to planners and economists. It is only when we fully understand how they react to events and what choices they are likely to make that we can suggest appropriate policies.

Schultz (1964) articulated with clarity and vehemence the growing opposition to the then entrenched argument that peasants were poor because of their conservatism and unwillingness to adopt new practices. For Schultz, poverty did not imply inefficiency. As happens to most polemical arguments, his rectification soon became an overstatement. It was Lipton (1968) who rephrased it in more realistic terms; farmers are reasonable, he clarified, but they cannot always be efficient, given the uncertain nature of their world.

[1] The first part of this chapter has already appeared in *Man*, 14(1):64–80, 1979. I am grateful to the editors of *Man* and the Council of the Royal Anthropological Institute of Great Britain and Ireland for their permission to include a section of the already published article in this chapter.

177

AGRICULTURAL DECISION MAKING:
Anthropological Contributions to
Rural Development

Those of us who have been interested in examining decisions to gain a better understanding of the dynamics of peasant production have been as concerned as Lipton with the constraints imposed by uncertainty. There are several ways of incorporating uncertainty into decision models to examine the consequences it has on choices: to treat decisions as a game against nature; to consider the decision process as a series of separate but linked sequential choices; and to consider decisions as a complex set of parallel processes. Anthropologists are familiar with the first approach as illustrated in the work of Barth (1967). They are less likely to be familiar with the second as illustrated in Gladwin's (1975) examination of fish sellers in Ghana. In my own research I have considered decisions as complex protracted processes following the format and some of the insights gained from Shackle's (1961) contribution to the subject. There are many other ways of modeling decisions under uncertainty that have been suggested by economists and psychologists (Anderson 1979) that have not yet been used by anthropologists.

Whatever approach one follows, attention must be paid to the way farmers conceptualize prospects. It is not enough to sidestep the problem by agreeing with Estes (1976) that individuals are unlikely to have enough information to think about the future in probabilistic terms. A farmer may, after all, be able to single out a future event as highly likely and convey such a belief in the form of a forecast. For example, although a farmer is aware that on rare occasions he may fail to find a buyer for his coffee harvest, he may regard such a prospect so unlikely that he will forecast the total sale of his coffee. Most evaluations of future prospects are, however, much more complex. The yield from a coffee plantation, for example, changes from year to year, none of the amounts harvested may stand out clearly as the most frequent type of yield. Forecasts are unlikely when information is complex and elusive. Yet we can expect a farmer to be able to construe some image of what is likely to happen. How prospects about the future are formulated will depend on how information is structured for retention and used to think about the future. In this chapter, I examine both processes with reference to a group of peasant farmers in Colombia. The findings are in line with the results of a survey research about expectations of small farmers in Iowa.[2]

The farmers interviewed for this study are the same as the ones

---

[2] During the 1940s and 1950s, several agricultural economists interviewed farmers in the hope of gaining some understanding of decision processes—some of the information gathered relates to the ability and techniques used by farmers to forecast and arrive at statistical estimates (see Johnson *et al.* 1961; Williams 1953).

described in my book *Uncertainties in Peasant Farming*. Some of them were extremely poor and, having a small allotment of land sufficient to grow only the bare necessities, had to depend on wage labor for their cash earnings. Others had managed to accumulate enough land and to retain rights over valuable coffee land, so as to concentrate most, if not all, their effort on their farms. Few of them could actually envisage expansion or capitalization of their farms (animals for the pasture land, more efficient cane presses for their sugar production, etc.). Some of the farmers were quite well off and able to compete with the population of recent White settlers. All of the Indian peasants had limited political influence and access to the economic resources tied to power or patronage. All of the Indians wanted very much to acquire land suitable for cash crops and to expand cash activities, but only as long as it would not conflict with subsistence production. They were hampered in their estimates and decisions by uncertain availability of inputs and uncertain profits and yields. The basic production strategies were to concentrate cash production efforts on crops that had well-established markets, to avoid selling food crops whenever possible, and to plan ahead of time only for inputs in cash crop production.

By 1976, 15 years after the initial study, new roads had opened the area to greater commercial activity. Although the income from coffee and sugar had risen considerably, some of it had been absorbed by an inflationary spiral that seemed to be more pronounced in this area than in Colombia as a whole. Information on new varieties of coffee trees and on fertilizer now is available more readily, yet is still painfully inadequate. Low interest loans, though difficult to obtain, are now a more probable option. Tourist industry also has made an incursion into the area, which is listed as a brief stop in one of Swan's tours. Indian peasants have made use of all these new opportunities and, to some extent, profited from the changes. But the economic impact has been small, so the old strategies did not have to be revised drastically.

Among the surviving farmers, eight were selected for interviews since they had at least .5 ha of coffee and/or land suitable for another cash crop. My previous knowledge of their holdings and capital investments allowed me to interpret what, at first, may have appeared to be loose responses; it also made it feasible to relate responses to changes I knew (from 1961 and 1976 data on farming activities) had taken place in their farms. Two White farmers, whose holdings approximate those of wealthier Indians, were initially included in the study to check whether yields and price estimates were directly defined by cultural values. As their answers did not differ in style or content from Indian farmers, I concluded that cultural differences are either too small or irrelevant to the task at hand compared to relevance of problems that all of them must face.[3] Indians and Whites belong to separate legal and cultural communities; all of them speak Spanish, and numerical calculations always are carried out in that language. Farmers were asked to describe present prices and yields for each crop and to recall past information on the subject. Furthermore, they were urged to estimate future prices and yields either by citing a most likely price, set of ranked prices—a forecast—or by indicating what range of prices they believed to be likely—a range of expectations. Farmers were urged to react to step-by-step increases and decreases in their standard income from coffee to measure the full range of expected values.[4]

The Processing of Information

The Perception of Price and Yield Information

The price and the yield that a farmer perceives are not just numbers on a newspaper but they are part of a deeply felt experience, perhaps too complex to dissect into sets of related elements. Nevertheless, if we are to understand the experience, it has to be analyzed in a piecemeal fashion: first, by noting how prices and yields are verbalized; second, by discerning how the statements about prices and yields are grouped into classes; and third, by noting the relationship between the classes.

To capture their mode of discussing prices, informants were encouraged to expand and explain their answers. Sometimes, prices (yields,

---

[3] I beg the reader not to construe this statement as an opinion that Indians are not exploited in the area. Quite the contrary. Whites find it morally easier to cheat an Indian of his land or in a transaction than to cheat a member of the White community. Local authorities show little concern for the rights of the Indians and do not always come to their protection. Whites have greater access to political power, to sources of credit, and to information. But the topic of this chapter is not the exploitation of an underprivileged minority, but the plight of farmers who face uncertain futures and must formulate some sort of expectation, if they are to decide rationally.

[4] Ms. Osborn first interviewed farmers in February 1976, with a set of questions designed to find out what strategies the farmers used, how they viewed new opportunities, and what range of outcomes they expected from various cash activities. The second interview schedule was designed to gather information on price predictions, as well as evaluations of yields, outputs, incomes, and costs, and this interview schedule was used when Ms. Osborn returned to the area in late October of the same year. The first set of questions was asked right after the major annual coffee harvest, when Indians were ready to plant annual cash and subsistence crops. The second set of questions was asked after the small—if any—coffee harvest and after the sale of any sugar produced; at that time, farmers can determine from their trees how good the next harvest might be.

more frequently) were expressed in quantitative terms (for example, the price was $400 the *arroba*), whereas at other times price information was conveyed by a qualitative statement. The form of the answer did not relate to how the question was asked. I am well aware that farmers' ways of talking about prices may reflect only some of the categories used to conceptualize prices, hence my conclusion can be only tentative. I did not, of course, entirely rely on questions and guided discussion to gain information on price encoding; I also used notes taken during my previous fieldwork when I spent a year listening to them talk about farming. Nevertheless, it is likely that formal tests may elicit yet another set of criteria. Further fieldwork using formal techniques no doubt would be extremely useful, though the results will be also subject to qualifications.

The following statements were recorded; I have sorted them into four exclusive categories according to meaning, as well as to indifference in their use (see Table 8.1).

Not everyone talked about prices in terms of the four categories listed. Three farmers described prices as either good or bad; six others used three categories to describe all prices experienced, and only one made use as well of "more than good" category. In fact, some informants answered price questions in terms of one qualitative judgment and one quantitative answer.

These conceptual categories represent only one way of encoding prices; the other system is a set of numerical categories. In other words,

TABLE 8.1
Informants' Evaluation of Prices and Yields

Bad price	Barely sufficient	Good	More than good
'Bad price' *Mal precio*	'Barely brings in anything' *Apenas rinde*	'Good sale' *Buena venta*	'Optimum' *Optimo*
'Does not bring in anything' *No rinde*	'Brings in a minimum' *Minimo que rinde*	'Good price' *Buen precio*	'Marvelous' *Maravilloso*
'Not enough' *No alcanza nada*	'Only covers costs' *Cubre costo solo*	'It pays' *Paga*	'My hope' *Mi esperanza*
'No use' *No sirve*	'Just barely' *Medio apenas*	'Worth it' *Asi vale*	'Very good' *Muy bueno*
'Cheap' *Barato*		'Just' *Justo*	

a farmer may answer a question concerning prices in several ways: "The price for coffee last year now seems cheap; it was $8 for a pound"; "*panela* [uncentrifuged brown sugar] is now $80"; "coffee sells now at a good price, but fat is now $15."

The numerical price given is not necessarily a true representation of the perceived price. For example, during 1975, the price of *panela* was not always $80; it varied greatly during the year and from buyer to buyer. To say that the price was $80 is to subsume a number of variations under the numerical category of $80. When farmers were asked to describe ongoing prices for major cash crops, as well as for subsistence items that were occasionally sold in the market, they answered with one or two prices or, at most, a range. When asked about ongoing coffee prices (the question was asked just before the major annual harvest, when prices are changing quite rapidly), seven of them answered with one price, one informant listed three ongoing prices, and two informants expressed the variation in terms of a range. The question was even more pointedly phrased for corn and beans, yet the types of answers were similar. Six informants gave a single price for beans, two again gave a range, and one listed three prices. Only a few informants sold some corn to buyers, hence, not many answered the question. Of those who did, once again the answers were in terms of one or two prices. The responses are striking because the price fluctuations are sharp and frequent during the time they are likely to sell, so that often various buyers will buy at different prices within the day—a fact of which they are well aware. The responses of the *panela* producers were similar. The above answers thus are not accurate descriptions of total reality, nor are all farmers equally imprecise. The numerical price, like the qualitative judgment of price, is the label for a category that includes a variety of prices observed. Prices other than the one used to label the category are likely to be forgotten and lost as material for concept formation, a point to be discussed later.

Although it well may be that actual prices and price judgments are integrated into separate structures,[5] they can easily be brought together so that, for example, the numerical equivalency of a good price is arrived at effortlessly. Nevertheless, one cannot discount the possible effect of separate classification systems on ability to forecast future prices, nor the possibility that the two systems are never integrated.

When prices are coded into qualitative classes, the criterion used is a judgment of the ability of a specific price to generate an income. Every

[5] Some researchers have suggested that digits and letters are separately stored in memory. The evidence is, at present, conflicting. See Posner and Snyder (1975:70–71).

class relates to every other class in its position within a continuum of hypothetical potential income, ranging from a negative to a positive pole. Economists would describe a set of such judgments as a discontinuous utility curve—each kink representing a utility judgment. Anthropologists, instead, are more likely to represent the system of qualitative judgment as a line:

/ bad /	/ barely enough /	/ good / better than good /
loss in income		gain in income

Whatever representation one chooses, one is dealing with a continuum of potentially experienced incomes, which is conceptualized as a set of distinct categories.

Information on Past Events

Judgmental as well as quantitative statements are used to describe past prices and yields: "The price of coffee last year now seems cheap; it was $8 a pound; last harvest was 6 *arrobas*; for three years it has been bad; now coffee trees are blooming; I think we'll harvest three loads; little less [the coffee harvest] in recent years." Information about the past is thus either retained using the same framework or at least can be integrated easily into the same framework. Farmers whose coffee plantations are large enough to sell most of it by the load can of course quote the total yield of recent past harvest in terms of *arrobas*. But this is not the way they normally talk about past events unless pressed on the subject by a prodding anthropologist. The point I want to make is not that past yields or prices are never retained in numerical form (though ongoing price variations often are subsumed under one numerical category), rather that, more often than not, past yields and prices are retained as qualitative judgments.

The judgments used to describe prices are of interest because they mirror the process of classification. Such judgments, in this case, imply that price is associated with certain other states: deprivation, plenty, and plans to be realized. Such associations were voiced by farmers when discussing strategies, prices, and yields. Past events are then classified according to obvious farming concerns. It may be that the categories ultimately can be reduced to a system of dialectical categories, but as this reduction can be neither proven nor refuted, it will be ignored. Furthermore, it would not bring us any closer to an understanding of how farmers formulate expectations of future prices. In this chapter, I suggest that if a coffee price, for example, is judged as

good rather than bad, barely enough, or very good it is because such a
price brings the farmer a particular state of well-being. His well-being
does not, of course, entirely depend on the income from coffee since
farmers also may earn money from the sale of other crops or from their
labor. Furthermore, the adequacy of his coffee income depends on basic
subsistence and production costs. If the farmer grows most of the food
he needs, he has more cash available for other purchases or for investment. If the cash required to maintain the plantation is proportionately
small, he likewise will have more money to spare. Price judgments are
commensurate with the acquisitive power of money, hence a price
judged as good a few years ago is no longer so regarded during this
inflationary period. Disparity in judgments is thus only apparent; the
differences are easily explained when the aforementioned criteria are
taken into account. For example, farmers with low-yielding plantations,
who have other income sources as well as an adequate supply of
home-grown staples, considered $600 for an *arroba* of coffee as a good
price; anything below that amount was regarded as barely enough.
Farmers with a viable but small plantation and no other source of
income regarded $500 for an *arroba* of coffee as good. Differences in
their judgments of what constitutes a bad price can be reconciled in the
same manner.

As the concerns of farmers are instrumental in how information is
systematized, farming problems must be outlined. Most of the Indian
farmers are no longer in a position to expand their holdings; they are
older men who have already allotted some of their land to a son, or are
working jointly with them, preparing for the eventual subdivision of
their holdings. Despite their ages, they are competent and active farmers. Although they have already made most of their investment decisions (a younger one, however, is planning eventually to buy ten cows,
one at a time, for the pasture land he has been acquiring), they are still
thinking about readjustments in their chosen strategy (for example,
how much of their land to fallow, and how much of the unused land to
use for further expansion of any of the various cash crops). They are also
thinking of new opportunities open to them: the possibilities of a loan
for improvements or investment in the new recommended variety of
coffee;[6] the rising price of sugar. Such considerations force them, as
they do most other farmers, to focus on a range of returns for each cash

---

[6] A new variety of coffee has been introduced to this region that does not require shade.
It can be planted closer together (an important consideration where there is a shortage of
land), and it is said to yield more abundant harvests, but it will require higher capital
investments in fertilizers.

crop, comparative returns of competing cash activities, dwindling returns from ongoing enterprises, and land and labor allocations. But unlike commercial farmers, they do not have to pay attention to the sequence of yields harvested from the same or very similar fields. As Murdock and Anderson (1975) indicate, the attributes to be coded and integrated are not all of the attributes present but are those related to the task the subject is asked to perform.

Hence, as a farmer awaits the ripening of coffee berries, he talks about ongoing prices and evaluates returns from observing how heavy his trees are with berries. This is an important calculation because this is the income that he will use to subsidize annual crops, capital investments, and sugar production, as well as to purchase necessities. But only rough calculations are necessary to decide whether he has enough cash or must attempt to borrow. Furthermore, it is easy to alter plans in case of miscalculations. Sugarcane, the next major cash crop, is a semipermanent crop; a field can be maintained in production for as long as 10 years, though each year it will yield lower profits. The farmer then has to keep a close watch on annual costs and returns of sugar production to decide when to let the field revert to pasture or fallow. Yields from past sugarcane fields are not very relevant, as they are unlikely to be reproduced in the new site; what is important is only a rough calculation of range of outcomes. It is only with the annual harvest that range and frequency of yields and returns are relevant since the same field is used in subsequent years, and one crop competes with others for land and labor. The decision, however, of how much corn, manioc, or potatoes to plant does not rest solely on yield. Furthermore, because they are basically used for subsistence, price is irrelevant. As noted in an earlier publication (Ortiz 1973), such decisions are made in the course of planting and undergo constant readjustments. Consequently, farmers are never sure of exactly how much it was that they planted or, as they harvest a bit at a time, what the total yield was.[7]

The integrative system that emerges from thinking about farming

[7] I was never able to verify exactly how much corn was harvested. When I lived in the area, I missed my chance of surveying the fields before the corn had ripened, and hence before any had been harvested for home consumption. The usual practice is to pick as needed, even before it is fully ready, so that final harvest represents only a share of total yield. The farmers' estimates for yields do not match calculations based on what is planted, proportion of seeds that are said to germinate, or cobs produced. There is one exception; 15 years ago one of the varieties of bean planted for subsistence gained in market value. Farmers soon began to plant special fields with this bean for the exclusive purpose of selling the product. It then became important to note yields, prices, and costs, which farmers did promptly and consistently.

must then consist of a minimal set of categories that are not hierarchically arranged:

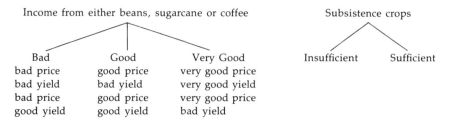

Income from either beans, sugarcane or coffee			Subsistence crops	
Bad	Good	Very Good	Insufficient	Sufficient
bad price	good price	very good price		
bad yield	bad yield	very good yield		
bad price	good price	very good price		
good yield	good yield	bad yield		

From what has been said about how farmers perceive, code, and structure information, it is not surprising that they manage to retain (*a*) prices, yields, and returns in terms of categories; (*b*) incomes as ranges rather than as sequential numerical lists; (*c*) prices and yields as judgments integrated into ranges and/or as numerical ranges; and (*d*) actual prices remembered only as an illustration of a judgment category or when the price has significantly affected the return.

Information is, of course, not likely to be perceived and retained only in terms of the suggested framework. Tulving (1972) and Estes (1976) suggested that also it is recorded automatically using a number of other frameworks. Informants, for example, are likely to note automatically the repetitiveness of events as well as their spatial and temporal connections with other events; semantic associations also are likely to be noted.[8]

It is hard to determine the impact of external associations on peasant problem-solving and forecasting activities. Information on the past frequency of yields and prices may have been automatically recorded, but only in terms of the coding categories (e.g., good, bad, or other). When asked about the frequency of a type of harvest, only four of the ten informants could answer: "Every so many years, I get a bad harvest"; "every 3 years I get a harvest like the last one." Others instead answered: "It really depends on weather"; "it is hard to tell." Contrary to what may be suspected, it was the better informed, more sophisticated farmer who gave the vaguer answer. The impact of automatic associative processes was discernible in discussion about seasonal price fluctuations, supply–demand relations and probable as well as viable income ranges. All farmers understood the price–supply–demand

[8] Although I discuss only automatic and intentional processes responsible for retention and retrieval, one must not neglect the effect of interest (Posner and Snyder 1975) and emotion (Broadbent 1971) on memory.

dynamics. They were also able to explain why certain strategies were best suited for certain market conditions.

The Formulation of Expectations

To imagine future states of the market or of a farmland, one must go beyond remembering past experiences. One has to determine, for example, which one of the experienced outputs is likely to be repeated, or by what measure a past outcome will change. For example, to forecast the future price of a commodity, a farmer must be able to determine from past information the frequency of repetitions, fluctuations, and the actual rate of change for each fluctuation. But as farmers do not memorize price sequences, they are only able to talk about the most familiar outcomes, note the general trend in prices, and give a few examples of past prices. For example, farmers knew that prices for coffee had been rising steadily but did not remember all price increments.

Just prior to the coffee harvest, farmers were asked to forecast the price of coffee beans in the coming harvest. The price of coffee was at the time $600; the previous year it had sold for approximately $450 in the major harvest, and the year before it had sold for $350. The most experienced farmers—half of those interviewed—answered with the ongoing price, although they were aware that prices fluctuate during the harvesting period. Five other informants suggested price increments of $25, $40, $70, and $90. The forecasts reflected the most recently experienced increments rather than an average of all past increments.[9] There was a reluctance to quote a forecast even when eventually they offered a guess. It was impossible to elicit forecasts for corn or beans.

Thus it should not be surprising that when asked for forecasts for a year hence—a much harder task—they all answered: "Coffee prices are things of the government." "It depends on the weather and how plentiful is the harvest." "It is hard to tell whether people will buy, and if they do not, then the price will go down again." Only indirectly was it feasible to calculate what coffee prices they considered as likely in near future. The suggested possible price reflected the most recent annual price increment rather than an average of past annual increments. This information was obtained by asking the farmers to comment on specific increases and decreases above and below their own estimate of a

---

[9] Williams (1953) interviewed 80 dairy farmers in Indiana, and noticed that, although past price trends were taken into account, farmers also relied more on most recent prices when estimating future prices.

standard income. In their answers, they conveyed that such a price was the highest likely price but not necessarily the most probable highest price. In other words, although they were able to formulate expectations of events for the future year, they were unable to forecast future events.

Although peasant farmers may be unable to arrive at probabilistic estimates or forecasts, they do not face the future without some notion of what is in store for them. They have expectations; they conceive some events as more likely than other events. Even better informed investors may find themselves in a similar position to that of a peasant farmer. Shackle (1955) was aware of how difficult it is to arrive at probabilistic estimates or even simpler forecasts, and so suggested that economic models should envisage decision makers as focusing on the most familiar or frequently experienced outcome as well as the highest expected gain and lowest expected return. In other words, their choice is not based on a single forecast or a carefully weighted set of probable outcomes but on a range of expectations, some of which are more prominent or familiar than others. The prominence of some outcomes is triggered by the recollection of extreme distress or satisfaction. If an outcome is recalled with familiarity, it is because it has been experienced frequently. In other words, expectations are summary statements of what has been experienced. It is with this knowledge that peasant farmers—who work in an uncertain world, who must rely on memory rather than on written records, who lack safe storage facilities, and produce a considerable share of what they consume—face the future. The decision process of peasants who cannot rely on forecasts is likely to be more flexible, to include a number of contingency strategies, and to limit comparative evaluations of options.

Expectations and the Farmer's Response to Uncertain Environments

Four general points have emerged from the discussion so far: Our memory of the past depends to a great extent on our intellectual concerns at the time of perception; these concerns, in turn, depend on information gathered through task-oriented or automatic processes; only under certain conditions can past recollections be used to estimate chance or forecast future events; most often individuals decide on the basis of a wide range of past experiences rather than on a vision of the future.

These findings are of relevance in the evaluation of some current arguments regarding the behavior of peasant farmers: the responsiveness of peasants to price incentives, their reluctance to behave as true maximizers, and their preference for risk-averting strategies. A discussion of each of these arguments will illuminate some of the hypotheses in their assumptions and will clarify the importance of the perspective developed so far. Since my findings are not necessarily pertinent to all peasant farmers—they apply to farmers who produce for their own subsistence, in addition to growing a crop exclusively for sale and who are dependent only minimally on credit—my qualifications and clarification cannot be automatically transposed to all other farming populations.

Price Responsiveness

Schultz' (1964) portrayal of peasant farmers as poor but efficient dealt another blow to the rather common practice of treating supply and demand in underdeveloped countries as physical quantities unaffected by price. In the sixties, the stage was set to question again the old argument with a new set of field and statistical studies measuring the supply response of peasant farmers. As new findings accumulated over old ones, however, the assumption of price responsiveness became as shaky as the contrary one. The factors that affect supply are many, and it is hard to hold any set of them effectively constant in quantitative models to reject hypotheses, however dubious. Furthermore, the tendency of economists to rely too much on statistical analysis confuses rather than clarifies their arguments, as Bauer had already warned (1957:12). Some of the confusion is due to limitations in the data: Official records do not always reflect accurately farm gate prices or acreage planted with a particular crop; economic events have to be deduced from lists of measures and values. Other confusions result from the assumption that what is produced is a reflection of the intent of the farmer. Hence, the inconclusiveness of studies on price responsiveness should not surprise us. No further comments would be required except for the fact that there is yet another hidden assumption which, unlike the previous ones, is rarely discussed: The decisions to plant or expand are based either on past prices, past average prices, or expected future prices.

Anthropologists are most familiar with Dean's (1966) study of tobacco producers in Africa. He assumed from conversations with farmers that planting and harvesting decisions were made on the basis of last year's price and was able to show, over several years, that changes in acreage corresponded to lagged prices. However, the correlation was not too strong.

Lagged prices as data appeal to economists who lack information on what prices are expected by farmers or how farmers calculate the next harvests' prices. It was Nerlove's (1958) study of supply dynamics in American commercial agriculture that made possible the considerations of expected prices in statistical studies of supply response. His representation of expected price was neat and simple. It was derived from his assumption that farmers do not bother to predict with accuracy; instead, they are satisfied with what, on the basis of past experience, they consider as normal prices. Normal prices do, of course, change; hence, farmers are constantly adjusting their conception of what they consider normal. "Each period people revise their notion of 'normal' price in proportion to the difference between the then current price and their previous ideal of 'normal' price [1958:53]." After a few more simplifying assumptions, he arrived at a formula of price expectations as a weighted average of past prices, assigning more weight to more recent past prices than to those further back in time. Nerlove was also aware that farmers are unlikely to make sudden and drastic output adjustments, as there are costs implied when their estimates turn out to be inaccurate. Instead, he expected farmers to increase or decrease output by a fraction of what would have been the desired quantity under the new expected price conditions.

Krishna (1963) was able to obtain a relatively good fit between predicted fluctuations in acreage for several crops and recorded estimates for farmers in Punjab after considering the effect of weather on amounts of grain planted. He did not entirely follow Nerlove's formula; instead he used last year's prices. He did, however, introduce Nerlove's adjustment coefficient reflecting partial and progressive response to changing prices.

Narain (as quoted in Askari and Cummings, 1976) did follow Nerlove more closely by introducing expected price in his study of changes in the acreage of several crops in various regions of India for the years of 1900–1939. He concluded that when a response to price could be discerned, it proved to be reasonable rather than perverse. Mubyarto (as quoted in Askari and Cummings) following the same steps, interviewed 560 Javanese farmers and came to the conclusion that although it was possible to note responsiveness to price fluctuations, the extent of the adjustment would depend on the farmer's reliance on cash to purchase farming inputs as well as food for the family.

In 1968, Behrman reviewed past attempts and contributed with his own study of farmers in 50 Thai provinces for the period 1937–1963. He examined changes in acreage for rice, maize, cassava, and kenaf, and using Nerlove's price expectation formula, concluded that, on the whole, responsiveness was positive but not homogeneous for all crops or for all regions. His findings support Mubyarto's contention that degree of commercialization affects the extent and the speed of adjustment to changing prices. Regional comparison allowed him to suggest yet other important factors: alternative opportunities opened to farmers, number of crops a farmer relies on for his cash income, and consumption patterns. Some regional variations, however, remained unexplained.

Nowshirvani (1968) examined the response to price fluctuations for rice, wheat, barley, sugar, and groundnut in 41 Indian districts for the period 1952–1964 and added yet other qualifications to the ones so far mentioned: yield variance and fluctuations in the cost of living. In a later article, Nowshirvani (1971) made another very important point that I shall take up later: that adjustments are made only to price trends.

There have been, of course, many more studies than the ones cited here—in fact, 260 of them—all of which have been clearly reviewed in Askari and Cummings' (1976) monumental survey. The proliferation of studies has not erased all doubts but has helped to point out the complexity of the response that seems to depend not just on price changes but on: yield variations, fertility of crop, availability of suitable land, income, the number of crops that may be harvested in one year, whether crop is annual or perennial, the competition of other crops for available inputs, the time needed to harvest the crop, the need of capital for expansion, whether crop is consumed or produced exclusively for the market, consequences of price fluctuations, size of holdings, forms of tenancy, government incentives, international price agreements, and literacy rates.

Such an enormous and no doubt still incomplete list of relevant variables should help bring the problem back to its original roots: the study of the decision process and the formulation of prospects. Helleiner admitted that "a positive correlation between price of a crop and sales to marketing boards thus tells the analyst very little about the nature of decision making amongst agricultural producers [1975:33]." The relevancy of price response to policymaking economists and the urgency of some large-scale test of the hypotheses was no doubt responsible for leading economists away from the appropriate framework for the study of price responsiveness: behavioral decisions models, and case studies of how farmers conceptualize future events. It is now time to heed Bauer's (1957) advice and turn our observations to farmers' behavior to unravel the arguments and counterarguments. I use my own data to examine some of the claims made about the decisions of farmers regarding how much to plant or to sell.

In the first part of this chapter, I argue that farmers do not automatically rank options on the basis of expected prices. In fact, only a few of them reluctantly ventured to predict a price or price range, and then only for the immediate future. In the second part of the chapter, I explain the reasons for the reluctance, which need not be shared by all peasant farmers. The explanation itself serves as a qualification for the generality of the statement: It applies only to farmers who still produce most of what they consume, make limited use of wage labor and credit, and are in no way encouraged or forced to outline their strategies as simplified cost accounting problems. Although such farmers represent only a sector of the small-farming population, they probably constitute a significant share of the population analyzed in the regional price response studies.

At time of planting, returns rather than price are considered by farmers in their decision of how much to plant. Return calculations are not set out as multiplication problems of expected yield range and expected price range. In fact, the problem that is kept in their minds is not how much they will get next year from an acre of corn, but what returns they have experienced so far, and which ones are more likely to be repeated. Farmers, as we have seen, do not keep in their minds the contribution of prices distinct from the contribution of yields to the value of past harvests; the two bits of information blend themselves in one qualitative category: good or bad harvests. Nowshirvani (1971) comes closer to the way such peasant farmers conceptualize their problem when he suggests that they must respond to price trends rather than to single price expectations. Farmers are aware that changes in price trends may, unless accompanied by inflation in cost of inputs or subsistence, alter the range of experienced returns, but the decision will be made on the changed range of returns (altered state of the range) rather than on altered price trends. The farmers discussed here do not average prices, and although recent past returns are kept more clearly in mind than more distant ones, returns that came close to the minimum bearable income or to the most satisfying level remain just as prominent.

As the populations examined by economists must encompass a fair proportion of subsistence—cash-cropping farmers as well as commercial small farmers, it is not surprising that the correlation in studies about price responsiveness remains inconclusive. Some farmers may use time-weighted averages, prices, or price trends in their calculation of returns. Other farmers, however, will (given their information-processing systems) react more slowly and only when price trends affect income and the share that a particular activity has over the composition of their income. For the second type of farmer, price policies are likely only to have a sharp effect when accompanied by other incentives.

Price policies, on the other hand, are more likely to have an impact in affecting decisions about how much of the harvest to market, or how much to retain for consumption (e.g., how much rice southeast Asian subsistence farmers may sell). These are short-term decisions; hence, farmers are more likely to estimate the likelihood of price fluctuations as well as to argue about the effect of ongoing prices. Responsiveness to price changes therefore must be studied as a decision problem (how much to plant, how much to harvest and to sell) rather than as a correlation problem.

Peasant Farmers Are Not Necessarily Efficient Maximizers

Although the assumption of maximizing behavior is a tempting one because it facilitates modeling and permits us to use some of the existing refined models, it is neither necessary nor an accurate representation of peasant behavior. There have been some attempts to avoid the assumption with safety-first models, focus gain–loss models, or satisficing models. Although these models lack the precision of approaches traditionally favored by most microeconomists, they are useful research tools when care is taken not to fall into the trap of rationalizing every bit of behavior. Some of the problems encountered when modeling decisions in uncertain situations, as well as the solutions so far suggested, were discussed in a conference on the subject and published in a volume *Risk, Uncertainty and Agricultural Development* edited by M. A. Roumasset *et al.* (1979). Papers by Roumasset, Anderson, and Boussard summarize the arguments for and against each approach.

Although the assumption of maximization is attractive, it has its complexities. Anthropologists are fond of pointing out that it may be all right to assume that people maximize, but the question remains as to what it is that they maximize. Obviously it is not always cash nor output; there are other goals that may be more important to the decision maker. What anthropologists often forget and economists have long ago captured, is that when individuals must choose among uncertain outcomes, they share the dilemmas of a gambler: to select an option on the basis of its intrinsic value and on the chances of being able to realize that value. In other words, a maximizing decision maker must keep in mind the probability distribution of each one of the possible returns as well as the value of the returns. Maximizing decision models thus make assumptions about forecasting as well as about ranking outcomes according to preference. Anthropologists, on the other hand, tend to disregard uncertainty when modeling decisions.

The assumption about forecasting that is most often implied in the models is that individuals have some idea of the probability of some outcomes or at least can rank them according to frequency. There are a variety of ways of incorporating subjective probabilities into mathematical models, but this is not the place to discuss them. Rather, it is important to note the limitation of such a behavioral assumption.

In order to widen the applicability of probability models, economists and psychologists have worked together to arrive at techniques of eliciting subjective probabilities that are not likely to bias responses. Slovic, Fischoff, and Lichtenstein (1977) review the many efforts to study how subjects arrive at probability judgments as well as the best ways of portraying such processes in mathematical terms. Dillon (1979) reviews some of the techniques selected by economists as most applicable to small farmers. The only one which could possibly come close to the way such peasants express the range of options is the technique of dividing the expected range of outcomes (as stipulated by the farmer) into intervals, and requesting the farmer to distribute a set of points (matches, stones, or whatever) among the intervals, according to the likelihood or familiarity of occurrence of each interval within the range. The difficulty is that, since the number of intervals and the range of each is determined by the analyst and represents his way of conceptualizing prospects rather than the farmer's way, the answers may not be consistent. Furthermore, it will be difficult to determine whether the points mark frequency judgments or judgments about the consequences or pleasure derived from the realization of each outcome. Nevertheless, we cannot discount the possibility that this elicitation technique is applicable to more highly commercialized farmers.

In *Uncertainty in Peasant Farming*, I describe how impossible it is for a farmer to outline clear maximizing strategies when planting food crops. Instead, some land is retained to make constant output adjustments as more information, need, or inputs are made available. I felt then that cash crop cultivation could be predicted using a variant of maximizing models. I have not attempted a formal test of my assumption, but my more recent examination of how expectations are formulated add a cautionary note (see Ortiz 1979). The farmer's vision of future outcomes and potential returns, as we have seen, is not expressed by single points but as ranges of expected returns that are not necessarily ranked clearly according to the likelihood of incidence. Farmers are aware, however, what is the lowest likely return and its relation to minimum bearable income (focus loss point). The most likely desired outcome is estimated in terms of highest yield experienced at the *ongoing price* rather than at an expected price. In other words, what we would call the maximum expected outcome is in reality the most satisfying outcome so far experienced with relative frequency by the farmers. Furthermore, the farmer cannot always distinguish whether the prominence of an outcome is due to its frequency of incidence or to the satisfaction derived from it. He compares prominent outcomes of various options as well as the concomitant possible loss point, choosing those which offer the best and most enticing solutions. But his choice does not imply that he can determine the relative utility (i.e., combined value and incidence) of each option and decide accordingly. He can strive, but he cannot maximize; he lacks the information and process-solving approach that would allow him to do so. One possible way of representing a decision process of this sort is to use Shackle's (1961) notion that individuals focus on single outcomes that they evaluate both as pleasant and as least surprising. Boussard and Petit (1967) have introduced Shackle's notion to mathematical models, an attempt that seems promising. But as Berry (1977) points out, we still have to resolve the problem of how preferences are ranked, as well as what is an acceptable focal loss level. To my mind, the more difficult problem is how a focus gain is determined.

To avoid some of these problems as well as the difficulties encountered when eliciting subjective probabilities, researchers have found alternative means of depicting decision processes. Tversky (1972), for example, describes decisions as processes whereby unsuitable options are progressively eliminated, rather than as acts of choosing the most suitable options. Gladwin (1975) has successfully applied the model to predict the marketing choices of fish sellers in Ghana. According to her study, fish sellers categorize the state of supply and demand into "good market," "spoiled market," "little fish," and "plentiful fish." On the basis of these four categories, they can calculate what their profit will be *once they are told the supplier's price* of fish. The model is neat if information is available. I suspect the model is more applicable to short-term marketing decisions and might be suitable as well to determine how much of the harvest a peasant farmer will decide to market rather than to consume. Planting decisions are fraught with greater uncertainties, hence they are more complicated. Whatever modeling solutions are adopted, the point to keep in mind is that poor and ill-informed farmers will be striving farmers who can neither offer the statistician a nice set of probabilities, nor point to single maximizing strategies.

Risk Preference and Risk Aversion

As indicated in the previous section, many of the models used by microeconomists portray farmers as choosing options (e.g., maize versus sugarcane, x acreage of maize plus y acreage of sugarcane, m of maize plus n of sugarcane) not only in terms of the value of each option, but also in terms of the likelihood of each return. In other words, value (in monetary terms) and probability distributions are equally important to farmers. The term *utility* is often used to express the combined function of value and probability. In the language of some economists, the efficient rational farmer is expected to choose the option with highest utility, given a certain set of conditions in the economic environment. Of course, farmers do not always abide by microeconomic rules.

There are two obvious reasons why farmers do not behave according to the ideal model of the efficient rational farmer. In the first place, farmers are neither well-informed nor are they necessarily competent statisticians. In the second place, utility decision models themselves are gross simplifications of reality. To bring greater reality and hence greater degree of predictive accuracy to the models, economists have made use of subjective probability distributions (see page 194). They also have considered a number of other variables that may affect the slope of utility curves as well as possible differences that may exist between points that appear to have similar utilities. It was found that different curves could be drawn depending on whether one focused on short-term returns, or on long-term investments, and whether profits or income are considered. Economists also became aware that it was not enough to expect efficient rational farmers to maximize utility; they must consider as well the possible welfare and economic consequences of some options. By incorporating the implications of disaster into formal microeconomic models, a new dimension was added to theoretical discussions on behavior of farmers and a new set of models—safety-first and focus loss–gain models—emerged.

Yet, despite the sophistication of present-day microeconomic models, they do not perform all that well when used to predict the behavior of peasant farmers. It has been assumed that the reason for the poor performance does not rest solely on the model, but on the peasants' preference for certain probabilistic distributions, a preference that typifies him as a risk averter. The previous discussion on expectations should convince us that such a judgment of farmers' behavior is often premature. We must first determine whether, in fact, farmers are able to determine the probability of outcome, which is not always the case, and only then determine whether they assign chance weight to prospects or whether, instead, they only try to rank them in terms of degree of belief. In the latter case, the variance in the probability of outcomes is likely to be undervalued. Thus we cannot conclude, when a farmer prefers option A to option B, that he has a preference for the variance and utility typified by A as he may not have captured the range as perceived by the analyst. Instead, his preference may be based on his subjective evaluation of the likelihood of only two critical outcomes: returns below a minimal income, and value of most familiar returns. In fact, Rothchild and Stiglitz (1970) argue that it may be more appropriate to talk about risk as a chance of loss than as a probability distribution; furthermore, that risk aversion is what risk averters will forego to avoid certain outcomes.

The notions of "risk preference," and "risk aversion," as well as the measure implied with the "risk aversion coefficients" (all of which have been introduced to improve the performance of formal models) are very confusing. At times, economists use them as expressions of fundamental formal decision rules; for example, the rule that income must be above a certain level to allow for the reproduction of the enterprise, or that indebtedness must not surpass a certain level, or that choice depends on level of wealth. But at other times, they are used to indicate the personal character and mood of the investor (Boussard, 1979:65). The choice of words has tended to obfuscate the distinction between the formal constraints of the models and their unreality. The confusion is responsible for the inappropriateness of some of the questions asked by social scientists about risk aversion. Binswanger (1979), summing up his thoughts on a conference on risk and uncertainty, encouraged sociologists, anthropologists, and psychologists to outline institutional arrangements or personality constraints that may account for attitudes toward risk. What he is asking for, in fact, is for us to provide him with an explanation that can be used to connect two noncomparable models: formal evaluations and farmers' perceptions. Both are gross simplifications of economic reality. In each case—one based on the analyst's perceptions, the other on the actor's perception—the procedure, assumptions, and rules used are different, hence the final comparability is tenuous. Sociological and psychological variables affecting preferences should, of course, be introduced, but in the modeling process rather than as a clause in the last stage of an economic argument about the performance of peasant farmers.

Questions about the sociological root of certain imputed attitudes, nevertheless, have found a receptive audience among sociologists and anthropologists. Cancian (1972), for example, has tried to relate risk taking with rank position in a stratified society. Nash (1965) makes the connection between behavior and an outlook that either emphasizes freedom of want or risk of success. Fogg (1965) illustrates the same point when discussing Ibo attitudes toward innovation as reflecting cultural attitudes to failure: A farmer who tries a new technique and fails will lose a great deal of prestige in his community. Foster (1967) adds the suggestion that such an outlook may be rooted either in cultural visions or in the nature of the activities pursued by peasants (e.g., potters are more conservative because low profit margin encourages a cautious personality). Other social scientists have elaborated on Atkinson's (1966) work relating personality and motivation to avoid failure by outlining what they believe are the institutional arrangements that have contributed to the development of certain personality characteristics. The work of Hagen (1962) and McClelland (1961) illustrate these aforementioned theoretical developments. Rural sociologists examine risk aversion in the context of innovation and draw some general conclusion about institutional arrangements that foster certain risk-averting strategies. Similar concerns are shared by Rogers when he suggests that the peasant's subjective evaluation of the minimal beara-ble income level must and will change as he increases his motivation for gain. Such an attitudinal change will be affected by increments in return that are above the motivational threshold (Rogers and Shoemaker 1971:143). Whatever the validity of particular sociological insights, to be useful, they must be introduced at the relevant point in the formal description of decision processes rather than be mechanically applied to rationalize the poor performance of mathematical models.

Summary

Categories used to retain price information and to think about future prices or price trends are not static, nor can they be defined only by semantic structures. Psychologists have pointed out that the processes used to store and relate information are multiple: Some are automatic, whereas others are an integral part of ongoing thought patterns. In this chapter, I have focused on the latter and have illustrated how farming concerns affect the format and content of information retained. Informa-tion about the past serves as data for thought about future events or ongoing trends. How the data are organized and verbalized shapes how allocation decisions are made. The options selected for evaluation when uncertainty prevails are likely to be few. The criteria used to compare them are more likely to conform with satisfying decision models than maximizing ones. In other words, it is unlikely that attempts will be made by farmers to rank options according to their probability; instead they are likely to focus on whether or not there is an overlap between the ranges of competing outcomes and the likelihood of having to face disaster or great satisfaction. Under conditions of uncertainty, decisions are likely to be flexible and leave room for a number of contingency strategies.

These findings are in line with my earlier suggestion (Ortiz 1973) that we must analyze decision making as a process that sometimes has determinable points of closure whereas at other times it extends itself through the period of enactment. The findings also correspond with those from surveys of information used by farmers.

The observation that at least some farmers keep ranges in mind and think about prospects in terms of alterations to trends should caution researchers about the limitations of the use of statistical decision models. If farmers favor an outcome, it is unlikely to be because of the combined value and probability of the particular outcome, but because it is the most satisfying of all familiar outcomes. Peasant farmers, like the ones examined here, cannot determine the probability of events nor rank them according to chance of incidence. Hence probability decision models will not perform adequately when used to predict decisions made by such farmers. Their failure should not be rationalized with clauses about preferences for particular probability distributions. Instead, the failure should be taken as a warning of their limited applicability. New models should be designed to capture more accurately the decision process of the peasant subsistence–commercial farmers. In so doing, we may either follow Shackle's (1955, 1961) lead or some of the more recent suggestions by Tversky (1972).

More realistic models are of fundamental importance to policymakers who need reliable tools to evaluate the impact of proposed incentives and regulations. For example, a better understanding of the impact of price policies would have been gained if the problem of price responsiveness had been studied as decisions about allocations, using models that portray the behavior of the set or sets of farmers in question.

Acknowledgments

I wish to thank Ms. Ann Osborn for her help in interviewing informants as well as for her comments and discussions. Ms. Osborn visited the region in 1975 to record changes in commercial activities, transport networks, and availability of capital resources. She returned the following year to interview the same farmers I had once studied.

References

Anderson, Jock R.
  1979 Perspective on Models of Uncertain Decisions. *In* Risk, Uncertainty and Agricultural Development. J. A. Roumasset and J. M. Boussard and Inderjit Singh, eds. pp. 39–63. New York: Agricultural Development Council.
Askari, Hossein, and John T. Cummings
  1976 Agricultural Supply Response: A Survey of Econometric Evidence. New York: Praeger.
Atkinson, John W.
  1966 Motivational Determinism of Risk Taking Behavior. *In* Theory of Achievement Motivation. John W. Atkinson and Norma T. Feather, eds. pp. 11–31. New York: Wiley.
Barth, Fredrick
  1967 Economic Spheres in Darfur. *In* Themes in Economic Anthropology. Raymond T. Firth, ed. pp. 149–173. London: Tavistock.
Bauer, Peter T.
  1957 Economic Analysis and Policy in Underdeveloped Countries. London: Cambridge University Press.
Behrman, Jere R.
  1968 Supply Response in Underdeveloped Agriculture. Amsterdam: North–Holland.
Berry, Sara S.
  1977 Risk and the Poor Farmer. Economic and Sector Planning Division, Technical Assistance Bureau, AID, Washington, D.C.
Binswanger, Hans P.
  1979 Risk and Uncertainty in Agricultural Development: an Overview. *In* Risk, Uncertainty and Agricultural Development. J. A. Roumasset *et al.*, eds. pp. 383–398. New York: Agricultural Development Council.
Boussard, Jean Marc
  1979 Risk and Uncertainty in Programming Models: a Review. *In* Risk, Uncertainty and Agricultural Development. J. A. Roumasset *et al.*, eds. pp. 63–84. New York: Agricultural Development Council.
Boussard, Jean Marc, and Michel Petit
  1967 Representation of Farmers' Behavior Under Uncertainty With a Focus Loss Constraint. Journal of Farm Economics 49:869–880.
Broadbent, Donald Eric
  1971 Decision and Stress. London: Academic Press.
Cancian, Frank
  1972 Change and Uncertainty in a Peasant Economy. Stanford: Stanford University Press.
Dean, Edwin
  1966 The Supply Response of African Farmers. Amsterdam: North–Holland.
Dillon, John L.
  1979 Bernoullian Decision Theory: Outline and Problems. *In* Risk, Uncertainty and Agricultural Development. J. A. Roumasset *et al.*, eds. pp. 23–38. New York: Agricultural Development Council.
Estes, William K.
  1976 The Cognitive Side of Probability Learning. Psychological Review 83:37–64.
Fogg, Davis C.
  1965 Agricultural and Social Factors Affecting the Development of Smallholder Ag-

riculture in Eastern Nigeria. Economic Development and Cultural Change 13:278–292.

Foster, George M.
1967 Tzintzuntzan; Mexican Peasants in a Changing World. Boston: Little, Brown.

Gladwin, Christina H.
1975 A Model of the Supply of Smoked Fish from Cape Coast to Kumasi. In Formal Methods in Economic Anthropology. S. Plattner, ed. pp. 77–127. Special Publication American Anthropological Association (4).

Hagen, Everett E.
1962 On the Theory of Social Change: How Economic Growth Begins. Homewood, IL: The Dorsey Press.

Helleiner, Gerald K.
1975 Smallholders' Decision-Making: Tropical African Evidence. In Agriculture in Development Theory. L. G. Reynolds, ed., pp. 27–52. New Haven: Yale University Press.

Johnson, Glenn L., H. R. Halter, H. R. Jensen, and D. W. Thomas
1961 A Study of Managerial Process of Midwestern Farmers. Ames, Iowa: Iowa State University Press.

Krishna, Raj
1963 Farm Supply Response in India–Pakistan: a Case Study of the Punjab Region. Economic Journal 73:475–487.

Lipton, Michael
1968 The Theory of the Optimizing Peasant. Journal of Development Studies 4:327–351.

McClelland, David C.
1961 The Achieving Society. Princeton: Van Nostrand.

Murdock, Bennet B., Jr., and Rita E. Anderson
1975 Encoding, Storage, and Retrieval of Item Information. In Information Processing and Cognition. The Loyola Symposium. Robert L. Solso, ed. pp. 145–193. New York: Wiley.

Nash, Manning
1965 The Golden Road to Modernity. Chicago: University of Chicago Press.

Nerlove, Marc
1958 The Dynamics of Supply: Estimations of Farmers' Response to Price. Baltimore: Johns Hopkins University Press.

Nowshirvani, Vahid F.
1968 Agricultural Supply in India: Some Theoretical and Empirical Studies. Unpublished Doctoral dissertation, Massachusetts Institute of Technology.
1971 Land Allocation Under Uncertainty in Subsistence Agriculture. Oxford Economic Papers 23:445–455.

Ortiz, Sutti
1973 Uncertainties in Peasant Farming. London: Athlone Press.
1979 The Effect of Risk Aversion Strategies on Subsistence and Cash Crop Decision. In Risk, Uncertainty and Agricultural Development. J. A. Roumasset et al., eds. pp. 231–246. New York: Agricultural Development Council.

Posner, Michael I., and Charles R. R. Snyder
1975 Attention and Cognitive Control. In Information Processing and Cognition. The Loyola Symposium. Robert L. Solso, ed. pp. 55–82. New York: Wiley.

Rogers, Everett M., and F. Floyd Shoemaker
1971 Communication of Innovation. New York: Free Press.

Rothschild, M., and J. E. Stiglitz
  1970   Increasing Risk: A Definition. Journal of Economic Theory 2:225–243.
Roumasset, James A., Jean Marc Boussard, and Inderjit Singh
  1979   Risk, Uncertainty and Agricultural Development. New York: Agricultural De-
         velopment Council.
Schultz, Theodore W.
  1964   Transforming Traditional Agriculture. New Haven: Yale University Press.
Shackle, G. L.
  1955   Uncertainty in Economics. Cambridge: Cambridge University Press.
  1961   Decision Order and Time in Human Affairs. Cambridge: Cambridge University
         Press.
Slovic, Paul, Baruch Fischhoff, and Sarah Lichtenstein
  1977   Behavioral Decision Theory. Annual Review of Psychology 28:1–39.
Tulving, Endel
  1972   Episodic and Semantic Memory. In Organization of Memory. E. Tulving and W.
         Donaldson, eds., New York: Academic Press.
Tversky, Amos
  1972   Elimination by aspects: A theory of choice. Psychological Review 79:281–299.
Williams, W. F.
  1953   An Empirical Study of Price Expectations and Production Plans. Journal of Farm
         Economics 35:355–371.

Chapter 9

Management Style: A Concept and a Method for the Analysis of Family-Operated Agricultural Enterprise

JOHN W. BENNETT

Introduction

This chapter summarizes the theoretical and methodological aspects of one section of a forthcoming book, *Of Time and the Enterprise: Management as an Adaptive Process in the North American Agrifamily*, a product of the Saskatchewan Cultural Ecology Research Program.[1] The key concept for this part of the book is *management style*: an experiment in pulling together various behavioral, social, and economic aspects of management decisions through the history of family enterprises, so that the approaches of individual farm operators to the problems of variation in resources and markets can be distinguished clearly. The conceptions of management style held by the farm operators themselves were used as the starting point of the analysis. Our general objective was to comprehend enterprise management as a form of social behavior in the

[1] The Saskatchewan Cultural Ecology Research Program is a continuing study of the social and economic development of a 7000 m² region in western Canada, funded by the National Science Foundation, National Institute of Mental Health, and Washington University. The Program has published 5 books and about 25 articles. Some key publications are the following: Bennett 1969–1976; 1967a; 1967b; 1968; Kohl 1976; 1971; Kohl and Bennett 1976. The present chapter is based on a book entitled *Of Time and the Enterprise: Management as an Adaptive Process in the North American Agrifamily,* to be published by the University of Minnesota Press.

AGRICULTURAL DECISION MAKING:
Anthropological Contributions to
Rural Development

context of family and community relations, and not merely as a set of more or less rational financial or economic decisions.

Before we examine the concept in detail, I wish to convey some of the background considerations that led us to use this concept, and which are explored at length in *Time and the Enterprise*.

The most general issue is the contribution anthropologists might make to the study of farm management in a thoroughly entrepreneurial economy. There is reason to believe that the distinctive methods employed by cultural anthropology are not well-suited to the study of economic pursuits governed by external markets, since anthropologists are mainly concerned with local-level phenomena, particularly internal exchange relationships. Their approach contains few specific tools with which to examine local and external power and resource interactions. The anthropological concept of "cultural broker," or the mediating agent in these interactions,[2] is of some use, but it is largely descriptive and lacks an analysis of key elements like opportunity cost, trade-off, alternative choice, and other features of socioeconomic behavior that govern the accessibility of the entrepreneur to the broker and their ability to make use of his services.

Despite the paucity of concepts suitable for the analysis of full-scale entrepreneurial agriculture, the anthropologist has a role to play in the study of these systems. Farmers the world over, and at all levels of national development, have much in common. All must make decisions about production in relationship to available resources; all must balance opportunities against constraints; all must cope with uncertainty, and its measured component, risk; all must deal with the "outside," however this may be defined. And all must make a living to support themselves and kin. These are all elements of the instrumental behavior of agriculturalists, and anthropologists have a task here: to analyze such behavior in its full social context.

Within the anthropological discipline, the question rests on the definition of subdisciplinary specializations. In the present context, this is the field of economic anthropology. This specialization has explored various facets of instrumental activity at different levels of subsistence, but it has not evolved a consistent or nuclear paradigm. In fact, most themes seem to be borrowed from other fields: formal analysis from economics; decision-making frames from management science, and resource utilization from ecology. This is not necessarily a defect, and

---

[2] Barnes (1954) and Wolf (1956) appear to have contained the first statements of the "cultural broker" concept in cultural anthropology. Barnes was concerned with resource acquisition; Wolf, with cultural influences.

the eclectic, descriptive aspect of anthropological work in the economic field is a useful undertaking. Moreover, the other disciplinary approaches, useful as they may be for analytic work, commit the user to a theory of human behavior limited to a particular universe: the decisions or judgments of producers in a rational–variable market system.[3] Their applicability is ideal–typical in the sense that the analyst assumes that *if* judgments are made in accordance with the quantitative specifications of the models, the producer will benefit. On the other hand, the anthropologist is really concerned with a different problem: why producers may follow these specifications on some occasions, but not on others.[4]

Once this "why" is asked, the focus of the inquiry shifts toward empirical behavior, and not formulae and idealizations of behavior. The anthropologist is concerned with variation in the whole system, and not simply with particular outcomes or methods. Moreover, the anthropologist studies the things the farm economist considers to be "informal" or "uncontrolled"—which are "givens" or simply irritating disturbances in the systems the economist analyzes or models. If this is so, it suggests that the most useful role of anthropologists in the study of productive behavior in entrepreneurial societies may be not to copy the methods of economists, but to study the phenomena the economist cannot study: the variable behavior of producers as people in a multidimensional social system.

Another important issue concerns the influence of time. There is a tendency in farm management research, and in most anthropological studies of agricultural decisions, to conceive of a decision as a one-time, unitary event. This occurs, but in our research, we saw it as only the tip of the iceberg, so to speak. Behind this event lay a history of thought

[3] By rational–variable market system, we mean a system of entrepreneurial production in which the rules are standardized, clearly formulated, and known to the participants, but in which the costs and gains fluctuate and are subject only in part to prediction. That is, gain cannot be realized with stable costs and output: Fluctuation is necessary.

[4] This is a crucial problem in agricultural economics as well, since the producers make the expected decisions in the economic game only part of the time. That is, they seek to *maximize* gain when they choose to do so, not necessarily when conditions are right to do so. Agricultural economics is a relatively conservative field, largely because the intense drive for production in North American agriculture has perpetuated neoclassic maxims as primary concepts. Other fields of economics have accepted greater variation in behavior, including the "satisficing" type of objective, and have incorporated these variable producer behavior patterns in their quantitative analysis. (For a history of the field of agricultural economics, see Case and Williams (1957); Black, Clawson, Sayre and Wilcox (1947) is a text representing the professional views prevailing during the period of our research.

and circumstance, and after the event, there stretched an indefinite sequence of consequences. Decision is a process, not an event. Although we made studies of particular decisions at particular points of time, the thrust of our analysis was on the consequences and antecedents of the decisions, and the way these formed patterns.

The entrepreneurial-farming system of North America—and its close neighbors in Europe, contains a built-in universe for study due to the unusual pervasiveness of the basic institutions. The relationship of the producers to the market, the mechanisms of acquiring the resources— physical and financial—needed for production, and the social-structural relationships among the members of the production team are remarkably similar over thousands of square miles. This system evolved under the tutelage of the capitalist system, and its producer members, have assimilated, by and large, its basic frame of reference. Two qualifications: first, the modeled rules of procedure, including the profit motive, maximum output at lowest cost, and other features, cannot always be followed by producers due to the variations in their personal and community situations. Hence they are, on the whole, aware of the limits of applicability of the pervasive frame. Second, the application of the rules is always subject to cultural modification depending on local and regional styles and values, and likewise for differing modes of production. Thus, grain farmers in the West follow a different style of operation from farmers in the Midwest—among other things, they take more risks. And out West, ranchers and farmers differ considerably in their subscription to the idea of farming as a pure business undertaking; many ranchers are more willing to restrain enterprise investment to indulge in outdoor hobbies that express cultural meaning and solidarity.

These variations in the application of the commercial model of management seem to be one domain for anthropological effort. However, there are, conceivably, two different objectives here: The anthropologist may wish to investigate the variant patterns in entrepreneurial agriculture as an ethnographic undertaking: description for its own sake. Or, he may wish to study those features of variable strategies that have policy and action significance. That is, how do these personal and cultural variations influence the conduct of the business, and the volume of production? Both objectives are respectable but there is an added incentive to choose the latter, or practical, objective when the anthropologist becomes concerned with making a greater contribution to a crucial issue: the conditions of food production.

Another impetus for anthropological study of entrepreneurial agriculture derives from the most common social basis of this form of produc-

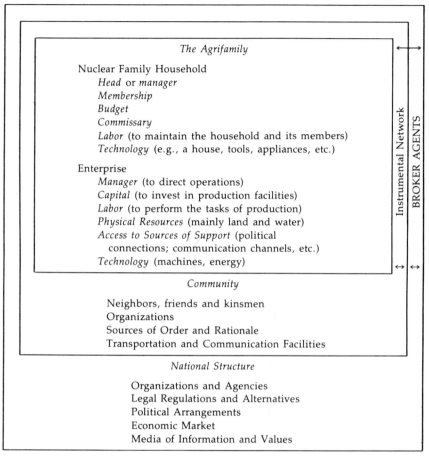

FIGURE 9.1. The agrifamily and its subsystems.

tion: the "family farm," as it has been called in North America.[5] This refers to the operation of an agricultural enterprise by a nuclear family with children, plus other relatives, when the occasion permits or demands. We refer to this arrangement by the term *agrifamily system* and its elementary structure is visible in the accompanying Figure 9.1. The core of the system is the nuclear family, which provides the personnel for two components: the household and the enterprise. Each member of the family plays roles in both components, and often these roles may

---

[5] For discussions of the family farm in Canada and the United States, see the following: Barkley (1976); Hedlund and Berkowitz (1977); National Farm Institute (1970); Ferris (1963).

conflict. The basic social matrix of decision making in North American entrepreneurial agriculture (excluding corporate enterprise farming, of course) is this complementary role structure, in which the family members cooperate, resolve conflicts, compromise, forego pleasures and needs, and generally strive to keep both household and enterprise on an even keel. This core unit of the system must function within a community that provides resources, but also provides constraints on the operation of the enterprise and the well-being of the family. Each family enterprise also must develop its own instrumental network to link it to the community. Finally, family, enterprise, and community must receive their basic resources and opportunities from the national structure and its many institutions and agencies. Broker agents bring the national structure into interaction with the local.

The structural elements of the agrifamily system are worldwide. Even in countries where the North American type is modified by different styles of kinship, community, settlement, and national institutions, the basic relationships seem to prevail, and these are becoming more, not less, common, as rational–variable agricultural management and production becomes the norm through the international economic development process. Consequently, the study of the agrifamily system and its parts offers the possibility of a central paradigm, which the anthropologist can use to provide a link to larger universae and generalizations. This is not a rigid paradigm, but it does contain a limited array of crucial processes: among them, the coping mechanisms that must evolve for the system to survive in a variable price and cost system; the need for political manipulation to secure basic resources by the producer and the community; and the compromises and trade-offs necessary to maintain a family and an enterprise in situations where the needs of one may be infringed on by the other. Much of the work of anthropologists in the decision-making field pertain to these processes, although they are sometimes not clearly identified as universal, and their reference to a possible central paradigm is neglected.

The Concept of Management Style

As previously noted, this chapter focuses on just one aspect of our effort to convey the larger context of family entrepreneurial management: The way individual farm operators developed distinctive styles of decision making and instrumental action, and how these styles might change due to internal (family and community) and external (markets, government bureaus) forces. Management style is an essentially anthropological concept insofar as it attempts to view managerial behavior as a holistic undertaking. The farm economist views management from a unitary standpoint: economic and financial considerations are his major concern. However, since the predictions of specific managerial decisions one might make from this limited frame are often wide of the mark, there is a need—and particularly in Third World agrarian development work—for including data on the "soft" or informal aspects of the managerial process.

Management style is an amalgam of such factors as the rate and number of innovations: economic performance variables, attitudes and practices in relation to uncertainty and risk, particular strategies of balancing prices and costs, sense of the future and its relationship to investment, and other factors. These factors are set within a known environment of physical and economic resource availability, and of family and community. That is, the *economic* response to opportunity and constraint is seen to be governed by such factors as prevailing attitudes toward life and social obligation, the number of kin that the enterprise is called upon to support, social customs pertaining to inheritance, community styles of cooperation and risk-spreading, and other factors. Our own work on this problem extended over a number of years in a 7000 m² region of the northern Great Plains, called "Jasper" after our pseudonym for the principal town, and we were able to observe such behaviors, and their causes, in a variety of ethnic and occupational groups, and in different modes of production. This approach included standard anthropological fieldwork methods, emphasizing the intellectual integration of observational data; quantitative economic data; and the computerization of key aspects of all forms of data.

Since the study extended over a period of 12 years (in fact fieldwork has never really ceased) it was possible to ascertain change in the management style and general social and economic postures of the agrifamily. Our basic conception of management style is dynamic; we see it as an adaptive mechanism that is susceptible to modification as the operator and his family experience change and growth. In this sense, it is not a classic cultural datum with static, one-time-slice implications, but a behavioral stance that can change as circumstances alter. That is, the change in style is not merely a matter of short-term adjustments to take advantage of the fluctuating structure of constraints and opportunities, but can display grand or secular changes during the lifetime of the operator, or over the duration of the enterprise. Thus management style is not geared only to personal, economic, and cultural factors, but also to cycles of family and enterprise. By extending our observation of the same samples of agrifamilies over a number of years, we were able to see these changes take place.

The Folk Categories of Management Style

As noted, our basic approach in the determination of management style was to combine the ethnologic with the objective. The former is the classic integrative–observational approach; the latter is the analytic examination of quantitative expressions providing measurements on standard or universal frames of reference. To study enterprises with the ethnologic approach, one becomes concerned with the "emic" depiction of these units by the operators themselves.[6] All operators in entrepreneurial economies have standard criteria for judging the quality of their own and others' enterprises. Management procedures can be judged as good, bad, energetic, progressive, conservative, and so on; likewise, the condition of the enterprises can be considered as good, bad, well established, or building-up. No instrumental culture lacks a vocabulary to describe its behavioral and organizational components.

These judgmental and behavioral depictions are soft in the view of the economist, since they are expressions of a complex orientation to reality, with its usual capacities for distortion. However, the "distortion," from an anthropological viewpoint, is part of the system, not outside it. Moreover, it can be shown that such elements have valid and typical relationships to objective economic facts, policy considerations, and resource imperatives, because they are part of the holistic adaptive process of management. That is, they are a response to the differing and often conflicting aspects of management behavior in a social setting. For example, a man may characterize his neighbor's operation as "not much good," when in fact it is average by some statistical criterion, such as returns to scale. However, the man's judgment is not wrong when we consider that he is judging the enterprise in comparison to some other and better one; or that he is considering a longer time interval than the immediate moment; or that he is implying that "that place could be a lot better than it is." All such judgments can be defined objectively and

---

[6] There are some echoes here from the ethnosemantics approach in cultural anthropology and linguistics, but no specific source in that field of study has been of particular influence on the analysis. We view the procedure of collecting informant expressions of the problem and its substance as a routine empirical matter, not as a specialized theoretical issue.

given quantitative dimensions if one views management as an adaptive social enterprise.

Our first step toward a description of management behavior and qualitative assessments of enterprises was taken in the first field season (1962), when we ran the first format of our omnibus Regional Schedule on a sample of 30 agrifamily households—the beginning of a stratified sample that eventually grew to a total of 216 enterprises, and 166 agrifamilies (many operated more than one enterprise), with a core group of 90 enterprises and associated households studied longitudinally over a decade or more. Within this longitudinal sample, there existed a special management sample of 30 cases, studied *very* intensively each year of the decade, and with detailed historical data for most cases, going back to the founding of the enterprise. During the interviewing in the first season, it became apparent that Jasper operator–respondents and other family members had definite concepts about the nature of their own enterprises and about those of their neighbors. We began to collect the language used to express these opinions and categories. The work continued through the research program.

As this work proceeded, it became clear that the most useful and consistent set of folk categories was one describing what we first called the "production orientation" of managerial behavior. As we analyzed these data, it became evident that they fell into a matrix, which we represent here in Figure 9.2. At the same time, we were collecting economic and performance data of a routine type, so it became possible to compare actual management performance data and data on the economic condition of the enterprise, with interview material of a qualitative nature. By 1973, it was possible to make a synthesis of the two kinds of information, although we became aware of the main patterns as early as 1963.

Figure 9.2 displays four cells, produced by two interrelated sets of criteria. These criteria—the degree of enterprise development under way, and the degree of establishment in the sense of adequate investment to sustain desired production—are obviously related to another array of data on the cycles and stages of enterprise change and development, more of which later. Although Figure 9.2 gives an impression of excessive order produced by the armchair analyst, there is no doubt that the matrix existed in the minds of our respondents, as we ascertained by checking it with them on numerous occasions. There has been, of course, a certain amount of economy and clean-up: Actually the terms used were more numerous, and it took work to determine shades of meaning. But once the keys were learned, it was relatively easy. For example, when an operator called his neighbor a "sitter," he

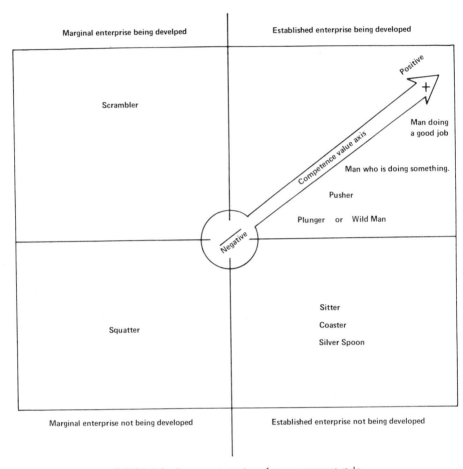

Marginal enterprise being develped   Established enterprise being developed

Scrambler

Positive

+

Man doing
a good job

Competence value axis

Man who is doing something.

Pusher

Plunger   or   Wild Man

Negative

Sitter

Coaster

Squatter

Silver Spoon

Marginal enterprise not being developed   Established enterprise not being developed

FIGURE 9.2. Jasper categories of management style.

meant that he was doing nothing to a well-established and reasonably productive enterprise. If he called him a "squatter," he meant that the man was doing nothing to an enterprise that was touch-and-go, or just barely surviving, and which might or might not be susceptible to further development. Moreover, it was capable of supporting the family at only a very low level of consumption.

Some of the variant terms were as follows: a synonym for squatter was "dirt-floor type" (recalling the early homestead level of development); a synonym for "man who is doing something" would be "a pretty good feller"; and sometimes the "plunger" type was called a "wild man." "Coasters" were sometimes "holding operations (with emphasis on the enterprise)." The terms selected for the diagram are a mixture of colorful local terms expressing the meaning clearly, and somewhat simplified terms to indicate shades of meaning (particularly in the difficult upper right-hand cell).

If we examine each cell separately, we see that the lower right-hand cell contains the term sitter, already defined. It also contains two other terms, coaster and silver spoon. Coaster was applied to people who were, for the time being, doing nothing with an established, productive enterprise, but who were likely to do something in the future; silver spoon to the sons of rich operators who inherited a good enterprise and felt no need to develop it further. All of the terms in this cell refer to operators of relatively profitable enterprises. The terms also imply negative value judgments, although the strength of these will vary with individuals. The silver spoon appellation is perhaps the most negative, since in the postfrontier culture of Jasper, strong value was placed on doing things by yourself, and if you received an enterprise without working hard for it, and then did nothing further to improve it, you could not expect much "social credit" (our term for the dominant pattern of status allocation in Jasper society—for the men, management style was a major criterion). This negative status assignment accorded to well-off persons on good or at least productive enterprises is one of the curvilinear relationships between management style and social factors we shall discuss in the final section of the chapter.

The left side of Figure 9.2 contains two terms: squatter, already defined; and "scrambler." The latter was used by Jasperites to describe the operators of undersized enterprises who were trying hard—but not always coherently—to establish a production base to pay off debts. It implied engaging in different types of production, each on a small scale, in an effort to realize the advantages of diversification—a dubious strategy in an era when specialization paid off. The value connotation was disapproval, but also doubt and ambivalence: "You'll have to wait and see with fellas like that—he's a hard worker, you can say that for him."

Between the top two cells is the type "land grabber": a manager who does something to develop his enterprise, but concentrates on a particular strategy—land acquisition. He often pursues this interest apart from any other productive strategy, and the buying and selling of land can become an obsession. The occasional ferocity of his activity level, his frequent success in creating a productive enterprise, combined with the ambiguous tenure of the enterprise, which he was likely to sell to the highest bidder, led Jasperites to think of him as a special case. We eventually excluded him from the computer analysis, since there were only two cases in the sample. However, we have dealt with him in other contexts.

The upper right-hand cell is the most complicated: It contains operators who persisted in following development strategies regardless of their income level, or the productivity the enterprise was capable of sustaining. In most cases, this was associated with people whose family consumption levels were on the rise for various reasons. (In general, this also was true for the scrambler, but his efforts were devoted to an undersized or otherwise submarginal enterprise.) A variety of terms were used by people to refer to these cases, and considerable comparative analysis took place before the pattern was clarified. We believe that the diagonal vector we call the "competence value axis" describes the pattern. This axis has its positively valued pole at top right, and the ambivalent–negative at the bottom left (and the negative, of course, meanders through the other cells as well). Thus, social credit was not given automatically to any operator who did something, but rather on the basis of what and how much, how energetically it was pursued, or how intelligent the striving. *Very* energetic, single-minded pursuit of economic gain was not highly valued (plungers, pushers); such people were faulted for their reckless disregard of prudent strategies and their heedlessness of risk in this uncertain environment. More cautious, but still steady effort received the highest value: a man who is doing something, or a "good job," and a *"good* man." It is significant that the highest accolades were not sobriquets or personal labels, but understated or indirect terms. Considerable familiarity with the Jasper style of avoidance of obvious or fulsome praise was required before these could be identified as the most approving labels.

Although we did not make psychological investigations, there was little doubt in our minds that characterological patterns underlay some of these types, and sometimes this was articulated by our respondents: "He's that kind of a fella—he'll push and push no matter what." "It seems like he don't have the right kind of drive to be a farmer." Our decision to avoid personality assessment was based on our growing knowledge that management style was no automatic reflection of personality, but was for most people a behavioral strategy influenced by situational factors. This became clear as the study proceeded through the years, and changes in management style were found for many operators. A plunger might be a sitter on the third-year visit, but such extreme transformations were unusual over such short periods; more common were degrees of the same behavior, in response to shifting opportunity, and to the turns in the family and enterprise cycles. Characterological tendencies probably operated mainly at the extremes: people who never did anything, and people who did a lot, all the time.

From the standpoint of the neoclassic ideal type of entrepreneurial behavior, it is important to emphasize the negative values and ambivalence associated with maximization strategies (maximizers, were of course, the plungers, etc.). This represented the caution bred of long years of adaptation to marginal resources and fluctuating economic conditions. But the do-nothing operators also were valued negatively, or at least viewed with ambivalence and reserve. The approved strategies were the careful building procedures that were based on a realistic assessment of a resource potential that could be developed without going too far into debt.

These local judgments also had an important ecological component. Jasper agricultural operators knew the capacity of the physical environment for degradation, under heedless use of soil and water. People who were heedless, even if their long-range objectives might include conservationist practices—"slowing down" this might be called—were assigned negative social credit values. The scrambler as well as the pushers and plungers were cases in point. However, a number of these people moved toward more conservative styles of management as the decade proceeded, and the possibility that a man might do this was known by Jasperites, as we have suggested already in reference to the personality issue. However, they also were aware that there was always a risk of failure, and of resource abuse associated with these strenuous methods. These standards applied to both grain farmers and cattle operators: The main problems were lack of adequate summer-fallowing or tillage methods in the former and inadequate care of pasture in the latter. Overused and poorly maintained irrigation works were also important.

The Jasper folk matrix of management style is contained within the frame of the entrepreneurial system of North America, itself based on neoclassic, capitalist economic philosophy. All of the operators were entrepreneurs in the fullest sense. However, there was no single standard of judgment: It was a matter of *how* the entrepreneurial mode was utilized. The viewpoint was adaptational: In the long run, the "best" manager, from the standpoint of the local culture, was not one who followed economically ideal maximizing management styles, but who *adapted* such standards to his own operations and to the constraints of the larger system. For "taking less" in this fashion, he was accorded prestige by the community. This type of behavior, and the community's evaluations of it, do not seem to be greatly different from the "limited good" behavior and attitudes described by some anthropologists as typical of "peasants." At least, it is only a matter of degree.

The dynamics of the judgments themselves are not reported in detail in this paper, but a few comments are in order. The management-style characterization of an operator was based on a number of criteria, not just one or two. Or rather, our assemblage of the criteria into a major variable of level or degree of intelligent development conceals the multidimensional basis of this judgment on the part of the respondents. The manager's perseverance, sincerity, ability to juggle conflicting objectives, as well as knowledge and skill, were used along with his production, income, condition of his machinery, fields, fences, or stock. This integrative aspect of the folk labels contrasts with the economist's tendency to apply single factors as judgmental conclusions (e.g., returns to investment). Thus the local people had a sophisticated conception of the management process: They knew it was a matter of trade-offs or compromises, and they applied this to the judgments of managerial performance as well. A Jasper farmer who might be poor at machinery maintenance might be good at livestock management, and his low credit mark on machinery might be balanced, in the judgment of his peers, by his high mark on cattle. In his case, the net judgment might earn him the label of a "pretty good fella, at that," somewhere in the middle range of the upper right-hand cell's competence value axis.

Criteria for Management Style and the Making of a Scale

We have noted that our fieldwork included both ethnographic and objective data collection, and that the categories of management style emerged simultaneously on both fronts; *but*, that in another sense, the objective criteria were used to check on the folk categories of style. This was an interactive methodology; it is probably impossible to be ruthlessly explicit about the order of steps in the procedure. At any rate, after intensive fieldwork tapered off in 1973, we engaged in a final cross-checking and validating operation. We shall turn now to this phase of the analysis.

To see if the folk labels were suitable as a basis for a scale, we had to perform several tasks:

1. To determine the precise criteria the respondents were using to make their choice of labels to apply to particular people.
2. To test these criteria by reviewing our own observational and objective data on our populations of operators. Our sample hierarchy evolved through the years on the basis of a mode of production-plus-income stratification scheme derived from the

Canada census of agriculture, but refined for our more detailed breakdowns. As previously noted, we compiled a total regional sample of 216 enterprises and households; within this, a longitudinal sample of 90 cases studied repeatedly over the 12 years of intensive fieldwork; and within this, a special management sample of 30 enterprises. For any particular objective, we could draw on one or more, a mixture, of these samples.

3. The third step involved careful comparison of the data from our samples with the folk labels applied to specific cases.
4. The final step was the production of an analytical classification of management styles, or really a refined, validated, simplified folk scale.

The whole operation was facilitated by the fact that we found very little disagreement between respondents on the labels applied to particular individuals. We usually asked each respondent to give us his label for all the people he knew best; names were always introduced conversationally; it was not a "test."

Continual analysis along these lines led to the following list of criteria used to determine management style. These criteria are presented in Table 9.1 in professional, not folk language, and the classification includes a mixture of folk criteria and technical elaboration.

We next used these criteria to develop our own ratings of management style for the management sample of about 30 cases, plus as many more from the other and larger samples as we had confidence in. We also impaneled three collaborators in Jasper, one of them an agricultural specialist, the others ranch and farm operators, and asked them to assign the folk labels to the sample cases on the basis of the criteria. Our own ratings used most of the folk labels, but simplified them and reduced the number. We called these the "analytical labels." We then showed the analytical label classification to the panel, and asked them to go through the procedure once more, using our classification. They found it possible to do this without difficulty. In most cases, it meant simply removing value judgments that had influenced the placement of individuals in particular categories, whereas on objective grounds—or rather, on the basis of the criteria—they might belong in another one. This entire procedure was done twice—once in the mid 1960s, and in 1971.

We now had several sets of ratings. The percentage of agreement among all of these was very high, and the amount of divergence between the panel's original folk label assignments and the analytical labels was consistent with our own similar work; that is, when

TABLE 9.1
Criteria for Project Ratings of Management Styles[a]

1. *Frequency of significant decisions over at least a 5-year period.* In general, if an operator made about two significant decisions a year, defined as actions that risked or spent a sum equal to 10% of invested capital, his style could be described as "active"; that is, somewhere in the upper riqht-hand established–developing cell, and above the pusher and plunger or scrambler categories. If such decisions were larger in number, or risked more capital, the operator would be likely to be considered one of the more active two just mentioned.

2. *Willingness to go into debt, and the size of the debt.* Anything above 20% of invested capital could be considered risky or daring, and would drive the category into the pusher–plunger domain, or the scrambler. Anything below, in the 5–10% range, was in the more cautious but still innovative field.

3. *Qualitative magnitude of decisions.* Aside from the quantitative factors, it is necessary to include a criterion of social and moral magnitude because Jasperites include these in their assessments. Possible future consequences to the enterprise and the family were the general factor here (that is, shifting from one grain crop to another may not involve much capital at risk, but it could have very large consequences in the future, if the price ratio between the two crops changed radically). This could affect seriously the family as well as the farm.

4. *The manager's own description of his business and his style of operation.*

5. Consideration of a number of factors and variables taken from our protocols and summary analyses of enterprises in our regional and longitudinal samples. The most important of these are as follows:

Planfulness:	the capacity of the operator to conceive and carry out a plan.
Carefulness:	a related factor: degree of caution in managerial actions.
Resources:	the nature of the resource base of the enterprise and the intelligence of the plans and projects in relation to it.
Support Load:	a quantitative variable concerning the total number of persons the enterprise was required to support.

[a] Cutting points on quantitative factors were flexible, since all ratings were worked out with qualitative factors as well.

NOTES:

1. It is clear that these variables interrelate and overlap, but this is precisely the way Jasperites made their judgments, and we saw no need to attempt to compile rigorously separated criteria—when, in fact, these do not really exist.

2. As can be seen, the main emphasis was on defining the differences between the categories in the crucial upper right-hand cell. Classification of operators in the various sitter-type categories, or in the squatter, were much easier, and in effect, an operator who "did none of the above" was automatically placed in those boxes (after, of course, taking into consideration the known social aspects of his operation).

operators were moved, for instance, from one category into another on the basis of a look at the objective criteria, both teams independently made the same change in the great majority of cases. The judges had a very high consistency among themselves: an internal divergence of only about 8%. These experiments convinced us we were dealing with a consistently variable universe. It also convinced us that whereas the folk labels did not always conform to the analytical in all cases, the reasons for not doing so were completely comprehensible in light of known value perspectives.

The folk terms that involved the most effort on our part were those in the upper right-hand cell of Figure 9.2. These were graded on a value continuum as well as an activity scale, but the subdivisions seemed fine-grained, and often a toss-up; hence often dependent on personal prejudices or predilections. Our work on these categories led us eventually to the global concept of a "developer" type of manager, a label that could be used, if one so wished, to include all of the cases in the upper right-hand cell. However, once we began to rate developers carefully with the criteria, we found all sorts of distinctions that the respondents always had not observed. For example, we noted the existence of a "blocked developer," or an operator who wanted to build his enterprise but could not do so for some special handicap not his own fault—like a tyrannical father who refused to give up control, or some bad luck with a bank loan. We also found "consistent developers," "intermittent developers," "conservative developers," and "frantic developers." We eventually found that, although these categories could be useful for individual case study analysis, they were not really necessary for the computerized analysis of the relationship of management style to social and economic factors. We ultimately distinguished only developers and conservative developers, and for some purposes (the condensed scale in Table 9.2) had simply one category: developer.

The final terminology in the experiment of moving from folk labels to more analytically based categories is shown in Table 9.2. *Both* the analytical labels and the condensed scale ratings were computerized, and identical cross-tabulations and correlations with social and economic data run for both. The final results convinced us that the condensed scale did as good a job as the other. Consequently one salient finding emerged: The folk classification was, by and large, and aside from all the complex value inclusions, basically a simple activity scale. It distinguished between people on the basis of how active they were in trying to make their enterprises produce at a more efficient, or higher, level, both for the present and as a measure of stability for the future.

We shall pass over the complications of this whole operation. Each folk label is a story in itself. For example, a scrambler might show the same investment criteria as a developer of some type, but he was distinguishable from the latter on the basis of the inadequate resource base of his enterprise. This "hopelessness" of his situation was reflected in a quality of behavior and in particular attitudes that the respondents and the expert panel were all responsive to—and ultimately, so were we. The scrambler thus survived all tests, although at

TABLE 9.2
Folk and First–Second Stage Analytical Managerial Style Categories

Folk labels	Analytic labels (1)	Scale of activity (1)	Condensed scale (11)
Plungers and pushers	Plunger Risk taker Pusher	1 2 3	1. Very active
Scrambler	(Scrambler)	(Not in scale) (7)	
Doing a good job and man who is doing something	Developer (of several subtypes) Conservative	4 5	2. Moderately active
Sitters; squatters; coasters; silver spoons	Sitter	6	3. Inactive
Land grabber	(Land grabber) (specialized)	(Probably not in scale) (8)	

several points, we were inclined to include him in the generalized developer category. The silver spoon occasioned similar argumentation, but ultimately we did not feel that his inheritance position was sufficient grounds for keeping him as a separate inactive type. The resulting classification, like the folk, though somewhat less so, was a conglomeration of criteria, with somewhat different weightings of particular factors. This will offend methodological purists who seek quantitative rigor and hence identical weightings and factors, but we were striving for acculturally or behaviorally based classifications, and in real life, such classifications always are mixed.

Quality of Enterprise

Before we review the data on the management style analysis, we need to examine our second-most important qualitative rating scheme. On the whole, judgments about enterprise quality were much simpler than those of management style. Of course, a number of criteria were involved, but the basic folk classification was mainly a good–average–poor nominal scale, with the usual details added for the particular enterprise under discussion. Value judgments were not as important as they were in the management style scheme.

Most complications in the interview data on enterprise quality arose because of the time factor. For example, a scrambler might take over a small, but well-developed, enterprise by shouldering a heavy debt. In such a case, the operator's frantic behavior to service this debt would earn him the scrambler sobriquet, but the enterprise itself, at the time of takeover, would be "good." But these cases were not numerous, and each could be handled on its own merits.

We followed essentially the same procedure with the quality variable as we did with management style. We did our own ratings, simplified the categories, proceeded to let the judges develop theirs, and finally asked the judges to rework their ratings on the basis of our list of criteria. This list is presented in Table 9.3.

The panel found it easy to use these, since most of them are already in common use by farm-management specialists, and can be found, often in somewhat different proportions and terminology, in farm-management research studies. The results of the panel and the project ratings were again very consistent, with only about a 10% disparity. All of the disagreements were over enterprises that were undergoing a significant transition at the time of ratings. Some judges and project personnel would rate the enterprise on its past performance, some on its present, changed basis. The judges themselves disagreed on about 7% of all cases they rated.

TABLE 9.3
Criteria for Project Ratings for Quality of Enterprise

---

1. *Consistency of performance* over period of at least 5 years, as shown by interviews. Consistency: sustained output of whatever crop contributes most income, plus frequency of favorable outcomes of projects (4 out of 5, roughly).

2. *Conservationist posture.* Application to resource sustained-yield policy in developing enterprise. It means that resources are not pushed to the limit; maximization strategies are not followed, in general.

3. *General condition of equipment and buildings.* Not necessarily "polished"—good repair was major criterion.

4. *Appropriate balance of machinery to returns.* That is, not over or undermachined. Determined by actual amount of investment in machines, which should not be over 10% of capital. (Local standard)

5. *General condition of fields,* etc. (also, see 2).

6. *Rate of return on invested capital:* for most ranches, from 3.5–5% for most grain and livestock farms (either grain–livestock or livestock–grain), 1.5–3%.

7. *Income:* gross returns from sale of agricultural products; and net, after costs.

NOTE: Rating will be influenced by utterances of manager, and by the amount of consistency between these utterances, as expressive of a "philosophy of management," and his actual performance. Stage of enterprise is also considered.

Once again we felt we had tapped some sort of judgmental consensus—an agreement between observational and objective data. In itself, this validated our awareness that the agricultural operators themselves were very much part of the larger system they worked within and manipulated for their economic survival.

Frequencies: Management Style; Quality of Enterprise; Mode of Production

Table 9.4 presents the frequencies of management styles in the two principal scale classifications, with the scrambler group added as a solitary type, for both the 1960 and the 1970 periods (1960 means roughly 1960–1964; 1970, 1969–1973). The data show that the inactive management style diminished during the period, and the very active dropped somewhat. The moderately active middle range gained at the expense of the extremes. The total number of cases in the two time samples differs somewhat, due to attrition: Jasper lost about one-fourth of its agricultural enterprises in the 12 years, although many of these were simply merged.

The general findings accord with the expected results due to the rapidly improving conditions of western Canadian agriculture: Increased opportunity invited some inactives to resume activity, but since results could be obtained without frantic movement, there was some

TABLE 9.4

Management Style Frequencies: 1960 and 1970 Condensed Scale; Analytic Labels (Modified Folk Labels); Regional Sample

Concordance with condensed scale	Analytic labels	1960		1970	
Inactive	Sitters	41		19	
			C29.3[a]		C17.3[a]
Moderately active	Conservative developers	43		37	
	Developers	20		25	
			C45.0[a]		C56.4[a]
Very active	Maximizers	20		17	
	Risk takers	6		3	
	Plungers	4		3	
			C21.4[a]		C20.9[a]
Scramblers	Scramblers	6		6	
			C4.3[a]		C5.5[a]
	Totals	140		110	

[a] C = Column percentage.

loss of very actives. The swell in the moderate group suggests a kind of "regression to the mean" associated with improving opportunity and greater ease of acquiring results.

Table 9.5 displays the *changes* in management style in a smaller sample of operators—this was the intensive management sample plus additional cases supplied from the longitudinal sample to make a total of 71. The data show that about half of the inactives moved into an active style by 1970, and that the 1960 very actives spread out, indicating that most of them were moving their enterprises onto a stable, less active plateau after their development activity in the 1950s and 1960s—thus, management style is influenced by the cycle of the enterprise. Many of the former inactives were in a redevelopment stage not only because of increased opportunity, but because they were preparing to transmit the enterprise to a successor son, or to sell it. These explanations were derived, of course, from our intensive observational data on these cases.

Table 9.6 provides frequency data for the enterprise–quality variable. The scale used here is a 5-point version, later compressed into a 3-point

TABLE 9.5
Management Style Changes in Same Operators: 1960–1970 (Longitudinal Sample)

Management style 1960	Management style 1970				1960 Style totals
	Inactive (sitters)	Moderately active	Very active	Scramblers	
Inactive (sitters, etc.)	8	4	—	—	12
Moderately active (developers)	5	41	1	—	47
Very active (pushers, etc.)	2	4	2	—	8
Scramblers	—	3	—	1	4
1970 style totals	15	52	3	1	71 available cases

scale as previously noted. In general, the table shows that the two categories of poor decreased over the decade; the category of very good increased substantially; and the group of good–average also increased. Here again, one is faced with the results of a combination of influences from improved economic opportunity plus changes in the maturational cycle of the enterprises studied (see later discussion, and Figure 9.3, p. 230).

The objective data on economic changes, not included here, show similar patterns. Gross incomes of the enterprises and associated households during the decade also regressed toward the mean (i.e., fewer very rich and very poor enterprises, with an increase in the middle range). Other measures, like return on investment, had the same pattern, but amount of indebtedness increased in 1970, showing that the seizure of opportunity included greater willingness to go to the banks or to the government to get the money to develop or to expand.

TABLE 9.6
Quality of Enterprise: 1960 and 1970 (Regional Sample)

Quality of enterprise	1960	Percentage 60	1970	Percentage 70
Very good	24	15.6	33	24.6
Good	55	35.7	59	44.0
Average	29	18.8	23	17.2
Poor	38	24.7	14	10.4
Very poor	8	5.2	5	3.7
Total available cases	154		134	

These economic variable arrays correlated highly with the style and quality variable patterns—about .8.

Table 9.7 combines the style and quality variables. The data suggest that the style of the operator's management, and the quality of his enterprise are, to some extent, the same factor, but that time and changing opportunity do make a difference. Thus, whereas active management tends to produce good enterprises in both time periods, changed economic conditions can result in improvements on quality in other management categories as well: the sitters improve in 1970; also the scramblers.

Table 9.8 presents management style in relation to mode of production (the previous tables pooled all modes). The data suggest that mode of production is an important correlate of style, particularly in the context of a changing economy. In both the 1960 and 1970 periods, the

TABLE 9.7
Quality of Enterprise in Relationship to Management Style: 1960 and 1970
(Regional Sample)

| Management style | Quality of enterprise 1960 | | | |
	Good	Average	Poor	Totals
Inactive (sitters, etc.)	R31.6 12 C19.0	15.8 6 25.0	52.6 20 55.6	38
Moderately active (developers)	61.5 32 50.8	28.8 15 62.5	9.6 5 13.8	52
Very active (including scramblers)	57.6 19 30.2	9.1 3 12.5	33.3 11 30.6	33

| | Quality of enterprise 1970 | | | |
	Good	Average	Poor	Totals
Inactive	42.1 8 11.4	15.8 3 16.7	42.1 8 44.4	19
Moderately active	74.1 43 61.4	18.9 11 61.1	6.9 4 22.2	58
Very active (including scramblers)	65.5 19	13.7 4	20.7 6	29

TABLE 9.8

Change in Management Style in Relationship to Mode of Production: 1960–1970 (Regional Sample)[a]

Mode of production	Management style						Totals
	Inactive		Moderately active		Very active (including scramblers)		
	1960	1970	1960	1970	1960	1970	
Straight grain	R15.9 / 7 / C17.9	11.4 / 5 / 27.8	29.5 / 13 / 21.3	20.4 / 9 / 15.8	9.1 / 4 / 12.5	13.6 / 6 / 20.7	44
Grain–livestock	11.3 / 6 / 15.4	5.7 / 3 / 16.7	35.8 / 19 / 31.1	17.0 / 9 / 15.8	28.3 / 15 / 46.9	1.9 / 1 / 3.4	53
Livestock–grain	11.8 / 6 / 15.4	3.9 / 2 / 11.1	27.4 / 14 / 22.9	21.6 / 11 / 19.3	11.8 / 6 / 18.8	23.5 / 12 / 41.4	51
Livestock	25.0 / 17 / 43.6	10.3 / 7 / 38.9	16.2 / 11 / 18.0	33.8 / 23 / 40.4	4.4 / 3 / 9.4	10.3 / 7 / 24.1	68
Balanced	15.0 / 3 / 7.7	5.0 / 1 / 5.6	20.0 / 4 / 6.6	25.0 / 5 / 8.8	20.0 / 4 / 12.5	15.0 / 3 / 10.3	20
Subtotals	39	18	61	57	32	29	236

[a] Total available cases 1960: 132
Total available cases 1970: 104

mixed-grain and livestock enterprises were the most active, and of course these represented the sphere of production in Saskatchewan that was expanding most rapidly, consisting mainly of former grain-emphasis farmers who had been making a transition to livestock-emphasis since the 1950s. The least active modes were straight grain and straight livestock: the two most specialized modes which had little capacity in, or in most cases, need for change. The "balanced" category (50% of the income produced by each mode, grain and livestock) showed no change, and these enterprises in the samples consisted of the older and well-established ones. (An "occupational ethnicity" factor was associated with cattle operation to be summarized in the concluding section.)

However, the straight-grain and the straight-livestock enterprises do increase their level of management activity slightly in 1970, again showing the results of increased economic opportunity. Thus, the table seems to show the combined effects of three influences on management style: *mode of production, changing economic opportunity*, and *stage of enterprise maturity*. (Each of these factors was also included as a variable, and was run against style and quality—some of the results are summarized later.)

Table 9.9 shows the relationship of management style to quality of enterprise. The data suggest that: (*a*) there was no strong relationship between the two variables in the grain–livestock category, but a bias toward the good side in the livestock–grain[7] (the enterprises that as noted, had received the most development during the period); (*b*) the relationship was definitely skewed toward the good side for both the straight grain and straight livestock, reflecting the age and maturity of the enterprises—also noted previously (the specialized modes were, in this period, more economical to operate, since they were simpler, required less overhead, and hence enjoyed more favorable ratios of income to cost); and (*c*) the great profitability of livestock (the highest poundage prices for beef on the hoof in history) meant that more investment capital was available for development in enterprises with livestock as the dominant mode—although the actual investments were relatively modest because the enterprises themselves were mature and stable.

We consider that this methodological experiment demonstrated that qualitative judgments can be made reasonably precise and valid, when measured against quantitative economic criteria. We are not implying absolute objectivity here; obviously there was some overlap between

---

[7] Grain–livestock means 51% or more of the gross farm income is derived from grain; livestock–grain is the reverse proportion.

TABLE 9.9
Quality of Enterprise in Relation to Mode of Production: 1960 and 1970 (Regional Sample with a Predominance of Longitudinal Cases)

Mode of production	Quality of enterprise				Quality of enterprise			
	Good	Average	Poor	Totals	Good	Average	Poor	Totals
Straight grain	R57.9 / 11 / C15.1	26.3 / 5 / 17.9	15.8 / 3 / 7.3	19	66.7 / 12 / 13.3	22.2 / 4 / 17.4	11.1 / 2 / 11.8	18
Grain–livestock	35.7 / 15 / 20.5	26.2 / 11 / 39.3	38.1 / 16 / 39.0	42	75.0 / 9 / 10.0	0.0 / 0 / 0.0	25.0 / 3 / 17.6	12
Livestock–grain	64.0 / 16 / 21.9	12.0 / 3 / 10.7	24.0 / 6 / 14.6	25	67.7 / 21 / 23.3	22.6 / 7 / 30.4	9.7 / 3 / 17.6	31
Straight livestock	66.7 / 28 / 38.4	16.7 / 7 / 25.0	16.7 / 7 / 17.1	42	74.1 / 43 / 47.8	18.9 / 11 / 47.8	6.9 / 4 / 23.5	58
Balanced	21.4 / 3 / 4.1	14.3 / 2 / 7.1	64.3 / 9 / 22.0	14	45.4 / 5 / 5.5	9.1 / 1 / 4.3	45.4 / 5 / 29.4	11

the sources of information. It would be impossible to completely exclude knowledge of economic trends from judgments of style and quality, but we believe that our methodology did everything possible within these limitations to develop the qualitative or ethnographic approach independently of the objective data. We turn now to a brief summary of how our management-style variables fared when compared to purely social aspects of the agrifamily.

Social Causes and Correlates of Management Style: A Summary

Our theory of the agrifamily is based on the simple structural diagram, Figure 9.1, but this diagram does not convey the essential processes associated with the agrifamily. These are the cycles or "temporal rhythms" of its various components, and the important ones are described in Figure 9.3. Although these cycles are tied to one another (e.g., an aging operator will "slow down" and bring his enterprise into a "maintaining" stage—they preserve a certain autonomy—for example, the same operator may have to resume development of the enterprise to prepare it for takeover by a successor). A careful study of the diagram will suggest that cycle points among the various horizontal components can intersect in various ways—sometimes to the benefit of the family, sometimes to the enterprise but with hardship for the family, and so on.

Our study of these cycle intersections was accomplished with both of our general types of data: integrative analysis of our ethnographic observational and interview materials, plus the quantified and scaled items given to the computer for variable correlation and association. Some findings were derived from ethnographic materials alone; others from the computerized analysis, but in all cases the latter were checked against the ethnographic information by simply identifying the particular enterprises and households in the tables and ascertaining the reasons for their presence in a particular cell or datum. The latter procedure was particularly important: Among other things, it overcame the major defect of computer analysis of complex holistic information by bringing the analysis back to the original real units.

Details of our findings on cycles as influences on management style, and other aspects of the problem, can be found elsewhere.[8] Some of the highlights can be summarized as follows:

[8] See, for example, a forthcoming paper: Bennett (1981), which summarizes the findings on how management style was influenced by the family cycle; and also the book in preparation mentioned in Footnote 1. Preliminary results were summarized—in rather different language—in *Northern Plainsmen* (Bennett 1969–1976, especially Chapters 4, 6, and 7).

Component and functions	Cycles	Process
Enterprise manager Decisions and tasks—accomplishment	Bachelor starters — Establishes — Develops — Slows down — Retires → New operator: develops	Aging
Enterprise Economic and technical	Establishment phase — Development phase — Maintaining phase → Redevelopment phase	Developing
Nuclear family household Reproduction and socialization	Courtship — Marriage — Birth and training of children — Transmission of headship → Courtship marriage	Expanding and contracting
Instrumental network Reciprocal exchange	Wife's family added ---- (various events will influence) ---- Offspring's family added	Ramifying and attenuating

FIGURE 9.3. Temporal Rhythms of Local Components of a Typical Jasper Agrifamily System

1. The chief influences on—causes of—management style were:

   a. *The number of persons in the household and kin group the enterprise was required to support* (e.g., the larger the number, the more active the style—with qualifications).
   b. *The cycle stage of the family household* (e.g., the early stages induced more active styles; but resumption of activity was fairly common in the advanced stages when an operator prepared to transmit the enterprise to this successor–son).
   c. *The general condition of the agricultural economy* (e.g., the better the conditions, the more active the management, in response to improved opportunity).
   d. *The quality of the enterprise as measured by objective economic criteria* (e.g., the better the enterprise, the more active the management style—with some qualification).
   e. *The financial and physical resources available to the enterprise* (e.g., the better the resources, the more active the style—with some qualification).

2. Several of the summarized findings have qualifications. Most of these can be described as a basic bimodal or curvilinear set of relationships.

   a. *The enterprises with the highest and the lowest incomes were both likely to display inactive management styles.* In the first case, this was due to income satisfaction; in the second, to inadequate resources and a kind of hopelessness factor, plus in some cases, an ethnic factor.
   b. *Many enterprises with very large numbers of people to support, and many of those with very few, were likely to display inactive styles.* In the first case, this was due to the hopelessness factor plus ethnicity in some cases. In the latter, the income-satisfaction influence was the main cause.

3. These bimodalities are of interest because they validate fictional or belles lettres themes concerning the effects of poverty or wealth in capitalist society, and indirectly challenge some of the assumptions of economic behavior models, where success is seen as a constant behavioral trait. That is, as many Jasper operators achieved economic stability and a desired level of income, they ceased to "push," in local parlance, and often neglected opportunity and further development or needed repairs to their soil and water resources. Their sometimes lackadaisical behavior resembled that of the very poor and resources-deprived, who saw no use in "pushing" since the resource base was

absent. This similarity sometimes was carried over into social relations, as when, for example, big ranchers habitually consorted—"hung out"—with small or ineffective ranchers.

4. The ethnicity factor mentioned previously had some relationship to "production orientation," but a complex one. Simple correlations of management style and economic factors with ethnic background of parental generation showed that agrifamilies with western–northern European, or Canadian–United States origins were in general more commercially oriented and more likely to display active styles of management, than agrifamilies with eastern European or new world—French-colonial origins. The former group were all Protestants; the latter, Catholic. However, this simple relationship was deceptive, since other factors were operative. The eastern European *et al.* group were the last settlers in Jasper, hence got the poorer land. This functioned to constrain possibilities of expansion and development. In addition, though, many of these families had social customs that constrained development. The most important of these was divided inheritance of land and the enterprise: At the death of the first generation father, the property would be divided among the sons, creating a series of substandard enterprises. There were ways to overcome this handicap, but in general, it had the effect of limiting development in a large number of cases. In any event, these constraints were disappearing in the second and third generation operators, and by the 1970s, differences in management between the western–northern–European and United States–Canadian groups and the eastern-European–French-colonial families were barely perceptible.

5. Management style was found to be an important criterion in the assignment of "social credit," our term for the main type of personal status in North American agricultural society—*reputation* is a reasonable synonym. The active style of management was the approved type, but *not* the most active. That is, the plungers or wild men were looked at askance, as foolhardy and disorganized operators likely to fail. The approved form was the cautious developer type. Operators who manifested this type of management were given status in the community and were "listened to." (Wives had similar social credit statuses, assigned on the basis of their skills as managers of the household and, in a less consistent fashion, as helpers in the management of the enterprise.)

6. Management style—particularly the active styles—did not, on the whole, have a reliable relationship to educational achievement, as measured by grades completed. Education as such did not seem to explain success in the building of a productive farm or ranch, particularly the

ranch. However, education can be seen as a measure of the stability of residence in the community. Many farmers who had stayed put, so to speak, to complete elementary grades and some high school did tend to have active management styles. Thus stability of residence is an important factor in forging such a management style, and may, incidently, result in longer schooling.

7. The longitudinal study showed that managers were capable of changing their management style in response to the demands of the cycles, and also of external factors like economic opportunity. Over the decade of research, the sample showed a general movement toward active styles, as western Canadian economy progressed. However, the usual effects of enterprise cycle, resources, income and other factors were not completely overridden.

With reference to the micro level of decision making, we can summarize our findings concerning the particular kinds of management decisions and actions manifested by the enterprises at particular points in the family and enterprise cycles. In general, the choice of a technique or innovation was guided by two factors: (a) the cost of the activity, and the amount of money available—which later was simply what the household could forego supplemented by loans; (b) the needs of the enterprise at particular points in its developmental cycle. The results, among other things, reflected certain general problems associated with family entrepreneurial farming: Key management decisions are often deferred because of high cost and need to use funds for the household—which can result in crippling delays in development. For example, the most expensive investment for most farms and ranches during most of Jasper history was land. By not buying sufficient land soon enough, the enterprise was likely to move into a new high-cost era with inadequate resources. This was, in general, possibly the most pervasive reason for what agricultural experts in Saskatchewan often referred to as the "backwardness" of Jasper-region agriculture in the 1960s. The habit was marked among first- and second-generation operators, but the new third generation, coming into control in the 1970s, was much more aware of the needs and was more willing to take risks.

We also found expectable differences in the kind of specific strategies followed at particular times or in particular conditions. For example, the measures taken in the redevelopment phase preceding succession could be summarized by the term "modernization": buying the latest machines, seed, breeding stock, and the like—all of which would make the enterprise attractive to a progressive son. The farms and ranches of very large families usually showed a choice between buildings and machines: dilapidated barns and houses, but abundant, well-maintained machinery, the emphasis being placed on production instrumentalities.

These summarized findings, as already noted, are only highlights, and ignore many significant details and complications. However, the ones presented serve to demonstrate that management in these enterprises is not governed solely by economic factors or by the criteria used to define incentive by the economist. Management is a complex social undertaking and the fortunes of the enterprise are determined by the whole temporal system in which the operator and his family must function. In short, management can be found to be an adaptive social process.

One point made in the Introduction is worthy of final emphasis. If anthropologists are to make a contribution to the increasingly crucial problem of food production, it is necessary to understand the sources of variation in producer response to market incentives. Economists have attempted to deal with this problem for years, with mixed results. The extent that producers respond to the "right" signals, in order to provide desired production magnitudes, is a rough measure of the extent to which they have assimilated the very frames of reference used by the economic analysts of the agricultural market. However, this response is never perfect, and often it varies widely. The reasons for this variation are not to be found in purely economic factors, but in complex trade-offs between economic and social factors, strained through individual characterological differences. This variation is especially marked in the family entrepreneurial mode of production, since production is performed in small social units with many internal constraints and peculiarities. The objective is not necessarily to "control and predict," but simply to learn to expect variation because of the imperatives of family entrepreneuring. Moreover, since many of these sources of variation reflect inadequacies of the present market system, with its disbenefits to the producers, an understanding of the larger systemic nature of decision making might result in more intelligently conceived support and assistance programs for farmers. Certainly, many mistakes in agrarian development in Third World countries might have been avoided through a more realistic understanding of how the farmer functions in a complex local and external world. We believe that management style—and our other approaches to the problem of managerial variation to appear in *Time and the Enterprise*—may assist in this understanding.

References

Arensberg, Conrad M., and Solon T. Kimball
    1965   Culture and Community. New York: Harcourt Brace.
Barkley, Paul
    1976   A Contemporary Political Economy of Family Farming. American Journal of Agricultural Economics 58:812–19.
Barnes, J. A.
    1954   Class and Committees in a Norwegian Island Parish. Human Relations 7:39–58.
Bennett, John W.
    1967a   Hutterian Brethren: The Agricultural Economy of a Communal People. Stanford: Stanford University Press.
    1976b   Microcosm–Macrocosm Relationships in North American Agrarian Society. American Anthropologist 69:441–454.
    1968   Reciprocal Economic Exchanges among North American Agricultural Operators. Southwestern Journal of Anthropology 24:276–309.
    1969–1976   Northern Plainsmen: Adaptive Strategy and Agrarian Life. 1969 ed., Chicago: Aldine; Arlington Heights Il: AHM Publishers (1976 edition; slightly revised and supplemented).
    1976   Anticipation, Adaptation, and the Concept of Culture. Science 192:847–853.
    1981   Farm Management as Cultural Style: Studies of Adaptive Process in the North American Farm Family. Forthcoming. In Research in Economic Anthropology (Vol. 4). George Dalton, ed. Greenwich, Conn: JAI Press.
    In Press   Of Time and the Enterprise: Management as an Adaptive Process in the North American Agrifamily (In association with S. B. Kohl and G. Binion). Minneapolis: University of Minnesota Press.
Benvenuti, Bruno
    1962   Farming in Cultural Change. Assen, Netherlands: Van Gorcum Company.
Black, John D., Marion Clawson, Charles R. Sayre, and Walter W. Wilcox
    1947   Farm Management. New York: Macmillan.
Brunthaver, Carroll G.
    1975   Agricultural Economics as an Aid in Management Decision-Making. American Journal of Agricultural Economics 57:889–891.
Burchinal, Lee, and Ward W. Bauder
    1965   Decision-Making and Role Patterns Among Iowa Farm and Nonfarm Families. Journal of Marriage and the Family 27:525–530.
Case, H. C. M., and D. B. Williams
    1957   Fifty Years of Farm Management. Urbana: University of Illinois Press.
Coughenour, C. Milton
    1976   A Theory of Instrumental Activity and Farm Enterprise Commitment Applied to Woolgrowing in Australia. Rural Sociology 41:76–98.
Ferris, Diana
    1963   The Farm Family in Canada. Revision, Incorporating Recent Census Data, of an article by Helen C. Abell. In The Economic Annalist 1959, 29. Economics Division, Canada Department of Agriculture, Ottawa.
Foster, George M.
    1965   Peasant Society and the Image of Limited Good. American Anthropologist 67:293–315.
    1972   A Second Look at Limited Good. Anthropological Quarterly 45:57–64.

Gilbert, Howard A., *et al.*
　1971　Recognizing Personality Characteristics Related to Managerial Potential in Agriculture. Agriculture Experiment Station, Bulletin 584, South Dakota State University, Brookings.
Gilson, J. O.
　1959　Family Farm Business Arrangements. Bulletin No. 1, Agricultural Economics. Faculty of Agriculture, University of Manitoba, Winnipeg.
　1962　Strengthening the Farm Firm. Bulletin No. 6, Agricultural Economics. Faculty of Agriculture, University of Manitoba, Winnipeg.
Grove, Frederick P.
　1928　Our Daily Bread. New York: Macmillan.
Hedlund, Dalva E., and Alan D. Berkowitz
　1977　Farm Family Research in Perspective: 1965–1977. Multigraphed, Department of Education. Ithaca, New York: Cornell University.
Jameson, Sheilagh S.
　1977　Women in the Southern Alberta Ranch Community: 1881–1914. *In* The Canadian West. H. C. Klassen, ed. Comprint Publishing Company and the University of Calgary. Calgary, Alta.
Johnson, Glenn L., A. N. Halter, H. R. Jensen, and D. W. Thomas
　1961　A Study of Managerial Processes of Midwestern Farmers. Ames: Iowa State University Press.
Kohl, Seena B.
　1971　The Family in a Post-Frontier Society. *In* The Canadian Family. K. Ishwaran, ed. 1st. ed. Toronto: Holt, Rinehart & Winston of Canada.
　1976　Working Together: Women and Family in Southwestern Saskatchewan. Toronto: Holt, Rinehart & Winston of Canada.
Kohl, Seena B., and J. W. Bennett
　1976　Succession to Family Enterprises and the Migration of Young People in a Canadian Agricultural Community. *In* The Canadian Family. K. Ishwaran, ed. Revised ed. Toronto: Holt, Rinehart & Winston of Canada.
Mitchell, W. O.
　1960　Who Has Seen The Wind. Toronto: Macmillan.
National Farm Institute
　1970　Corporate Farming and the Family Farm. Ames: Iowa State University Press.
Pond, G. A., and W. W. Wilcox
　1932　A Study of the Human Factor in Farm Management. Journal of Farm Economics 14:470–479.
Rogers, Susan C.
　1975　Female Forms of Power and the Myth of Male Dominance. American Ethnologist 2:727–756.
Rolvaag, O. E.
　1927　Giants in the Earth. New York: Harper & Brothers.
Rosen, Harvey S.
　1974　The Monetary Value of a Housewife: A Replacement Cost Approach. American Journal of Economics and Sociology 33:65–73.
Sandoz, Mari
　1935　Old Jules. New York: Hastings House.
Sherman, William
　1967　The Germans from Russia. *In* Symposium on the Great Plains. C. C. Zimmerman and S. Russell, eds. Fargo: North Dakota State University.

Stefanow, Marlene
    1967   Changing Bi- and Multi-Culturalism in the Canadian Prairie Provinces. *In* Symposium on the Great Plains. C. C. Zimmerman and S. Russell, eds. Fargo: North Dakota State University.
Stegner, Wallace
    1962   Wolf Willow. New York: Viking Press.
Taylor, Lee, and Arthur R. Jones
    1964   Rural Life and Urbanized Society. New York: Oxford University Press.
Wilkening, Eugene A.
    1958   Joint Decision-Making in Farm Families as a Function of Status and Role. American Sociological Review 23:187–192.
Wilkening, Eugene A., and Denton Morrison
    1963   A Comparison of Husband and Wife Responses Concerning Who Makes Farm and Home Decisions. Marriage and Family Living 25:349–351.
Wilkening, Eugene A., and Lakshmi K. Bharadwaj
    1967   Dimensions of Aspirations, Work Roles, and Decision-Making of Farm Husbands and Wives in Wisconsin. Journal of Marriage and the Family 29:703–711.
    1968   Aspirations and Task Involvement as Related to Decision-Making Among Farm Husbands and Wives. Rural Sociology 33:30–45.
Wolf, Eric
    1956   Aspects of Group Relations in a Complex Society: Mexico. American Anthropologist 58:1065–1078.
Zimmerman, C. C.
    1967   The Rise of the Wheat Empire. *In* Symposium on the Great Plains. C. C. Zimmerman and S. Russell, eds. Fargo: North Dakota State University.

PART II

PATTERNS OF AGRICULTURAL DECISIONS

Chapter 10

Agricultural Business Choices in a Mexican Village

JAMES M. ACHESON

In the growing literature on decision making in peasant agricultural communities, emphasis has been placed on choices made by people who are already established in agriculture, or those who abandon agriculture completely and migrate to urban areas. It is widely known and accepted that peasant farmers often are engaged in more than one agricultural operation, and that they often combine agriculture with other occupations. Nevertheless, there is still a strong tendency to think of farming as a single business, and of farmers as being engaged in only a single occupation. Very little emphasis has been placed on business entry choices, particularly as such decisions influence the ways peasant agriculturalists combine various enterprises to earn a living. In Cuanajo, Michoacan, Mexico, such business combinations can scarcely be ignored, as a large proportion of the inhabitants combine agriculture with other enterprises over the annual cycle. If informants are to be believed, as land has become scarce and population has grown, the tendency to combine businesses has grown markedly. Only a small number of people are involved exclusively in one kind of agricultural enterprise any more. Moreover, there is a pattern in the way in which people combine various businesses. In this chapter, we will use some standard economic tools to analyze how people decide to enter a combination of businesses, and the characteristics of people who choose various agricultural business combinations. It is hoped that this analy-

241

sis will make some contribution to understanding a phenomenon, the importance of which goes far beyond the agricultural regions of central Mexico.

Cuanajo and its History of Business Complexity

Cuanajo is in the Tarascan area of the State of Michoacan, in the highlands of central Mexico. The pueblo is located about 12 miles southeast of the City of Patzcuaro. In 1972, the town contained an estimated 2731 people organized into some 565 households. The vast majority of the inhabitants are "Indians" who are bilingual in Spanish and Purepeche, the local Indian language.

This part of west–central Mexico is a land of lakes, rolling grasslands, and pine-covered mountains. It is a well-watered region, long one of the richest agricultural areas in the country. Cuanajo is about 7800 feet above sea level, in the *Tierra Fría*, 'cold country', where frosts occur about 120 days of the year. Annual temperature variation is small. Even in April and May, the warmest months, it is never uncomfortably hot. Winter nights, however, can be decidedly cold.

The pueblo itself lies in a small valley surrounded by mountains and hills. Virtually all the inhabitants live in the nucleated part of town, in houses that are lined up on streets laid out in the grid system. Close to the central plaza, where houses are very close together, there is no agriculture, although households commonly keep a few chickens in their patios. As one moves toward the outskirts of town, houses are farther apart, and most households have small vegetable gardens, plots of corn, or a few fruit trees growing behind their houses. The nucleated part of town is ringed with cultivated fields, planted in corn and wheat with smaller patches of barley and clover. These fields cover the whole floor of the valley. Beyond the fields, the mountains begin, their steep slopes covered with pine forests except for an occasional sprinkling of hardwoods. The base of the closest mountain is scarcely one-half mile from Cuanajo's plaza, and *Cerro de Condembaro*, over 10,000 feet and capped with snow 9 months a year, is only 8 miles to the east.

Like all other towns in the Tarascan area, Cuanajo is an agricultural community with a strong craft tradition. In the case of Cuanajo, that craft is woodworking. According to West (1948:58–59), the people of Cuanajo have made wooden products since at least 1789. Before 1910 and the start of the Mexican Revolution, all of the men in the community, according to very old informants, worked in agriculture for 6 or 8 months a year and made furniture out of local pine during the winter months. All agricultural produce and furniture were sold at the Patzcuaro market.

Three major changes have occurred in the past few decades. First, population has increased greatly. In 1910, there were an estimated 1000 people in the pueblo; in 1946, an estimated 1735 (West 1948:19); and in 1972, an estimated 2731. In all of this time, the amount of agricultural land has remained almost constant. This means that the average amount of land per capita has decreased 273% per capita in just 62 years. By the 1930s, the population had expanded to the point where many people in town did not have enough land to earn an adequate income. From that time on, the community has had a class of people forced to work as day laborers during some part or all of the annual cycle.

Second, Cuanajo has moved from a subsistence economy to a highly commercialized economic system. To be sure, many people still continue to produce a good deal of what they consume, and furniture can still be regarded as a cash crop for many households. But everything is bought and sold on the open market: corn, wheat, land, houses, machinery, clothing, and oxen. In addition, since 1960 businessmen in Cuanajo have begun to do business regularly with banks and other credit-granting agencies. Many of the carpenters in town, for example, have purchased motors and other machinery on credit from companies in Morelia or Celaya.

Third, in the same time period, a number of new businesses were established in Cuanajo. In the mid 1930s the first nonagricultural business was founded—a small store. By 1972, there were two large stores and ten small ones; six corn-grinding mills; five professional money lenders; and four butcher shops. There was still only one large grain dealer, although two other men (both store owners) bought up substantial amounts of grain after the harvest. The greatest changes occurred in the 1960s and 1970s as the carpentry industry rapidly expanded. In 1961, an electric power line reached the pueblo; in 1963, the first electric-powered woodworking shop was set up. In 1967, there were 23-mechanized shops in Cuanajo and 101-unmechanized shops; by 1970, there were 56-mechanized shops; in 1972 the number had expanded to 152-mechanized shops, and about 72-unmechanized ones. Between 1960 and 1972, carpentry had changed from a handcraft done in the agricultural off-season by perhaps 150 men, to a mechanized industry employing at least 250 men throughout the year and another 200 on a seasonal basis.

As carpentry has expanded, ancillary occupations have expanded as well. In 1950, there were only 2 or 3 furniture salesmen. In 1972, there were 31 such salesmen who transported furniture to cities within a

200-mile radius and sold it in the local markets to peasants and urbanites.

In short, the total range of business opportunities in the pueblo has increased greatly in complexity since the turn of the century, with the introduction of stores, mills, grain dealerships, and the rapid expansion of mechanized carpentry and furniture selling. Nevertheless, agricultural businesses still provide employment for most of the male population, and furniture making is still the prime secondary occupation for most of those involved in farming.

The Business Opportunity Map of Cuanajo

To understand the reasons people enter various agricultural businesses, we not only must know something about agricultural businesses but also about the total range of opportunities from which people can select. At this point, we will not attempt to describe the choices people actually have made, but rather the business options from which they could select. It is only against a background understanding of what people *can* do that a discussion of what they actually do makes any sense. In studying the business opportunity map, it is especially important to describe all of the various businesses, the amount of capital necessary to enter those businesses on a minimal level, and the kinds of returns one can expect on investments in each.[1] In this section, we will first delineate the various kinds of businesses in the pueblo, and second, compare the returns to investment in each by using the standard internal rate of return (IRR) method, which will be explained in the following sections.

Agricultural Businesses: General Information

At least 90% of the land in cultivation is devoted to growing corn and wheat. When we are talking about agricultural businesses, we are really speaking of the various ways of earning money from these two crops.[2]

---

[1] Detailed information on these businesses was obtained from depth interviews with key informants who have first-hand experience in each business. Most interviews were obtained from men since most entrepreneurs are men. Three women entrepreneurs contributed a great deal of information on moneylending, storekeeping, and land rental.

[2] Some tree crops such as apples, pears, peaches, and cherries are grown along with squash and lentils, but these are grown in small household gardens. They are not primary agricultural businesses. For all practical purposes, animal husbandry does not exist. There are a few sheep, goats, and pigs raised in the pueblo as well as 65 cattle and oxen, but most of the meat consumed locally is supplied by farmers in the surrounding hamlets.

Most agricultural work is carried out with the aid of steel-tipped plows and ordinary hand tools. Oxen are the most essential piece of capital equipment because they are used in every phase of farming. No chemical fertilizers or pesticides are used, and there are no tractors or other agricultural machinery.

All of the agricultural land in Cuanajo is owned privately by individuals in the pueblo. Since land traditionally has been inherited by all the children of a family equally, plots have been divided and subdivided. The typical landowner has several small pieces of land in scattered locations, rather than one consolidated field. In 1972, the 565 households in Cuanajo held approximately 924 ha of agricultural land and woodland (3.4 m²) or about 1.67 ha each. Although land is not divided equally, neither is it concentrated in the hands of a very few people. As Table 10.1 indicates, only eight men held more than 8 ha. The largest landowner admitted to owning 20 ha, which is only 2.1% of all the land available. Some 316 men had no land at all, but many of these were young men who some day will inherit at least a little land.

In 1972, approximately 480 out of the 752 men of working age in Cuanajo either owned land or worked on the land during some part of the year. Most of the men in agriculture do not have enough land to support their families. They are either day laborers or small subsistence farmers who supplement their incomes in other ways. Only the 161 men with over 3 ha of land can supply their own grain consumption needs and still sell sizable amounts of grain commercially, but many of these men also work at some other occupation during the agricultural off-season. Most of them work in carpentry. Only 10–15 men have enough land so that they can do nothing during the year except farm.

It is virtually impossible to buy land in Cuanajo. People are loath to sell the land they have, even though they might not have enough to make a living. As a result, general consolidation of farms has not taken

TABLE 10.1
Distribution of Agricultural and Woodland Land in
Cuanajo in 1972

Hectares of land owned	Number of household heads
None	316
Under 1 ha	126
1–2 ha	146
3–4 ha	135
5–7 ha	17
Over 8 ha	9

place, and moreover, it is impossible for an individual farmer to increase his own output by adding land to his production mix.

If we view grain farming in Cuanajo from the point of view of distribution theory, and ask about the kinds of inputs various farmers control, it is clear that there are at least six different kinds of agricultural businesses.[3]

1. It is possible for a farmer to control all of the inputs he uses. This is the stereotypic farmer—the man who earns his own living working on his own land, with his own family and oxen. The revenues he obtains then are a return to his own labor, his family labor, his land, the oxen and other capital equipment he has invested in, the seed he supplies, and so on. This type of agricultural business is very rare in Cuanajo. There are perhaps only 10–15 such farms.

2. It is possible for a man to supply only his own labor to the agricultural process. Such day laborers in Cuanajo are called *jornaleros*, and the money they obtain is a return only on their own labor. As we have seen, a very large number of men in the community are engaged in this business, on a full- or part-time basis.

3. Another type of agricultural business involves only land rental. In this case, the income received is a return on the land. In Cuanajo this type of arrangement is rarely used.

4. Sharecropping is a fourth kind of agricultural business. That is, the landowner supplies not only the land, but also half of the seed and fertilizer. In this case, the revenue the landowner receives is a return not only on the land, but on the seed, fertilizer, and other needs he contributes.

5. The complement of Case 4 occurs when a person contributes only his labor and oxen, and usually half the seed and fertilizer. In Cuanajo, these *yunteros*, owners of ox teams, work on land owned by someone else, and get half the proceeds. Sometimes they simply are hired by the landowner for a flat fee. In 1972, the going rate to hire a *yuntero* in Cuanajo was 28 pesos per day. In this case, the income received is a return on labor and the investment in the ox team.

6. One can make money in agriculture merely by investing capital. This is essentially what people are doing when they rent land to others. There is another variant on this possibility—the grain business. In Cuanajo, as in most agricultural societies, the price of grain is lowest at the time of harvest, and rises over the course of the annual cycle as

[3] According to neoclassic economic theory, income is distributed to individuals according to the factors of production they own and the returns on those various productive assets (e.g., land, labor, capital).

supplies decrease. Profits can be made then not only in growing agricultural products, but also by holding them off the market and selling them on the annual price rise. In Cuanajo, there is one very large grain dealer who buys as much grain as possible after the harvest, and sells it throughout the year to local people in his store or to an urban grain dealer in Morelia.

These six agricultural businesses are logical possibilities. There are people in Cuanajo who are involved in only one or another of these types of businesses. There are 96 day laborers, for example, men who do nothing but get a return on their own labor. There are some 15 men who do nothing but farm their own land and thus get a return on all inputs. On the whole, however, people are involved in more than one of these variant agricultural businesses or combine one or more types of agricultural businesses with other nonagricultural occupations. The way they do this demands considerable analysis.

Each of the several types of farming enterprises requires very different sets of resources. The person who is in the business of renting land to sharecroppers needs only the land. A sharecropper needs oxen, a plow, a little working capital, and a good deal of skill in agriculture. A man who farms his own land needs, besides the land, farming skills, a team of oxen, and working capital.

The most expensive asset is the land. The price of land varies tremendously according to its productivity, the distance from the pueblo, and the size of the parcel. In 1970, a hectare of good quality farming land near the pueblo cost a minimum of 15,000 pesos; medium quality land cost about 5000 pesos and a poor piece of land, high in the mountains, might cost as little as 3500 pesos. House lots in the pueblo itself are sold by the meter and cost considerably more than even the best farming land. Besides the land, the most important piece of capital equipment is a team of oxen and a plow. Teams of oxen sold for about 30,000 pesos, and a metal plow for about 800 pesos.

Given the high price of land, farmers have an enormous amount of capital tied up in their businesses. A small farmer with under 3 ha of good land and a team of oxen might have as much as 60,000 pesos in the business. Large farmers with over 10 ha can have assets valued at over 100,000 pesos. This means that even a moderate-sized farm involves a larger investment than any other business in the community.

Nonagricultural Businesses

All seven types of nonagricultural businesses are small family enterprises, physically housed in the homes of their owners. All involve some combination of the owner's labor, capital, and management. None require as much capital as is tied up in businesses involving agricultural land. There are, however, some very important differences among these businesses.

Stores and butcher shops are very similar. Both types of operations demand a good deal of close management and a relatively large amount of capital. In most cases, the owners of such businesses and their wives spend all of their time in the shop. The owners of the largest stores have approximately 20,000 pesos invested in store stock and the rooms housing the store. To have a viable-sized store necessitates at least 10,000 pesos in initial capital. The butcher shop owners have about 14,500 pesos invested in animals, equipment, and shops, and it is difficult to see how a viable shop could be operated on under 12,000 pesos in initial capital.

Mechanized and unmechanized carpentry shops are two very different businesses. The mechanized shops involve the owner, one to four hired men, and between 3500 and 6500 pesos in capital. Most of the capital is invested in a band saw, and a combination table saw–drill, which are run by 1- to 5-hp electric motors. The unmechanized shops consist of very little except a small shed and a few hand tools, costing a total of 1000–2000 pesos. Unmechanized shops are usually small, part-time operations, involving only the owner himself and perhaps a minor son. Such shops do not hire outside help. In unmechanized shops, revenues generated are a return only on the owner's labor and the small amount of capital he has invested. In mechanized shops, income is a return on the owner's labor, a larger amount of capital equipment, and the managerial skill of the owner.

Corn-grinding mills are in essence nothing more than a 10-hp motor attached to a table-sized grinding apparatus, which is operated by one man. The entire apparatus is housed in one room. The machinery and other equipment cost some 20,500 pesos minimum. All of the six mills in town are owned by storekeepers who have the mill housed next to their stores. The mills themselves are operated by hired men, who are supervised closely by the store owner or by his wife.

Furniture selling requires no investment in capital equipment or any hired labor. To enter the business, one needs about 1400 pesos to buy a load of furniture and finance a selling trip (i.e., bus tickets, food, hotel expenses). Success in this business does require a good deal of skill in marketing and unusual ability to get along in the *mestizo* world. Thus, income to furniture salesmen is a return on their own labor, working capital, and managerial ability.

The five moneylenders in the pueblo supply small amounts of money to local people to tide them over in emergencies. They are not a source of capital for long-term investments. At any one time, a moneylender might have loaned out 5000 pesos to some 10–12 people. The interest rate charged is a phenomenal 5–10% per month (60–120% annual rate), which reflects not only the scarcity of capital, but the high risks involved. It is very difficult to force a recalcitrant person to pay his debts even with a secured loan, and it is expensive to go to the law to force payment. Thus, a moneylender's income is primarily a return on capital and risk.

There are three factors that greatly simplify the business opportunity map of Cuanajo. First, there are no silent partners (a person who contributes some or all of the capital but does not actively participate in the day-to-day management of the business). Capital for improvements and long-term investments is very scarce, as the high rates of interest paid to moneylenders attest. People who have capital to invest generally either lend it out (moneylending) for short-term emergency loans, expand their own businesses, or buy land, considered a safe investment. They do not put it in a business owned and managed by someone else. Second, it is impossible to hire managers to run local businesses. Managerial skill is at a premium. Men who have the skill and ambition to do a good job managing a business generally have their own. Third, few businesses are large enough and complicated enough to involve hired labor. Virtually all the hired laborers are employed in some kind of farming operation, or in mechanized carpentry shops.

Returns on Investments in Cuanajo's Businesses

According to economic theory, the decision of a businessman to enter a business or combination of businesses is greatly influenced by an assessment of the returns to be earned. Other factors also enter into the calculus of decision making, but in Cuanajo, as in many other commercial cultures, the key to understanding business-entry decisions comes from an understanding of the way possible investments are ranked.

Economists and accountants have developed various intellectual tools to rank possible investment options. The most important of these are the net present value (NPV) method and the closely related concept of internal rate of return.

Local businessmen obviously do not rank investments using either set of concepts. They calculate their profit in terms of *ganancia*, which is essentially a short-term cash-flow concept. They simply add up all their cash expenses for a day or week (ignoring family labor costs and long-run expenses such as taxes and depreciation), and subtract this from cash revenues received in the same period. This concept is then used not only to calculate profit, but also to rank businesses. That is, the best business is seen as the one which gives the highest cash flow in the short run. This technique, which ignores all capital accounting, leads to an odd view of business opportunities in the pueblo and to innumerable investment errors (see Acheson 1972a). People discover viable investments really by the painful process of trial and error (i.e., watching who succeeds in business and who fails).[4]

Use of internal rate of return corrects the problems inherent in the use of *ganancia* since such calculations take into account not only short-term and long-term cash expenses, but also the time value of money.[5]

The formula for the net present value of any business or other investment opportunity is as follows (Weston and Brigham 1975:268):

$$NPV = \left[\frac{NCF_1}{(1+i)^1} + \frac{NCF_2}{(1+i)^2} + \frac{NCF_3}{(1+i)^3} \cdots \frac{NCF_N}{(1+i)^N} \right]$$

$$- C = \sum_{t=1}^{N} \frac{NCF_t}{(1+i)^t} - C$$

Here, *NCF* is net cash flow; *i* is the interest rate or the marginal cost of capital; *C* is the initial cost of the project; and *N* is the expected life of the project.

The formula for the internal rate of return is exactly the same except the equation is set equal to zero and solved for the interest rate involved. In the real world, internal rates of return for any given investment option are solved by a trial and error method in which various rates of interest are used. Calculating the internal rates of return figures for each business option in Cuanajo is a laborious job. However, it needs to be stressed that these techniques are the accepted way of ranking business options among economists and accountants. Use of less sophisticated techniques brings severe criticism from those who work with investment problems on an every day basis.

[4] No invidious comparisons are being drawn here. Many small businessmen in the United States and other western nations do not know much accounting. Moreover, all entrepreneurs make investment errors—even those who have the finest accounting skills and other technical expertise at their disposal.

[5] When a businessman invests, he usually is committing himself for a long period. Revenues produced by a business now are worth more than those produced far in the future, since money received now can be invested and made to earn interest. One forfeits that interest on money received in the future. A good explanation of these concepts is contained in Weston and Brigham (1975:268–274).

The figures for internal rate of return cannot be calculated unless a number of problems are recognized and solved.

First, the internal rate of return can only be calculated if we have figures on net cash flows, a concept used by accountants to measure the flow of money through a business. It ordinarily is calculated by subtracting from gross revenue (sales X price received) all business expenses and taxes. One can obtain figures on net cash flow from *ganancia* by subtracting from *ganancia* figures all long-run cash costs (i.e., taxes, family labor, etc.) that are not taken into account by local businessmen. Net revenue, a measure of profit can be calculated from figures on net cash flow by subtracting out all costs of depreciation of capital equipment.[6] In Table 10.2, we will report the figures for the *ganancia* and net revenue in each business. Net cash flows will be used in calculating figures on interest rates of return for each business.

Second, the model assumes that the investor will be in the business for a specified length of time and then sell the business. The object, after all, is to be able to compare investment options in the same time frame, and this cannot be done if one investment lasts 20 years and another 6 months. In this chapter, we will assume all investments are made for 5 years.

Third, there are two factors that make it difficult to calculate net cash flows in Cuanajo. One is the lack of information on long-term costs. This lack of data causes problems especially in analyzing the flow of cash produced by the termination of a business. Accountants always have problems estimating future salvage costs and revenues under the best of circumstances, but the relative lack of information on long-run costs makes this problem even more acute in Cuanajo. To handle some of these problems, we will always assume straight-line depreciation in calculating salvage costs. Thus, if a piece of equipment lasts 10 years, we will assume that the owner will get back half of his initial investment when he sells the business in 5 years.

Fourth, local businessmen do not take into account their own labor or family labor. In some businesses, there are no labor costs of any kind (for example, moneylending or renting land); in other businesses, such as mechanized carpentry, labor is hired and paid for, and influences calculations of profit. In still other businesses, for example, storekeeping, the business is a one-man or one-family operation in which no

---

[6] The units of time used locally are short and not uniform. Day laborers are paid by the day; carpenters talk of "profit" per week; furniture salesmen think of profit per selling trip, etc. In calculating the net cash flow and IRR we will translate all these figures into years.

TABLE 10.2
Summary of Income and Return Figures for Cuanajo Businesses

Business	Ganancia[b]	Annual net revenue	Annual net cash flow	Initial investment	NPV at 20%	IRR
Agricultural businesses						
a. Land rental	18 pesos/day, 5600 pesos/year for 10 ha farm	490 pesos/ha	490 pesos/ha	5000 pesos/ha	−1324/ha	11%
b. Runs farm but hires man with oxen	28 pesos/day, 8400 pesos/year for 10 ha farm	740 pesos/ha	740 pesos/ha	5000 pesos/ha	− 575/ha	16%
c. Runs farm and supplies all inputs	32 pesos/day, 9600 pesos/year for 10 ha farm	810 pesos/ha	810 pesos/ha	5600 pesos/ha	− 480/ha	18%[a]
d. Grain business[c]	7500 pesos/year total, 1800 pesos/year from one part of business	1360 pesos	1360 pesos	6000 pesos	+ 120	23%
e. Day laborer[d]	6–7 pesos/day	1170 pesos				
Furniture salesman	15 pesos/day, 375 pesos/month	500 pesos	4500 pesos	1400 pesos	+5443	147%[a]
Unmechanized carpentry	6.6 pesos/day, 160 pesos/month	1780 pesos	1920 pesos	1000 pesos	− 163	14%[a]
Mechanized carpentry	13 pesos/day, 320 pesos/month	340 pesos	3840 pesos	3500 pesos	+ 713	28%[a]
Store	19 pesos/day	6900 pesos	6900 pesos	20,000 pesos	+1002	22%[a]
Corn grinding mill	12 pesos/day	1460 pesos	2160 pesos	20,500 pesos	−1771	16%[a]
Moneylending	10.3 pesos/day, 3100 pesos/year	3100 pesos	3100 pesos	5000 pesos	+5728	58%
Butcher shop	31.75 pesos/day, 375 pesos/week (1972 prices for 2 men)	16,400 pesos	16,600 pesos	14,500 pesos	+5002	31%[a]

[a] Imputed wage rates for proprietors are used in assessing the Internal Rates of Return in these businesses. Thus, the IRR figures reflect only returns to capital, management, and hired labor in all cases.

[b] Ganancia is expressed in the units the people of Cuanajo actually used, and then converted into longer units, such as the month or year.

[c] There are three aspects to the grain business. Figures are presented for only one. The 7500 pesos is the ganancia earned by all three aspects combined. All the rest of the figures on the grain business concern only the one aspect of that business on which we have accurate data.

[d] Since no capital, land, or other assets are used, it is impossible to calculate the internal rate of return or the net present value for day laborers. Men who earn their living in this way simply do not invest.

252

labor costs are paid. Our estimates will be thrown off badly if the proprietor's labor is taken into account in one case and not in another. This problem will be solved by figuring internal rates of return using an imputed wage rate for the proprietor's labor in those businesses where such labor is one of the major inputs.

Fifth, estimating the costs of entering a business is made difficult because many people do not buy all the assets they own. Farmers, for example, usually inherit much of the land they farm. In estimating initial costs, we will derive data from cases where people bought all the assets they use.

Sixth, we will assume a 20% discount rate. Businesses in the United States generally use a discount rate of 8–10%, but it generally is recognized that discount rates in underdeveloped countries are higher due to increased risks. Businessmen in Cuanajo paid 3% per month to banks or 24% per year; and of course local moneylenders earn much higher interest rates. Under these conditions, a 20% discount rate does not seem high.

To demonstrate the way this formula is used to calculate internal rate of return and net present value for one business, let us consider the case of a man who rents 1 ha of land to a sharecropper. We will assume that the hectare rented is of medium quality, which is used to produce corn and wheat grown without irrigation and with only minimal amounts of manure. Six informants estimated that 1 ha of such land would produce about 14 *hanegas* of corn per year or about 6.2 *cargas* of wheat. Since the average annual price for corn was 54 pesos per *hanega*, and the price of wheat was 125 pesos per *carga*,[7] the gross revenue from this hectare would be 1143 pesos. Since half of this money would go to the land owner, and half to the sharecroppers, each would receive 571 pesos. If the only additional costs per year were 10 pesos for seed and 70 pesos for taxes, then the net revenue and net cash flow is 490 pesos per year.[8]

The initial costs of entry into this land rental business are about 5000 pesos, since this was the cost of a hectare of such land. Land has been rising steadily by 2 or 3% per year, so that salvage costs would be somewhat higher than the initial costs.

In calculating the net present value of investment in this land rental business we will assume (a) the investment will be terminated in 5 years; (b) the interest rate or discount rate is 20%; (c) the net cash flow is

---

[7] These were the February prices in 1967 and appear to be about average over the annual cycle. A *hanega* of corn weighs about 65 kilos; a *carga* of wheat approximately 156 kilos.

[8] Since none of the capital involved depreciates, there is no difference, in this case, between net cash flow and net revenue.

490 pesos per hectare and will remain constant for 5 years; (d) the initial cost of entry is 5000 pesos, and the salvage value will be 5500 pesos, assuming that land appreciated 10% over the course of the 5-year period. Under these conditions, the net present value is:

$$NPV = \sum_{t=1}^{N} \frac{NCF_t}{(1 + i)^t} - C$$

= present value of 490 pesos received for 5 years *plus* present value of 5500 salvage value received in 5 years *minus* 5000 pesos initial cost.
= 1465 pesos[9] plus 2211 pesos[10] minus 5000
= −1324 at 20%

As was pointed out earlier, the internal rate of return is calculated by a trial and error method. Since the internal rate of return is the interest rate that makes the net present value formula equal zero, the internal rate of return is calculated by calculating the net present values of this kind of agricultural business for different interest rates until an interest rate is tried that reduces the equation to zero. There are no tables by which the internal rate of return can be calculated.

In this case, we know that the net present value of this agricultural business is −1324 pesos at 20%, which indicates that the internal rate of

---

[9] Since the 490 pesos is a constant amount received every year, it can be treated as an annuity. The formula for the present value of an annuity is

$$A_n = Pmt \times PVIF_a,$$

where:

A_n = present value of annuity
Pmt = payment
$PVIF_a$ = present value interest formula for an annuity.

The $PVIF_a$ can be found in a table for the present value for an annuity. In this case, the $PVIF_a$ for an annuity in 5 years with a 20% discount rate is 2.991. Thus, to calculate the present value of this annuity, one multiplies 2.991 ($PVIF_a$) by 490 pesos annual payment to get 1465. For a good short explanation, see Weston and Brigham (1975:245).

[10] The 5500 pesos will be received only once by the owner, at the end of the 5 year period when he goes out of business. The present value of a single payment to be received some time in the future is:

PV = $FV \times PVIF$, where
PV = present value
$PVIF$ = present value interest formula
FV = future value (the amount to be received in the future).

In this case, the FV is 550. The $PVIF$ at 20% for five years is .402. Thus, the PV of 5500 pesos to be received in 5 years at 20% is (.402 × 5500) or 2211 pesos.

return is lower than 20%. If we do the calculation assuming the interest rate is 10%, we get a net present value figure of 272. If we run through the same calculation using 12% rather than 10%, then the net present value is minus 120. Thus, we know that the internal rate of return is a little higher than 10% but not as large as 12%. Under these circumstances, it is reasonable to assume that the internal rate of return is about 11%.

Although internal rate of return and net present value are calculated from the same formula, the concepts differ slightly in application. Internal rate of return gives a figure analogous to an interest rate, a very familiar concept. Net present value, however, gives a cutoff point on investment.

The net present value and internal rate of return were calculated for every agricultural and nonagricultural business in Cuanajo. Each business examined involved a different set of figures and some unusual features that had to be taken into account. Nevertheless, the techniques and the formula used are exactly the same as in the case outlined. The results of all these calculations are summarized in Table 10.2.

Business Entry Decisions

An accountant looking at these figures would be struck immediately by the wide range of returns offered. Businesses like moneylending, furniture selling, butcher shops, stores, mechanized carpentry, and the grain business all give internal rates of return over 22%. The internal rate of return figure for moneylending is 58%, whereas furniture selling is a phenomenal 147%. By way of contrast, the figures for all agricultural options are relatively low (i.e., 11% and 16%), with the exception of the grain business, which gives an internal rate of return of 23%.

Net present value figures are used in a slightly different way to assess investment options. The general rule is that one should invest when the net present value figures are positive and not invest when they are negative. Given this rule of thumb, there are four poor investment options in Cuanajo: unmechanized carpentry shops, corn-grinding mills, and all forms of agriculture, except for the grain business. There are, of course, no figures for internal rate of return and net present value for day laborers, but their earnings are some of the lowest in the pueblo.

The critical question this raises is: Why is agriculture in Cuanajo not abandoned as a business and businesses like moneylending, butcher shops, mechanized carpentry, and grain sales attract all the business people? Why is it that the vast majority of people in the pueblo are still in agriculture and are loath to sell land under any circumstances?

To some extent, these questions are unfair because some of the people of Cuanajo *have* moved into high return businesses. The number of mechanized carpentry shops has expanded very rapidly, and virtually every farmer is in the grain business in that he typically holds some of his grain off the market in the fall to get the higher price later in the annual cycle. But on the whole, there has been no general move away from agriculture.

In the first analysis, one might assume that the unwillingness to sell land and move into other businesses could be traced to the fact that Cuanajo businessmen do not have concepts like internal rate of return, and thus really do not perceive the opportunities in some of these businesses. After all, local people measure the desirability of an investment in terms of *ganancia*, and there can be little doubt that this is a very poor indicator of investment options (Acheson 1972a).

To be sure, the local accounting system confuses the pattern of decision making, but this is not the fundamental factor influencing business decisions. Businessmen have to deal with hard economic reality. If one makes too many mistakes, one goes out of business. This is true regardless of what the initial perceptions of business options might be. Under these conditions, if people are remaining in agriculture, and are not entering other businesses, the reason is apt to be something far more compelling than perceptions arising from the local accounting system.

There are two sets of reasons why people tend to remain in agricultural businesses in large numbers despite the low returns. The first relates to the total set of business opportunities in the pueblo, and the second to the positive value placed on land ownership.

In Cuanajo, there is no real opportunity to invest a large amount of money in a business or combination of businesses that will give high returns. Businesses such as moneylending, butcher shops, stores, furniture selling, and mechanized carpentry give very high internal rates of return and positive net present value figures. Both these facts indicate that they should be good investments. The problem is that these businesses can absorb very little capital, and consequently very few additional people could move into them. There is, for example, only a very limited amount of money that can be loaned out at 60–120% per year. People borrow only to cover short-term emergencies, and repay the moneylender as quickly as possible. Since there is only a limited amount of money that can be loaned out at these rates, the earnings to moneylenders are limited as well. The owners of existing large stores and butcher shops earn high returns on investments, but the market for such services is strictly limited. There may be room for one or two more butcher shops in Cuanajo with the expanding demand for meat, and perhaps one more large store could be established and survive. But larger numbers of people could not move into those industries without saturating the market. Demand for store products is clearly close to saturation now. Everyone who entered storekeeping on a small scale in the past few years (early 1970s) has failed.

There is one large grain dealer in Cuanajo, and he gets virtually all the business. There might be room enough for one, two, or three more such dealers, but certainly no more, since there is only a limited amount of grain grown. Business that additional sellers gained would almost certainly come at the expense of the established dealer.

The market for corn-grinding mills also is saturated now. The net present value to owners of these mills is negative, indicating that such mills are a very bad investment. Undoubtedly earnings to owners of such mills was higher in the recent past when there were only two or three mills in the pueblo. When the sixth mill was added, however, the owners of the mills agreed to operate on alternate days so that only three mills were open on any single day. When the seventh mill was added in 1970, the market was so bad that the owner could not make enough to meet his monthly payments, and he went out of business within a year.

Mechanized carpentry is the only business in the pueblo that offers high returns and is capable of absorbing a large amount of capital in aggregate. A very large number of men have moved into mechanized carpentry and the closely related business of furniture selling. These are the only businesses that offer a general solution to the people of Cuanajo, and certainly the local business community has been very responsive to opportunities in this area. Again, the number of mechanized shops has increased from 23 in 1967 to 156 in 1972.

However, it should be noted that the amount of money that any single individual can earn from mechanized carpentry is strictly limited. Once one has invested about 6500 pesos in machinery and has hired four or five employees, there are no other assets that can be added to the production mix to increase income. In economic terms, after one has about 6500 pesos invested and about five employees, the marginal productivity of both labor and capital fall to zero. The question needs to be asked, why could one not purchase two or three such carpentry shops and thus greatly increase the amount of money invested at a high return? In reality, this is impossible, and management is the limiting factor. Mechanized carpentry shops demand a good deal of day-to-day supervision and management by the owner. When the owner is absent for long periods, tools disappear, routine maintenance is not carried out, wood is wasted, and production slows. Moreover, it is impossible to hire managers for such shops.

Much the same is true for furniture-selling businesses. Once one has invested 2000 pesos in working capital, there is nothing more that can be purchased that will greatly increase revenues. Like mechanized carpentry shops, furniture-selling businesses demand the full-time attention of the owner himself.

In summary, then, a businessman can invest small amounts of money at high returns in mechanized carpentry, furniture sales, or the grain business. But since the amounts of money that can be invested are small, the net revenues earned are limited as well (see Table 10.2). Anyone with a lot of money is forced to invest it in businesses such as stores, corn-grinding mills, or land. If these businesses give low returns, they at least provide an opportunity to invest.

There is only one business in Cuanajo capable of absorbing a great deal of capital—namely agricultural business involving land. The 3.4 m^2 of agricultural land is worth at least 5.4 million pesos or some \$450,000. A very few people might convert some of the money invested in land to mechanized carpentry shops, or other businesses, but the vast majority of the people in Cuanajo have no other option but to keep their money invested in land regardless of the fact that it is a poor investment (i.e., negative net present value figures, and low internal rate of return). Of course the fact that people must keep their money tied up in such businesses as agriculture reflects the general lack of economic opportunities in the pueblo, and gives some insight into the difficulties of earning a living in rural sectors of countries like Mexico.

It is not just the absence of many economic opportunities that stops people from leaving agriculture and selling their land; there is a positive value placed on land ownership. When a sample of some 52 people were asked if they would sell their land, *no one* said he would sell it except if forced to by some emergency. Several people said they would not sell it even then, but would rent it out and get income in that way. When pressed to elaborate on the reason for their unwillingness to sell land, people tended to give a variety of incomplete answers or mumble something about "paper money," or "security." Several people, however, were able to articulate the reasons for their reluctance. There were, in essence, four reasons. First, they think of land as a secure investment. Land not only has been holding its value, but generally has appreciated in the past several decades. Local people, however, are not holding their land for speculative purposes (i.e., making a profit by selling when the price rises) because they do not plan to sell the land regardless of how valuable it becomes. Rather, they see land as an investment that will return income to its owner if he is ill or incapacitated. It is the Mexican peasant's analog of the insurance policy.

Second, Mexican peasants are fully aware that the annual rate of inflation has ranged from 9 to 30% in the years between 1967 and 1978. This means that if a person sold his or her land, the money received would quickly lose its value. Mexicans have known something we in the United States are just beginning to learn—namely, that in times of rapid inflation, land, gold, and other tangible commodities are to be valued over paper money.

Third, Cuanajo residents clearly see money put into land as providing a place where they and their family can put their labor. If one has enough land, one has a steady job for life. Although it would be difficult to prove, some men might be willing to purchase land even if it gave no return to capital, simply to have a permanent occupation and the income from their labor on the land. Fourth, it must be recalled that in Cuanajo, there are few opportunities for silent partners. One cannot simply invest money in a business; one must put in one's own labor and managerial skills as well. A man whose only skills and experience are in agriculture might be forced to think of increasing his income in terms of buying additional land, or at least keeping the land he has. Given his skills, there may be no other business he can enter, and no other investment he can make. For him to sell his land would be to sell the only means he has of making a living.

Finally, though few in Cuanajo can articulate it, the ownership of land is valued for purely symbolic reasons. Land ownership links one to the pueblo in ways that people in the United States find difficult to understand. Even people who have migrated to distant cities and have remained there for decades try to retain ownership of a little family land to maintain some ties with their natal pueblo. In addition, land ownership is associated with prestige. People with the highest prestige in the pueblo are the large landowners, those with the least, the landless day laborers. In the eyes of many people, to sell land is to divest themselves of one of the things that grants them standing and full membership in their home pueblo.

For all these reasons, the very high value placed on land is not completely irrational. However, land is not the best investment one can make, and one wonders if the security and prestige associated with land are worth the great sacrifices sometimes made to keep it.

Agricultural Business Combinations

The vast majority of the men in Cuanajo are involved in some combination of businesses, and most of these involve agriculture in one form or another, since most of the capital in the pueblo must remain in the agricultural sector due to lack of opportunities elsewhere. Agriculture, however, cannot be combined with every business. There are some definite rules concerning (a) the ways businesses can be combined; and (b) the kinds of people who can take advantage of these options.

On the whole, the ways businesses may be combined with each other is determined by the managerial requirements of the businesses rather than by the returns. The exact way businesses are combined is strongly constrained by the fact that storekeeping, butcher shops, mechanized carpentry, and large-scale agriculture demand most of the time of their owners. There are four types of business combinations involving agriculture.

Business Combination 1: Small-Scale Agriculture and Unmechanized Carpentry

As one can see from Table 10.1, page 245, the vast majority of the people in Cuanajo have under 4 ha of land. This is simply not enough to supply an adequate income for a family. Even people with 4 ha of land would earn a net revenue of only 740 pesos per ha, or 2960 pesos per year. This is only 8 pesos per day, which, even in 1970, would scarcely be enough for survival. Clearly, the men with under 4 ha have to have some other source of income. Fortunately, small-scale agriculture has one asset: It is combined relatively easily with other businesses. Small-scale farmers do nothing in the winter, and even during the agricultural season, they have a good deal of free time since they must hire a man with an ox team to do many of the most important agricultural tasks for them (plowing, harrowing, weeding). Virtually all of these 407 small-scale landowners work on their own land part of the time and earn a good part of their income in other occupations. About 15 of these men are house carpenters or stone masons; another 15 combine agriculture with furniture-selling trips; and some 28 of the owners of the very smallest plots own mechanized carpentry shops and are only in farming on a marginal basis. The largest number, however, about 230 of these men, work in furniture-making shops as hired laborers, or else have their own small, unmechanized shops. Many who combine small-scale farming with work in their own unmechanized shop also work as day laborers in mechanized shops owned by other men. We have very little information on the remaining 119 men. Many of them are very young and do not really do much of anything; others pick up any kind of work they can during the annual cycle. Many of these men remain in Cuanajo for part of the year, and then find work in the cities of central Mexico or in the United States. A lot of these men were very vague about their occupations, and this is scarcely surprising given the unplanned, haphazard way they change jobs and the fact that many do not want to talk about their illegal trips to the United States.

Business Combination 2: Small-Scale Land Rental and Mechanized Carpentry

Unlike unmechanized carpentry, which can be combined nicely with agricultural pursuits, a mechanized carpentry shop demands the full-time attention of the owner throughout the entire year. About 110 of the 156 owners of mechanized shops have no land at all, along with most of the day laborers who work in all shops. These people do nothing but work in carpentry. There are, however, some 46 owners of mechanized carpentry shops who have some land, although most do not have enough to earn a living. Most of these men make no pretense of farming this land themselves. Rather, they rent it out on a sharecropping basis, keeping only a small patch or two to farm themselves.

Business Combination 3: Land Rental, Storekeeping, Moneylending, and Corn Mill Operations

The economic situation in Cuanajo makes it very difficult for those few people who have a large amount of money to invest it locally. Complicating the picture is the fact that large-scale agriculture, in which the owners provide all inputs, is a full-time occupation. Three very successful businessmen in Cuanajo have managed to invest large amounts of capital successfully by combining land rental, store owner-ship, moneylending, and ownership of corn-grinding mills. There are two advantages to this combination of businesses beyond the fact that they can absorb a large amount of capital. First, such a business combi-nation can be operated by one man from behind the counter of his store. The corn-grinding mill and store are housed in the same build-ing. The moneylending business and the rental of land take great business acumen, but little time on the part of the owner. Second, one of the few ways to obtain land in Cuanajo on a regular basis is to foreclose on unpaid debts in which land was used as collateral. Thus, moneylending and land rental complement each other. By entering moneylending, one not only can lend small amounts of money at very high interest rates, but the business also periodically results in acquisi-tion of land—the only business where one can invest very large amounts of money with a high degree of security.

Business Combination 4: Large-Scale Agriculture and Sharecropping

In Cuanajo in 1970, there were 21 ox teams. Ten of those teams were owned by some of the few men who had such a large amount of land they could afford to do nothing but agriculture throughout the year. The other 11 teams were owned by men who had some agricultural land, but not enough to supply an adequate income throughout the year. These men combined work with their own team on their own agricultural land and rented additional land owned by other men on a sharecropping basis. Some of them also earned money by plowing and harrowing small plots of land owned by other people. Although these men did nothing but farm throughout the year, they clearly are involved in at least two separate agricultural businesses.

Choices of Agricultural Businesses

The lack of opportunities outside agriculture and the inability to purchase land greatly diminish the choices any single individual has. For all practical purposes, the business or business combination one enters is determined, in great part, by the amount of agricultural land one inherits. A young man who is fortunate enough to inherit a lot of land has several different options open to him. He can enter full-time agriculture in which he provides all the inputs, or he can rent his land and enter another business that takes a good deal of time (e.g., Business Combination 3). He can sell the land and use the proceeds to finance entry into another business requiring a good deal of initial capital, such as the combination of storekeeping and corn-grinding mills. No other people, and this includes virtually everyone else in the pueblo, have this range of choices. If a young man inherits a small amount of land, earning a living by full-time agriculture is virtually impossible, unless he is one of the very few who can manage to rent a good deal of additional land to work on a sharecrop basis. Most of the men in the pueblo have some land and are faced with the necessity of combining farming with some other occupation such as furniture making (i.e., Business Combinations 1 and 2) or working as day laborers.

The growing number of landless men have very few choices, since most agricultural options are closed to them. A very few young men inherit their father's stores or butcher shops, but they are a tiny percentage of the total number in the pueblo.

In the recent past, these men have had no choice but to become day laborers in agriculture or carpentry. The mechanization of carpentry shops has greatly increased opportunities for these landless men, and many of them have taken advantage of them. Moreover, the expansion of furniture making has increased opportunities elsewhere. It has created opportunities for furniture salesmen and people in mountain villages who cut logs. In addition, men who own mechanized shops have rented a good deal of their own agricultural land to sharecroppers, which has given several farmers access to enough land to greatly increase their own incomes.

Summary

In studying business decisions in Cuanajo, it is the constraints that need to be stressed. Whereas local businessmen are responsive to new business opportunities when they occur, there are unfortunately few good investment options in Cuanajo. A study of all the businesses established in the pueblo reveals that there are six different agricultural businesses and seven other well-established businesses. None of these investment options is capable of absorbing a great deal of capital at high returns. The only business options capable of utilizing a great deal of capital are agricultural businesses involving ownership of land. Unfortunately the internal rate of return on such businesses is very low, and the net present value of such agricultural businesses is negative. Both these factors indicate that agricultural businesses are poor investment options. Nevertheless, the vast majority of people in the pueblo are involved in some business or combination of businesses involving agriculture.

There is nothing else most people can do, given the general lack of investment opportunities. Moreover, land is one of the few secure investments one can make in an economy riddled with inflation. Control over enough land guarantees one a job and security for life.

Given the fact that most people in the pueblo must remain in agricultural business or in some combination of businesses involving agriculture, what factors influence choice of agricultural businesses? Two factors strongly influence such decisions

1. It is virtually impossible to buy land. This means it is difficult to enter most agricultural businesses unless one is already established, and it is very difficult to increase revenues by adding land to one's production mix.
2. There are strong constraints on the ways businesses may be combined with one another, given managerial requirements. There are four patterns of business combinations in Cuanajo, involving vir-

tually all men in the pueblo. The option selected depends on the amount of agricultural land inherited.

References

Acheson, James M.
 1972a Accounting Concepts and Economic Opportunities in a Tarascan Pueblo: Emic and Etic Views. Human Organization 31(1):83–91.
 1972b Limited Good or Limited Goods? Response to Economic Opportunity in a Tarascan Pueblo. American Anthropologist 74(5):1152–1169.
West, Robert C.
 1948 Cultural Geography of the Modern Tarascan Area. Smithsonian Institution Institute of Social Anthropology Publication No. 71. Washington, D.C.: U. S. Government Printing Office.
Weston, Fred, and Brigham, Eugene
 1975 Managerial Finance (5th ed.). Hinsdale IL: Dryden Press.

Chapter 11

Agrarian Reform and Economic Development: When Is a Landlord a Client and a Sharecropper His Patron?[1]

KAJA FINKLER

Introduction

Various anthropologists concerned with rural agrarian societies have applied Chayanov's theory of peasant economies to the anthropological study of peasant and primitive societies (Wolf 1966; Sahlins 1972).

The object of this chapter is to examine peasant decisions respecting deployment of land and labor from the perspective of Chayanov's theory. It will be shown that peasant behavior fails to conform to Chayanov's model when land is scarce and subject to agrarian reform laws.

Chayanov, a Russian economist working in the early decades of this century, challenged the validity of applying a capitalist model of production to peasant economic behavior on the grounds that peasants do not operate a business: They manage a household. He focused his study on "a peasant family that does not hire outside labor, has a certain area of land available to it, has its own means of production, and is sometimes obliged to expend some of its labor force on nonagricultural crafts and trades [1966:51].[2]" Thus, following Chayanov, peasant households,

---

[1] This chapter is a revised version of an article published in *Economic Development and Cultural Change* 27:103–120.

[2] "Crafts and trade" refer as well to seasonal nonagricultural work (see Glossary in Chayanov 1966:271).

AGRICULTURAL DECISION MAKING:
Anthropological Contributions to
Rural Development

unlike capitalist firms, employ only family labor, which cannot be converted into monetary terms. Inasmuch as peasants fail to pay themselves a wage for their labor on the family farm, the family's operations cannot be subjected to a cost accounting. Chayanov argues that, for all these reasons, a peasant household unit has greater survival value than a capitalist farm venture, especially under adverse economic circumstances. Therefore, a peasant family operation is well suited to compete with a capitalist farm enterprise.

Moreover, Chayanov advances the thesis that the peasant family's major objectives, in contrast to a capitalist undertaking, are, in his words, "determined by a peculiar equilibirum between family demand satisfaction and the drudgery of labor itself [1966:6]." From this follows that the amount of land peasants cultivate is subjectively defined and is dependent upon their subsistence needs. A peasant household's consumption requirements will thus correlate with the family's size and composition, and peasant household production will correspond to the family's developmental cycle. Hence, according to Chayanov's model, a peasant unit will expand its production as the family grows larger, and it will diminish its agricultural output as family size decreases as children marry and leave the household.

To a great extent, Chayanov's model hinges on the presupposition that land is readily available to a peasant household for the purpose of expanding its operations as needed; he deals only briefly with the problem of land scarcity. In his succinct words: "In farms greatly short of land, on the other hand, the concern to meet the year's needs forces the family to turn to an intensification with lower profitability [1966:7]." Having done fieldwork in an agricultural region in Mexico (Finkler 1973, 1974), where size of landholdings are small and predominantly fixed owing to the extant legal restraints under Mexico's land reform laws, its *ejido* system, I posed the following question: What economic strategies do peasant households follow when land, as well as other factors of production including capital, are scarce, and what are the ways in which a peasant family intensifies its labor output?

In my study of Itel[3] village (Finkler 1973, 1974), a peasant community situated in the Mezquital Valley, Hidalgo, where land reform has been instituted following Mexico's 1910 revolution, I observed that numerous peasants worked as share-tenants and also hired themselves out as day laborers. The fact that peasants hire themselves out as wage workers on a daily basis is consonant with Chayanov's theory when he states: "We have established that the peasant family without enough

[3] "Itel" and personal names used here are pseudonyms.

land and means of production at its disposal for the complete use of all its labor in the agricultural undertakings puts its surplus in another form of economic activity (crafts and trades) [1966:113]." However, I also noted that numerous families with small-size plots and in varying stages of their developmental cycle turned over their land to share-tenants. Although my experience in the region tended to support Chayanov's thesis that peasants aim first and foremost at subsistence, the fact that they entered into share-tenancy agreements was puzzling. Even more perplexing was the finding that, contrary to Chayanov's theory of peasant behavior, the household decreases its income by turning over its land to a share-tenant. The decision to seek a share-tenant to enter a sharecropping agreement rested with the landholder. This of course was contrary to the usual references to sharecropping in relation to landless peasants seeking access to land by working plots owned by absentee landlords (Paige 1975; Redclift 1977; Whetten 1948). Thus, the question that emerges is: What prompts peasants in this region to relinquish their meager land plots to another to be worked on a sharecropping basis? This chapter will examine the various circumstances that lead peasants to this decision and the socioeconomic consequences that ensue from it. With land scarce and subject to bureaucratic control, in addition to a lack of labor and credit, individuals emerge whose business-like operations, involving multiple sharecropping arrangements, place their enterprises in the growing group of business firms, rather than as peasant households working for a livelihood. These individuals gain access to *ejido* and privately owned lands, which they work for profit rather than for subsistence. Unlike peasants, these men emerge as "agricultural entrepreneurs" both because they maintain a labor force of their own and because their production is geared for profit. Moreover, as we shall see in an interesting reversal of the usual relationship, there are cases in this system where the landholder has become dependent on the share-tenant. Thus an understanding of the sharecropping system permits us to observe directly a process of socioeconomic differentiation.

I will discuss first *ejido* land tenure, which is crucial to an understanding of the dynamics of the sharecropping system practiced in Itel. The sharecropping system will be discussed in turn, followed by a review of the household decisions prompting landholders to give their lands to share-tenants, that is, to individuals who work these lands on a share basis. Any discussion of peasant household economics within the context of agrarian reform requires examination not only of the availability of land, but also of all productive factors together as a single unit. Therefore, a discussion of the system of water allocation and the shortage of labor and capital within the community also will be presented. Economists' production functions will be used to demonstrate why sharecropping is profitable. It will also be shown that, under conditions of land scarcity, generated in this case by land reform, Chayanov's theory fails to explain peasants' behavior because of the institutional constraints created by the land-reform program. A brief discussion of the sharecropping system within the context of economic change will follow, and it will be suggested that agricultural entrepreneurs will quite likely form a future core of agricultural producers, while *ejido* holders will be forced to become proletarians, despite the aims of agrarian reform guaranteeing land to the peasantry. Finally, it will be emphasized that the anthropological approach provides us with the necessary insights into the various dimensions affecting peasants' economic decisions, dimensions that also lead to economic distinctions within the village community.

Mexico's *Ejido* System

Mexico's *ejido* system has been described by numerous scholars (Brandenburg 1969; Cline 1963; Eckstein 1966; Simpson 1937; Stanvenhagen 1970; Tannenbaum 1929; Tannenbaum 1950). Its history and formal organization are well known and need not be discussed at length. Briefly, the *ejido* system came into existence through direct government action on January 6, 1915. Venustiano Carranza, then president of Mexico, issued his now famous decree ordering the return of land to the villages. In 1917, article 27 of the new constitution provided the legislative foundations for agrarian reform (Tannenbaum 1929; Whetten 1948). Despite the existence of these laws, the expropriation and redistribution of lands was not implemented fully until the administration of President Lázaro Cárdenas (1935–1940), who expropriated and redistributed some of the most highly developed farm land in Mexico. Most importantly, during Cárdenas' era *ejido* tenure became a fundamental part of the Mexican national economy (Simpson 1937; Whetten 1948).

The legal size of *ejido* plots of land has varied over time. During Miguel Aleman's administration (1947–1953), the minimum grant was raised to 10 ha of irrigated land, or 20 ha of dry land. The measure was designed to reverse a trend toward the growth of *minifundia*. It also opened more lands for colonization rather than continuing the expropriation of land "into uneconomic units, the small and poor land resources inherited by the revolution [Cline 1963:215]." According to

Whetten (1948), the average allotment per recipient in the state of Hidalgo was 3.1 ha of irrigated land or 6.3 ha of nonirrigated land, whereas, in Itel, the community in which the data that follow were collected, the average size of the irrigated *ejido* plots is 1.11 ha.

The formal administrative structure of the *ejido* system is composed of national and local administrative bodies. On the national level, the *ejido* is dependent upon the Ministry of Agrarian Reform for such matters as the validation of land grants, confirmation of old communal titles, the issuance of certificates of title to individual holders, and the assessment of land boundaries. On the local level, the administration of the *ejido* is vested in two committees. The first is the *Comisariado Ejidal*, executive committee, consisting of three elected members with three alternates. This body is responsible for active management of *ejido* affairs. The broad functions of the committee are to represent the *ejido* before the administrative and judicial authorities, to supervise the division of plots, to designate successors to plots, and to call a meeting of the general assembly at least once a month. The general assembly is composed of all *ejidatarios, 'ejido* landholders'. The second committee operating on the local level is the Vigilance Committee, composed of three members and charged with watching over the executive committee. Knowledge of, and interaction with, these committees play an important part in promoting the careers of agricultural entrepreneurs, and, therefore, these committees are important in contributing to the economic and political influence of these people.

More important for our discussion is the fact that under the law of *ejido* tenure, land is a nonnegotiable resource. The peasant is provided with a small plot of land on the condition that the land is not sold, transferred, or mortgaged. The *ejido* holder can lose his right to the plot by failing to work his land himself for 2 consecutive years. In view of the small size of *ejido* plots, peasant households are usually faced with financial dilemmas requiring additional revenues to meet their daily subsistence. This being the case, *ejidatarios* have essentially one of two options: (*a*) seek wage labor outside the community; or (*b*) seek land to sharecrop. Based on my study of 1970–1971 in Itel (Finkler 1973; 1974) and subsequent observations during my field stays in the region, *ejido* holders prefer the latter course: to seek additional private or *ejido* lands to work on a share basis. From the *ejidatarios'* perspective, the conditional nature of *ejido* tenure leaves no *ejido* holder certain of his plot unless he is present in the community to defend his title to the land. *Ejido* landholders say that there is always someone quick to fabricate a rumor that a man's plot is not being worked and that he is therefore not entitled to retain it. When peasants were asked why they did not leave

TABLE 11.1[a]
Yields per Hectare, Ejido, and Privately Held Land

	Ejido—metric tons	Privately held land—metric tons
Corn	3.6	3.2
Alfalfa	7.5	7.1
Beans	1.9	2.1

[a] I wish to acknowledge Ing. Guillermo Garmendia of the Mexican Ministry of Agricultural and Hydraulic Resources for providing me with these data.

their community, the most frequent response was "I have my *ejido* here. I have my livelihood here." Thus, while *ejido* tenure guarantees the peasant a minimum level of subsistence, it also serves to tie the peasant to his plot and village and usually precludes his search for alternative and supplementary forms of livelihood outside the confines of the community or its immediate environs (see also Chevalier 1967). In sum, a peasant with a fertile *ejido* plot of land tends to prefer a secure agricultural existence. He does not wish to forsake his land parcel in favor of wage labor away from the community or its environs. Access to *productive*, that is, irrigated *ejido* land tends to mitigate against migrant wage labor.[4]

It is important to indicate that in this region, *ejido* lands and privately held plots are equally productive. The literature is replete with controversy regarding the degree to which *ejido* lands are worked as efficiently as privately owned lands (Barchfield 1978). It is often argued that *ejidatarios* operate at a reduced level of efficiency as a consequence of legal institutional factors including the conditional nature of *ejido* tenure. This, however, seems not to be the case, as can be seen from Table 11.1, which shows mean agricultural outputs for four cycles from 1976–1978. As these figures demonstrate, both *ejido* and privately owned lands obtain approximately the same yields for crops grown in the region, given the variability in soils, rainfall, and other ecological factors. Hence it cannot be argued that *ejidatarios* turn over their lands to share-tenants because *ejido* lands are less productive than privately

[4] On the other hand, as I demonstrate elsewhere (Finkler 1973; 1974), in a community where land is unproductive due to lack of rainfall and irrigation, virtually the entire male population leaves the village in search of wage labor. Some go to Mexico City, others to the United States as wetbacks, and some engage in petty commerce. For a similar view, see Stavenhagen (1975). Guillet (1976) has shown that agrarian reform in Peru has resulted in the return of migrants from Lima to claim rights to land that guarantees a secure subsistence, supporting the argument made here. However, further studies are required along these lines.

owned lands. In fact, a common strategy practiced by peasant house-holds in Itel and its environs is to enter a sharecropping arrangement, usually with fellow villagers.

The sharecropping system in Itel involves peasant landholders whose landholdings exceed what they can manage with the labor available in their households as well as that of individuals with insufficient land of their own. Landholders who seek share-tenants usually do so for at least one of three reasons: access to labor, cash, or water. As will be seen presently, close inspection of the sharecropping arrangements, using econometric techniques, reveals that, viewed from the share-tenants' perspective, the decision to enter into a sharecropping arrangement is economically sound and conforms to Chayanov's hypotheses. From the perspective of the landholder, however, the decision to seek out a share-tenant is not economically sound.

Methodology[5]

For the purposes of this investigation, data were collected by means of an extensive interview schedule administered to 50[6] household heads selected randomly from 116 cases of sharecropping families identified in the village in 1970–1971. The data thus collected were subjected to an economic analysis using econometric techniques.[7] This analysis was done for two reasons: (a) to ascertain the ways in which decisions to enter specific sharecropping arrangements were economically advan-tageous or disadvantageous to the household; and (b) to test the hypothesis that share-tenants who become entrepreneurs owe their success not only to their superior managerial skills and to their vertical

[5] The data presented here were gathered in Itel in three separate field trips in 1970–1971; May–September 1973, and June–September 1974. During my current stay in the region from 1977–1979 on another field project (supported by NSF Grant #BNS77–13980 A01), I was able to continue observations in the region and obtained the raw data for Table 11.1. The fieldwork from June–September 1974 was supported by NIMH small grants division No. 1RO3MH246688.

[6] Of these, 39 cases were used in the economic analysis. The remaining cases could not be included because of insufficient data in some instances; in other instances, it was not possible to include some cases because of the limitations of the Cobb–Douglas analysis. It cannot deal with extreme variation, which was exhibited in the data from a few house-holds in Itel.

[7] I also owe special thanks to Yung Chung Lee and Glen Johnson of the Agricultural Economics Department at Michigan State University, East Lansing, for their assistance in doing this analysis. It could not have been done without their help, nor without Jim Brandy's computer programming.

relations with bureaucratic networks, but also to the discrepancy between the traditional system of allocating shares, and the actual contributions of each input category to gross income based upon an economic analysis (see Finkler 1978).

The economic analysis was computed in three successive steps:[8] (a) Cobb–Douglas functions were computed that generated coefficients of each input category;[9] (b) from these coefficients, the marginal value product (MVP) of each input was calculated. The contribution of each productive input was then converted to percentages following Euler's theorem (Allen 1964; Heady and Dillon 1961); and (c) the final step allowed the production function figures to be compared with the traditional system of division of shares. Percentage contributions of the various input categories were thus derived for corn and alfalfa, and for the aggregate output by each household per annum. However, only the result for the aggregate output by each household is reported here because, according to economists, the figures derived from aggregate functions are the most representative and relevant to our discussion.[10]

The dependent variable or product output for the analysis of aggregate income was gross income in pesos from all crops cultivated by the household during the 1973–1974 agricultural cycle. Here, the regression analysis includes corn, wheat, barley, and alfalfa. The independent variables, or productive inputs, for each of the regressions were land in tillable hectares, labor in man-days, seed (cost in pesos), water (cost in pesos per tillable hectare), and transportation (cost in pesos).

Land and Land Use

There are 313 households in the village, of which 292 own or hold land. Members of 238 households hold title to *ejido* plots of arable crop

[8] For a more detailed explanation of how the analysis was done see Finkler (1978:27,108).

[9] The Cobb–Douglas production function is used widely by economists (Lee 1975). It is a production function model used to measure the value of various categories of inputs. In the application of this type of production to the analysis of farm data, gross income is the dependent variable. The independent variables are groups of inputs that generate gross income or yield. The marginal value product of an input is the addition to total value product attributable to the increase in output from the use of an additional unit of the input, other inputs remaining unchanged.

[10] Glen Johnson and Yung Chang Lee, Department of Agricultural Economics, East Lansing—personal communication. Heady and Dillon also state: "The fitting of a separate function for each product assumes that the productivity of a resource relative to a specific type of output is uninfluenced by the level of resource use associated with the other products produced by the firm [1961:277]." But where there exists a gross interdependence between various products, as is certainly the case in Itel, the single aggregative functions with all outputs pooled are the most reliable.

land whose average size is, as already noted, 1.11 ha. Another 168 households own private land, and 168 households hold both *ejido* and private land. Private holdings of irrigated crop land vary in size with the majority of households (125) holding 1 ha or less, one man owning more than 17 ha, and the remaining households owning between 1 and 10 ha. All private holdings are subdivided into scattered lots; even holdings of 1 ha are not normally represented by one plot.

As shown in Table 11.2, of the 295 households I visited in the village, 133 let land to share-tenants, and 129 had members working another's land as share-tenants. Twenty-nine of these simultaneously let land and worked as share-tenants for others. The discrepancy between the total number of households that work land on the share, and the total number of households that give their land to sharecrop is explained by the fact that several Itelanos sharecropped land in neighboring villages and a few individuals from elsewhere work some Itelanos' holdings.

The irrigation system of the Mezquital Valley makes possible the cultivation of a variety of crops by the villagers; among these are maize, beans, barley, oats, and alfalfa. Maize is the basic subsistence crop, and at least some is cultivated by every household. The major cash crops are alfalfa, wheat, and, more recently, oats. Villagers used to plant wheat in the winter (Finkler 1973, 1974); however, during the past three agricultural cycles they have ceased to cultivate it because of its unprofitability. Wheat and oats are winter crops; alfalfa is a perennial fodder crop. The Mezquital Valley's lands are highly productive relative to Mexico as a whole. In concrete terms, during the 1970–1971 agricultural cycle, the average yields in Itel for corn per hectare was 4829 as compared with 1200 kg/ha for Mexico as a whole (FAO 1970). (On a regional basis, the average yields in the Mezquital Valley are 3918 kg from the years 1971–1978). The reason for these high yields in this region is that the Mezquital Valley uniquely benefits from untreated sewage waters diverted from Mexico City for irrigation.

TABLE 11.2
Share-cropping Arrangements in Itel

	Number of households	%
Landholders letting land to share-tenants	133	45.1
Share-tenants	129	43.7
Let land to share-tenants; also sharecrop themselves	29	9.8
Not involved in any sharecropping arrangement	4	1.4
	295	100

Households with one landholding usually follow a strategy of planting maize, followed by wheat or oats. Households with more than one plot put one field to maize followed by a winter crop, and the second field to alfalfa. After about 3 years, the alfalfa is turned up and the cropping pattern is reversed. Generally speaking, the overriding consideration in crop choice is the landholder's understanding that crops must be rotated for the land to produce. Consequently, all villagers adhere to the crop-rotation cycle, which is also followed by the parties in a sharecropping agreement.

The Traditional Sharecropping System

When speaking of crop-sharing arrangements, Itelanos refer to five factors or inputs, which they call *puños*, 'fistfuls or handfuls': land, seed, water, labor, and traction (which refers to the use of tractors or animal teams for plowing and cultivation). They calculate the provision of these factors on the following basis. The provider of water is entitled to 25% of the crop. The remaining 75% is then divided among the other four factors, breaking down to 18.75% for each. The parties to an agreement can also agree to provide complementary fractions of any input except, of course, land. The actual agreement of division of these inputs can be any of a number of possible permutations. The provision of water and seed involves only the purchase of these components and entails no labor. The supplier of the labor input is responsible for the planting of the seed, the care and maintenance of the field, the operation of the irrigation system (the actual purchase of the water vouchers from the irrigation district administrators), and the harvesting of the crop except in the case of corn, for which the landholder is responsible. The landholder is, of course, responsible for the assessments on the land and the semiannual cleaning of the irrigation ditches, but these are not considered elements of the sharecropping arrangements. Table 11.3 displays the variety of sharecropping agreements entered into by share-tenants and landholders in 1971 and their frequencies. As we can see, a majority of the agreements entail the supply of the labor factor only, allowing sharecropping families to supplement their income by providing just this factor.

One can enter into a sharecropping arrangement by furnishing only the water factor. Prior to 1968, this was a fairly common arrangement between holders of small tracts and the *canalero*, the man in charge of opening and closing the sluice gates, who controls the irrigation supply. Currently, a similar arrangement has been established in which a share-tenant provides only the seed input. For example, the entre-

TABLE 11.3
Sharecropping Arrangements in Itel (1971)

Factors supplied by tenant				Tenant's share of production %	Frequency (N = 133)	Percentage of total number of arrangements
Labor	Traction	Water	Seed			
1				18.75	42	31.6
1	1	.50		50.00	34	25.5
	1			18.75	20	15.0
1	1			37.50	7	5.3
.50				9.37	7	5.3
			1	18.75	4	3.0
1		.50		31.25	4	3.0
.50	1			28.12	3	2.3
	1		1	37.50	2	1.5
	.33	1		50.00	2	1.5
1	1	.33		45.83	1	.7
	1	.50		31.25	1	.7
	1	.25		25.00	1	.7
1	1	1		62.50	1	.7
1	1		.50	46.87	1	.7
	.50			9.37	1	.7
.50	.50			18.75	1	.7
1	1	1	1	81.25	1	.7

preneurs have relationships with suppliers of alfalfa, and previously, wheat seed, who provide them with a credit-purchasing power not available to the landholders with whom they sharecrop. By purchasing the seed on credit, which is provided to the landholder, they obtain a share in the landholder's crop.

For the past 2 decades, this handful of five factors has included a sixth requirement, which has become a sine qua non to many sharecropping arrangements and, as such, should be considered here with the formal components of these agreements. The sixth requirement is that the share-tenant, in addition to supplying the agreed upon inputs, lend the landholder a sum of money in order to sharecrop his land. Several informants expressed a rule of thumb that, in any crop-sharing arrangement providing the share-tenant with 50%, it is almost certain that a loan exists. The loan is not interest bearing and is not repayable until such time as the landholder decides to discontinue the sharecropping agreement.

Interestingly, when this practice began, the amounts paid were between 100 and 200 pesos; in recent years, however, the figure has increased to between 1000 and 1500 pesos. There are two results of the competition for land to sharecrop: First, sharecropping arrangements have lost the stability that once characterized them. Examples abound. One landholder reported that in 1 year, his share-tenant loaned him 1000 pesos, and that the following year another offered to lend 1200 pesos. He explained that, inasmuch as he was in need of money, he transferred his sharecropping arrangement to the second man. After repaying the 1000 pesos loan, he had made a profit of 200 pesos. In fact, with the practice of extending loans, the landholder retains his advantage in a sharecropping agreement which otherwise is disadvantageous to him. Second, share-tenants have been motivated to increase their production. Share-tenants point out that they take especially good care of the crop to gain higher yields so that the landholder will not let his land to another.

At this juncture, it might be noted that the requirement of a loan, and the share-tenant's willingness to extend it, suggest that sharecropping is a lucrative endeavor. Moreover, in addition to the profit derived from agriculture by sharecropping, it must be noted that sharecropping is favored as an investment when an individual has some extra cash available, since other forms of investment opportunities are lacking. Moneylending could be an alternative; however, the risk is so high that moneylending is rarely, if ever, regarded as an alternative to sharecropping, where the risk is familar to every member of the community. Because moneylending is not regarded by the sharecropper as an alternative to sharecropping, the potential interest that a loan would bear is not incorporated in my calculations of the sharecropper's gains.

Factor Shares by Economic Analysis

We have seen thus far that the traditional system of allocating shares gives equal weight to all factors of production except water, which is allotted a 25% share. The economic analysis, however, suggests that the various factors of production contribute unequally to income for 1 year's agricultural cycle.

Table 11.4 shows the percentage contributions of each input category to the aggregate income when all crops cultivated by the sample households are included. Significantly, none of the percentage contributions derived from the program analyses corresponds to the traditional calculations of the shares of production. In fact, in every type of sharecropping arrangement the landholder contributes, in percentage terms, more to the output than he collects by the traditional division of shares. The discrepancies between the traditional system of allocating shares

TABLE 11.4
Contributions Made by Each Input Category to
Aggregate Farm Income (R² .80554)[a]

Input category	Percentage contribution
Land	38.09
Water	36.98
Labor	10.54
Seed cost	6.94
Traction cost	5.13
Transportation	2.33

[a] The multiple R–value indicating the proportion of variation in the output observations explained by the inputs is .80554 and for our 39 cases is significant at the .01 level. The remaining 20% variability must be attributed to differential soil fertility, rainfall, location of land plot with respect to irrigation canals, and other random environmental factors.

and that derived from the economic analysis is clearly seen by looking at the estimated gross income in pesos for 1973–1974 (see Table 11.5).

The estimated gross income per household in 1973–1974 was 8531 pesos. Table 11.5 displays the peso amount each input category yields when the two modes of allocating shares are compared. For example, with the traditional mode of allocating an 18.75% share to the land input, the contributing share equals 1562 pesos. Using the percentage contributions derived from the economic analysis, the land input's share is 38.09% or 3249 pesos. This suggests that the difference between

TABLE 11.5
Share Contributions by Traditional Calculations and by Programed Analysis (in pesos)[a]

Input categories	Returns based on traditional calculations	Returns based on programed analysis
Land	1562	3249
Water	2083	3155
Labor	1562	909
Seed	1562	582
Traction	1562	436
Subtotal	8331	8331
Transportation (shared equally)	200	200
Total	8531	8531

[a] The average gross income per household for the 1973–1974 agriculture cycle = 8531 pesos. In 1973–1974, one peso = 8¢ United States currency.

TABLE 11.6
Shares Provided and Received for Six Types of Sharecropping Arrangements (in Pesos)

Input category	Receives (traditional calculations) (1)	Provides (programed analysis) (2)	Difference between 1 and 2 (3)
	Landholder		
Land + water + traction + seed	6769	7422	653
Land + seed + ½ water	4165	5408	1243
Land + water + seed + labor	6769	7895	1126
Land + water + seed	5207	6986	1779
Land + traction + water + seed + ½ labor	7550	7876	326
Land + water + traction + labor	6769	7749	980
	Sharecropper		
Labor	1562	909	653
Labor + traction + ½ water	4165	2922	1243
Traction	1562	436	1126
Labor + traction	3124	1345	1779
½ labor	781	454	327
Seed	1562	582	980

the two sums is clearly disadvantageous to the landholder. Table 11.6, in fact, compares the contributions and returns to both landholder and share-tenant for six types of sharecropping arrangements when the division of shares is calculated in pesos.

Inspection of Tables 11.4, 11.5 and 11.6 reveals that the share-tenant stands to gain from any one type of sharecropping agreement practiced in the village. For example, assuming that the share-tenant retains a parcel of land for more than a year, which he usually does, he gains 15.85% above that which he provided in inputs at the 50% level of sharing.

Discussion of the Sharecropping System

Viewed from the share-tenant's perspective, the decision to enter into a sharecropping arrangement is economically sound and rational. In fact, as can be seen from Table 11.3, the majority of sharecropping agreements involve an 18.75% share, with the share-tenant providing his labor. This type of arrangement allows the peasant to supplement his income by seeking out one or two plots of land to sharecrop. Indeed,

84% of Itel households work one or two parcels of land and 3% work from 5 to 10 plots. The remaining 13% work 2–5 plots of land.

From the perspective of the landholder, however, the decision to seek out a share-tenant does not seem economically sound for any one type of agreement. As can be seen in Table 11.6, the landholder's contribution is greater than the amount of his return under the traditional system of dividing shares.

Nevertheless, the practice of letting land to share-tenants persists for several reasons. First, the difficulties Itelanos face in obtaining credit from national financial institutions can make sharecropping arrangements a very practical means of financing the seed and monthly irrigation costs of their crops. Second, a landholder in need of a large sum of cash will seek a share-tenant on a 50% share. The many arrangements in which share-tenants provide either the water or seed factor, and the recent practice of share-tenants providing substantial loans in order to obtain a 50% share of the crop, are both evidence of the financial need involved in many arrangements. Moreover, the decision to enter into a sharecropping agreement is based not only on financial considerations. An assurance of easy access to water is yet another reason for landholders, especially those with only one holding, to seek share-tenants.

Water is distributed to the peasants in a bureaucratic chain, at the end of which stands the *canalero*. When a peasant wishes to irrigate his land, he must purchase a water voucher at a payment station located some distance from the village. The voucher must then be presented to the *canalero*, who will place the user's name on a waiting list, and the user normally receives his water within 72 hr. This procedure was instituted in 1968 and greatly diminished irregularities, including the need for bribing the *canalero*, to obtain water.

Nevertheless, the power of the *canalero* continues to foster sharecropping arrangements. A *canalero* must be available if the user is to present his paid voucher and have his name placed on the waiting list, but in fact peasants complain that they must waste considerable time when trying to find the *canalero*. A number of landholders with only one *ejido* parcel told me that they became tired of searching for a *canalero* every time their fields required irrigation, and that they therefore sought a share-tenant who did not have similar difficulties.

Consequently, *ejido* holders may turn their land over to a share-tenant who enjoys favorable relationships with the *canalero* and to whom the *canalero* is easily accessible. Those individuals who can sway the *canalero* to give them water enter into sharecropping arrangements with peasants holding one or more hectares but lacking the means to sustain good relations with a *canalero*. Hence, a peasant's decision to enter into sharecropping arrangements often rests on this lack of influence with the *canalero*.

Finally, access to labor is another reason for landholders to seek share-tenants. Access to labor figures significantly in various types of share-tenancy systems. Absentee landlords prefer share-tenants to wage laborers to work their lands because share-tenancy arrangements are more economical than paying wages, as is the case in Ecuador, for example (Radclift 1977). In Itel, landholders are frequently short of cash at planting time, and a sharecropping agreement obviates wage payments. Moreover, contrary to the commonly held notion that peasant communities are overburdened with surplus labor, in Itel, many households experience a shortage of labor produced by the double-cropping of maize and alfalfa cultivation. As was noted earlier, households with more than one plot alternate their fields between maize and alfalfa. The labor demands created by the double-cropping cycle and by existing techniques of cultivation cannot usually be met by individual households with more than one holding. In fact, the interaction of crops, seasons, and techniques of production creates a seasonal labor shortage in Itel because holders of alfalfa land require a steady supply of labor, especially to maintain their schedules of monthly irrigation. Not only are day laborers frequently unreliable, but seasonal peak labor demands can even render them unavailable (see Boserup 1965; Finkler 1973, 1974).[11] Therefore, landholders with more than one holding seek share-tenants to assure themselves of a labor supplier who is both reliable and constant.

Effects of the Sharecropping System

The various circumstances that lead landholders to seek share-tenants have opened the way for several share-tenants to obtain high financial status in the community by virtue of their ability to organize and capitalize on multiple sharecropping arrangements. For these men, sharecropping offers the opportunity to prosper by functioning as agricultural entrepreneurs. The success of the men who sharecropped five or more parcels indicates that sharecropping is a definite vehicle for economic mobility. This is so because, for the share-tenant, a 50% share

[11] Moreover, Boserup demonstrates the relation between high labor demands and irrigation agriculture. In my previous study of Itel, I provide detailed descriptions of labor demands for each crop.

of a 1-ha field renders income almost 16% above that which he provided or, in 1973–1974, 1243 pesos. (See Table 11.6 column B3.) A man working five pieces of land simultaneously in 1973–1974, taking 50% of the harvest from each, accrues a substantial gain. Pablo's experience illustrates the potential profit in these multiple sharecropping arrangements. He worked five pieces of land simultaneously, taking 50% of the harvest from each. A calculation based on the figures displayed in Table 11.6 establishes his minimum net gain from these operations at more than 6215 pesos.

The agricultural entrepreneurs are able men. As noted earlier, they possess managerial skills for manipulating labor and cash resources. Moreover, unlike their clients who raise crops for consumption and then exchange their surpluses for cash to purchase products, they do not produce themselves—the entrepreneurs reinvest most, if not all, of their surplus earnings into their agricultural enterprises rather than in items of consumption. For example, one informant, Jesus, the most successful of the share-tenant entrepreneurs, was able to accumulate sufficient cash to purchase a tractor and a truck which, in turn, increased his ability to sharecrop multiple plots of land and to become an alfalfa wholesaler and merchant. In fact, through the years, these men have expanded their operations to include the purchase and sale of alfalfa, a highly lucrative business in the region.

Clearly, an important factor in explaining the rise of these entrepreneurs is their position in the village social structure. Two of these men are closely related to the wealthiest landowner in the village, a man who held important civil posts, including *comisariado ejidal* for 20 years, following the redistribution of Itel's lands. Another entrepreneur, who now works together with his three brothers, studied accounting in Mexico City and also worked as a bookkeeper there before returning to Itel to work in agriculture. (It is not uncommon for men to migrate to Mexico City or to other points of the Republic for a year or more and then return to Itel to take up farming.) The first two men have established a network of relationships with the water-distributing authorities as well as with other governmental agencies. For example, Jesus delights in recounting how he wines and dines *canaleros* and other bureaucrats. These activities without doubt are costly, if also enjoyable to him. Such expenditures, however, contribute to the success of his sharecropping operations.

More important, all of these entrepreneurs have an intimate knowledge of the government bureaucracy, knowledge which serves them well in their multiple enterprises. Their easy access to the water-distributing agents helps them obtain water without undue delay.

Moreover, knowledge of *ejido* matters enables them to assist the land-holders who, for one reason or another, may have difficulties retaining their *ejidos*. In more than one case, a widow was threatened with the loss of title to her *ejido* but, fortunately, her sharecropper had the necessary knowledge and connections to help her retain it. By safe-guarding the right to the *ejido* for the landholder, the share-tenant also retains his access to the land parcel.

Interestingly, however, in Itel one does not encounter consolidation of large tracts of land by share-tenants, such as was noted, for example, by Erasmus (1961) for the Mayos of northern Mexico. First, the share-cropping arrangements in Itel, based upon permutations of the five inputs, mitigate against such consolidation. The five-factor system al-lows for a degree of flexibility and opens various alternatives to land-holders from year to year. For example, it is not uncommon for a landholder to look for a share-tenant only for the year in which he is short of cash. Informants reported that in a year when the household was not burdened by medical expenses, they were able to work their land themselves; whereas in other years, when cash was necessary for medical expenditures, they would resort to sharecropping arrange-ments.[12]

Second, the sharecropping system optimizes ties within a larger extended family as family relationships are obvious in many sharecrop-ping arrangements. A full 40% of the share-tenants in 1971 were work-ing land which was titled to their parents, their in-laws, or their cousins.

Furthermore, even among nonrelatives, the sharecropper–landlord relationship is not solely an economic arrangement; socioeconomic obligations are implied as well. In fact, a landholder frequently becomes dependent on his share-tenant, and a type of symbiotic relation emerges from the economic arrangements. For example, in an emergency, individuals will borrow money or seek assistance from their share-tenant. Jesus described how he had paid the hospital bill of one landholder and had extended a loan to another. One landholding family related how their sharecropper–entreprenuer helped them meet their emergency medical expenses, and how he took them in his car to the doctor in another town. "He always does us favors which may or may not involve money." In yet another case, the share-tenant financed

---

[12] Two other usufructuary practices by landholders are to rent or pawn their land. Of these, rental was less common. Generally, landholders resort to pawning in an emergency when the household urgently needs a large sum of money. These emergencies are usually related to illness, death, or bailing a household member out of jail. Taking pawns is a common form of investment by individuals, including sharecroppers, who find them-selves with some extra cash.

the funeral expenses of a deceased landholder and continued to work the land for the widow. In fact, a widow's sharecropper is, in some fashion, her protector.

The social relations that emerge from the economic ties between the share-tenant and the landholder are admittedly to the advantage of the former. The few sharecropper–entrepreneurs who have succeeded in obtaining the greater number of plots to work no doubt encourage the relationship of dependency. Significantly, one of these entrepreneurs, after criticizing his fellow villagers—for spending their time in the cantina and letting their land to a sharecropper, ended by saying "I wish I had more of those."

One final point still needs to be made. It is a well-known fact the *ejido* tenure was initially designed to break up monopolistic control of land to create symmetrical social and economic relations among landholders. However, as we have seen, the sharecropping system leads, paradoxically, to economic differentiation, and is economically disadvantageous to the landholder while favoring a small group of men who are able to capitalize on the existing conditions. Ironically, too, while the *ejido's* hierarchical structure facilitates the vertical integration of *ejido* communities with the nation state, which ultimately retains control of the land (Carlos 1974; Finkler 1973, 1974; Stavenhagen 1975), it simultaneously creates opportunities for a few individuals to develop vertical links to the bureaucratic structure. The direct linkages of *ejido* communities with agencies representing Mexico's national sector, including the agrarian departments, the irrigation authorities, and the National Peasant League, have thus created opportunities for a few individuals to develop vertical networks not only in the community's interests, but for their personal benefits as well. In sum, contrary to the aims of agrarian reform, the sharecropping system has contributed to furthering socioeconomic differentiation within the village community.

Summary and Conclusion

In the course of this chapter, I have tried to demonstrate the ways in which individual *ejido* tenure enters into peasant assessments of their situation regarding the use of household labor and available land. The *ejido* holder tends to allocate his labor to agriculture rather than to wage-earning activities in urban centers because he aims at retaining his right to his *ejido* plot, providing the land is productive, as is the case in Itel. Availability of land alone fails to guarantee the peasant equal-life chances in regions such as the one described here. In addition to land, access to water and sufficient labor are critical productive resources, and individuals with access to water distribution authorities and in control of a small labor force gain economic advantages, above subsistence level. Hence, any discussion of peasant production requires that the various factors of production be treated as a unitary constellation. Moreover, examination of household production and of productive factors as a single unit clearly points to the socioeconomic consequences of different household strategies.

A peasant, by intensifying his efforts and allocating his labor time to sharecropping one or two plots, can indeed supplement his earnings; but those individuals with managerial skills, some capital, and an ability to mobilize a labor force of their own can greatly improve their economic position. Moreover, their economic enhancement is facilitated not only by their abilities and their position in the social structure but also by the traditional system of allocating shares in a sharecropping arrangement. It must be noted that the sharecropping practices described are widespread in the region and there is some evidence that the ascendance of sharecroppers occurs as well in other *ejido* regions in Mexico (e.g., Guanajuato).

There are two important conclusions to be drawn from this discussion. First, we have seen that when peasant households are subject to land reform institutions, peasant production strategies fail to conform to the theoretical model developed by Chayanov. Given the Mexican institutionalized system of land tenure, water distribution, and the peasants' limited access to credit institutions, households are forced to diminish rather than expand their operations to retain their small landholdings and to meet their subsistence requirements. Furthermore, contrary to Chayanov's thesis, peasant households are not actually fitted to compete with capitalist ventures such as those of the agricultural entrepreneurs. In fact, as we have seen, the peasant family is in a subordinate position to these men upon whose assistance it must depend.

Second, it might be argued that these agricultural entrepreneurs help to promote economic development. The men characterized as entrepreneurs, unlike the majority of *ejidatarios*, invest their profits in capital equipment rather than in consumption items. As was noted, these men maintain laborers in their employ and own agricultural equipment, such as trucks and tractors. In sum, they operate relatively large enterprises and accumulate capital, which they reinvest rather than dissipate. Additionally, with their greater economic resources they also obtain larger yields than the average *ejidatarios,* and these yields supply larger surpluses for the national market.[13]

[13] They produce the larger yields because they have the machinery to cultivate the soil

Thus, on the one hand, *ejido* tenure promotes political and economic stability by providing a sedentary peasantry with a minimal subsistence. On the other hand, access to water and scarcity of labor and credit in the region facilitates opportunities for individual entrepreneurship, which some view as instrumental in economic development (Belshaw 1967; Hagen 1962). In fact, Mexico is currently experiencing an accelerated pace of economic growth under an industrial capitalist economy, and the national economic organization favors the entrepreneurs who have formed a core of farmers in the region. During the past 3 years, production costs have increased disproportionately to the government supported price for corn and to the rise in value of agricultural products in general. Consequently, given the lack of infrastructural governmental supports for the peasant—other than the *ejido* land itself—fewer small holders can afford to work their land plots without the assistance from agricultural entrepreneurs to share in the cost of seed, traction, and labor.

Furthermore, recent industrial development in the region additionally strengthened the entrepreneurs' role for, during the last few years, the construction of an oil refinery and a hydroelectric plant have created a demand for day laborers within the region. Consequently, these governmental projects have siphoned off men from Itel, who are now commuting daily to their place of employment. These day laborers, hired on short-term contracts, are paid a relatively high weekly wage, averaging in 1973–1974, 500 pesos weekly, with weekends off. This sum exceeds the peasant's earnings from his *ejido* plot which he earns in close proximity to the community, averaging roughly 165 pesos weekly earnings, with no specific days off.

The data on the number of men who are working on these projects are not available, but the impression gained during recent field stays (summer 1975, 1977–1979) is that the number has increased considerably since 1971, at which time only a small number of men worked outside the village (Finkler 1973, 1974). The siphoning off of men from the community has contributed to an even greater labor shortage in the village, resulting in an increasing demand for share-tenants with an available labor force. Interestingly, too, the labor shortage has been promoted in yet another way. During the past several years, daily wage workers have become more scarce due to the diminished supply of juvenile labor. Whereas previously youngsters attended primary school till about the age of 12 and then joined the labor pool, at present many

---

at least twice before planting, and the labor for repeated weedings. They usually use pesticides; this cost is divided equally between the two parties to any sharecropping agreement. Because in 1974 it was not a common practice to use pesticides and because the cost is divided equally, I have not included it in the computed analysis.

youngsters go on to secondary schools within the region, including to the new high school built in Itel in 1975. Within the past several years, there has been a growing emphasis on secondary education, and most households send both sons and daughters to secondary school. Among the various reasons for this new trend in the region is the villagers' realization that there are no more *ejido* lands for distribution as well as a general demand for a better-educated labor force due to national industrial development. The increased emphasis on education has not only expanded the peasant household's daily consumption requirements for supporting youngsters in school but has also concurrently decreased the available labor supply in the household and village.

If present trends continue with a shortage of labor and continually rising costs of production coupled with newly created consumer demands for education and manufactured goods, *ejido* holders eventually may be forced to relinquish their *ejidos*. The result will be a total proletarization of the Mexican peasantry, whereas agricultural production increasingly will be taken over by agricultural entrepreneurs.

In sum, any discussion of peasant economics must take into consideration the institutional constraints generated by the bureaucratic structure produced by agrarian reform. To gain meaningful insights into trends here and elsewhere in peasant societies, it is necessary to focus on individual peasant household strategies and the decision-making process upon which such strategies are based. Furthermore, application of a multidimensional approach to the study of the production process illuminates the ways in which socioeconomic differentiation takes place, leading to significant changes in the future of the village community.

Acknowledgments

I wish to thank Peggy F. Barlett for her helpful comments and for suggesting the chapter title.

References

Allen, R. G. D.
    1964   Mathematical Analysis for Economists. London: Macmillan.
Barchfield, John, J. W.
    1978   The Mexican *Ejido* as Victim of Its Institutional Environment. Department of Economics. University of Southern California: Research paper No. 7711R.

Belshaw, Michael
  1967   A Village Economy: Land and People of Huecorio. New York: Columbia University Press.
Boserup, Esther
  1965   The Conditions of Agricultural Growth. Chicago: Aldine.
Brandenburg, F. R.
  1969   The Making of Modern México. Englewood Cliffs, N.J.: Prentice–Hall.
Carlos, Manuel
  1974   Politics and Development in Rural México. New York: Prager.
Chayanov, A. V.
  1966   The Theory of Peasant Economy. Homewood, IL: Richard D. Irwin.
Chevalier, Francois
  1967   Ejido y estabilidad en México. América Indígena 27:163–198.
Cline, Howard F.
  1963   México. New York: Oxford University Press.
Eckstein, Salomón S.
  1966   El Ejido Colectivo en México. México City: Fondo de Cultura Económica.
Erasmus, Charles J.
  1961   Man Takes Control. Indianapolis: Bobbs-Merrill.
Finkler, Kaja
  1973   A Comparative Study of the Economy of Two Village Communities in México with Special Reference to the Role of Irrigation. Ann Arbor: University Microfilms 73-14-375.
  1974   Estudio Comparativo de la Economía de dos Comunidades de México. México City: Instituto Nacional Indigenista.
  1978   From Sharecroppers to Entrepreneurs: Peasant Household Production Strategies under the Ejido System of México. Economic Development and Cultural Change 27:103–120.
  1970   Food and Agricultural Organization of the United Nations. Production Yearbook. (Vol. 24).
Guillet, David
  1976   Migration, Agrarian Reform, and Structural Change in Rural Peru, Human Organization 35:295–302.
Hagen, Everett E.
  1962   On the Theory of Social Change. Homewood, Ill: Dorsey Press.
Heady, Earl O., and Dillon, J. L.
  1961   Agricultural Production Functions. Ames, Iowa: Iowa State University.
Lee, Yung Chang
  1975   Adjustment in the Utilization of Agricultural Land in South Central Michigan with Special Emphasis on Cash–Grain Farms. Ann Arbor, Mich. University Microfilms.
Redclift, Michael R.
  1977   Reform of Tenancy Systems and Agrarian Development. Institute Agricultural Economics 6:44–56.
Paige, Jeffrey
  1975   Agrarian Revolution. New York: Free Press.
Sahlins, Marshall
  1972   Stone Age Economics. Chicago: Aldine.
Simpson, Eyler N.
  1937   The Ejido: México's Way Out. Chapel Hill: University of North Carolina Press.

Stavenhagen, Rodolfo
   1970   Agrarian Problems and Peasant Movements in Latin America. Garden City, NY:
          Doubleday.
Stavenhagen, Rodolfo
   1975   Collective Agriculture and Capitalism in México: A Way Out or a Dead End.
          Latin American Perspectives 2:146–63.
Tannenbaum, Frank
   1929   The Mexican Agrarian Revolution. Washington, D.C.: Brookings Institution.
Tannenbaum, Frank
   1950   México: The Struggle for Peace and Bread. New York: Alfred A. Knopf.
Whetten, Norman L.
   1948   Rural México. Chicago: University of Chicago Press.
Wolf, Eric R.
   1966   Peasants. Englewood Cliffs, N.J.: Prentice-Hall.

Chapter 12

Stratification and Decision Making in the Use of New Agricultural Technology[1]

BILLIE R. DEWALT
KATHLEEN MUSANTE DEWALT

Introduction

Social scientific contributions to understanding agricultural decision making generally have been characterized by relatively simplistic theories that attempt to use one or two key variables or sets of variables to explain how decisions are made. One variable that has received more attention than any other, particularly in trying to explain decisions to adopt new technological innovations, is economic inequalities. The relationship between stratification and innovation adoption, however, is not a simple one; at least four different theoretical models have appeared in the literature concerned with the relationship of these two variables.

The purpose of this chapter is to present these various models and then to test them using data on the adoption of two innovations in a Mexican *ejido* community. We find that none of the models can account for the relationships found in these two empirical examples. Further information on the two new innovations and economic realities in the region is provided to explain why different models apply and to further

[1] This chapter is a revised and expanded version of a paper originally published by Billie R. DeWalt in the *American Ethnologist* (2:149–68) under the title "Inequalities in Wealth, Adoption of Technology, and Production in a Mexican Ejido." Appreciation is expressed for permission to reprint some of the data and material from that article.

AGRICULTURAL DECISION MAKING:
Anthropological Contributions to
Rural Development

elucidate reasons why various decisions are made. We suggest that, if scholars are to begin to understand peasant decision making, knowledge of much more of the total context in which decisions are made is necessary. That is, our models and our analyses must become more complex to allow us more adequately to appreciate the processes involved in the way that peasants (and other people) make decisions.

Model 1—The Homogeneity Model

The simplest model of the relationship between economic position and adoption of new technology is advocated by those individuals who see little or no variation in either of these variables. Particularly in anthropological research, there has been a tendency for researchers to ignore the variability in all sorts of behavior and to present homogeneous descriptions that are supposedly characteristic of most of or all of the people (Pelto and Pelto 1975). Much has been made in these studies of the conservative, nonchanging character of peasants (e.g., Banfield 1958; Foster 1967a, 1972; Rogers 1969). Peasants are thought to be poor candidates for adoption of new ideas and techniques that will make them more modern.

Socioeconomic homogeneity also has been posited as existing in peasant communities. Often, anthropologists and others quote peasants who say "we all are equal here" as justification for presenting descriptions of them as a relatively undifferentiated, nonstratified mass (e.g., Iwanska 1971:58). Some anthropologists have spent time describing "leveling mechanisms" (Nash 1962, 1971; Wolf 1955) that supposedly help to keep peasants fairly equal in wealth or in a situation of shared poverty (see Huizer 1970:311).

On the basis of these and similar social scientific studies, it is possible to illustrate a model of the relationship between economic position and adoption of new cultural items as shown in Figure 12.1. The single point indicates that peasants are homogeneously poor and, further, that they are unwilling to change. Expressed in this way, we are sure that no researcher would subscribe seriously to this position. On the basis of those peasant studies that have ignored intracultural and intrasocietal differences, however, this unrealistic theoretical stance appears to summarize their position.

This kind of research has contributed unfortunately to the maintenance of stereotypes about what peasants are "really like." Perhaps the most significant negative consequences of these stereotypes have arisen in the context of planned change programs where expectations about

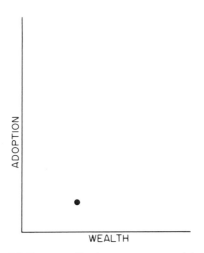

FIGURE 12.1. The homogeneity model.

peasants are created among change agents. On the one hand, this position has led to a feeling that the most important changes needed are in the values, attitudes, and motivations of the peasant. Several major theories of sociocultural change have reinforced such views (cf. Kahl 1968; Lerner 1958; McClelland 1961; Rogers 1969). Although there may be some credibility to these theories, they conveniently shift the burden of failure to modernize to the peasants, leaving the larger societal structure relatively free of criticism. A second unfortunate consequence of these stereotypes is that they provide a built-in excuse for economic development agents when their programs fail. Thus, reasons cited for the failures of development schemes are not the inability of the pottery cooperative to make money, or the failure of a new type of wheat seed to grow, or the fact that the chickens died, but rather the supposed conservatism of the peasants who refused to cooperate with the benevolent agents of socioeconomic modernization.

These statements about the expectations of change agents are not merely conjectural. Instead, they are based upon empirical evidence gathered in interviews with a number of development personnel in the region of Mexico in which our fieldwork was done. The assertion was commonly made that the people in the area are very resistant to change and that they do not understand the goals of the development agents. The "scientific validation" provided in the literature for the impressions and/or prejudices of these change agents is unfortunate. Perhaps as a result of this well-meaning social science research, an experimental school for adults was set up in the region. In addition to providing some training in modern agricultural and livestock-raising techniques, a large part of the month-long courses were designed explicitly to change peasants' cognitive orientations by presenting information that would, it was hoped, make them more cosmopolitan, more empathetic, more achievement oriented, and so on. Few local people were interested in this school, and this lack of interest was cited by change agents as another example of how peasants in the region were resistant to change and content with their poverty. Figure 12.1 also provides a fairly accurate description of local change agents' views of peasants (also see B. DeWalt 1979:1–3).

Model 2—The Linear Model

Another relatively simple model of this relationship is that for which Rogers (1971), Pelto (1973), and others find considerable support. It asserts that within any community the wealthier individuals (or families) are likely to be the first, or principal, adopters of new technology. Figure 12.2 summarizes this view. In effect, this model has the greatest support in the literature as many studies of technological adoptions are statistically oriented. The correlation coefficients presented in these studies express only the linear relationship between two variables. This view also has the most common sense support in the popular saying, "The rich get richer and the poor get poorer."

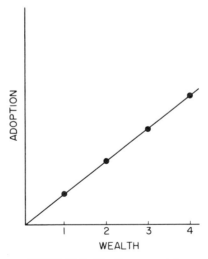

FIGURE 12.2. The linear model.

Model 3—The Middle-Class Conservatism Model

Homans (1961), on the basis of evidence from social psychological experiments, hypothesized that people of very low status and people of very high status were likely to be innovators or adopters of innovations. Although he did not directly apply his theory to peasants, he felt that low-status individuals were generally of low reputation and, therefore, easily could adopt behavior not in conformity with their groups. Upper-status people have a secure position and have nothing to gain by conformity and also will be able to adopt behavior not "normal" to the group. Only the middle class maintains close conformity to traditional or group behavior. Essentially, this is a position that alleges the existence of "middle-class conservatism." Homans' position can be diagrammed as the U-shaped curve found in Figure 12.3.

Model 4—The Modified Middle-Class Conservatism Model

Another view of the process of the adoption of new technology has been advanced by Cancian (1967). The Cancian model is very similar to the linear model just discussed, with the addition of some of Homans' ideas about middle-class conservatism. Cancian tested his model using studies from several different countries (and including peasants and nonpeasants) and found considerable support for it (1972:153–156;

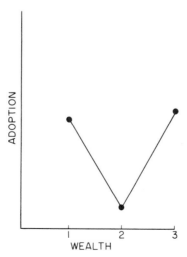

FIGURE 12.3. The middle-class conservatism model.

1977). The most important feature of his model is his feeling that the "upper-middle rank" (his terminology) is more unwilling to take risks than the "lower-middle rank." Cancian interpreted this as signifying a willingness to risk among the lower middle rank because of a greater desire for upward mobility in this economic group. The upper-middle rank, on the other hand, is unwilling to risk their relatively favorable socioeconomic position by investing in new opportunities that may fail. Cancian's model follows the linear model in regarding the lowest economic group as unable to invest in new opportunities, and therefore lowest in terms of adoption of new techniques. The wealthiest individuals are, as in the linear and middle-class conservatism models, regarded as the group most likely to adopt new cultural items because of their secure economic position. The "modified middle-class conservatism" model is that found in Figure 12.4.

Thus, we see that there have been four different hypotheses advanced about the relationship between stratification and adoption of innovations. These hypotheses will be tested after we present an introduction to our research community.

Historical and Ethnographic Background

The research site is an *ejido* community[2] (a minimum of 20 farmers organized into a group to receive and work lands expropriated from the great estates after the Mexican Revolution) in an area about 100 miles northwest of Mexico City. The *ejido* actually contains people from three separate, small, contiguous communities. Not every household has *ejidal* land; in fact about half the families are landless. When the *ejido* was first established (1933), nearly every family received land rights, but rapid population increase has resulted in many landless individuals.

The three communities are located in the *municipio* (roughly equivalent to an American county) of Temascalcingo and have had considerable contact with the town center of the same name for centuries. Only in this century has rapid cultural change begun, however. Although 30 years ago the communities still retained the language and other features of Mazahua Indian culture, today everyone speaks Spanish, wears clothing indistinguishable from the mestizos in the region, and Mazahua is spoken only by older informants.

[2] We prefer not to name the *ejido* in which this study was carried out. Some of the data collected could be of some damage to certain individuals. As a result, we believe that it is important to maintain the anonymity of our informants.

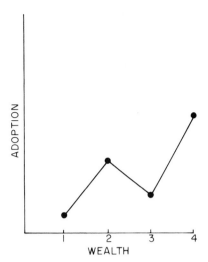

FIGURE 12.4. The modified middle-class conservatism model.

Of people in the region, the members of this *ejido* were in the fore-front of the fight for land after the revolution. As a result, they obtained landholdings slightly larger than other *ejidos* in the area. Also, the lands they obtained were among the richest since they are located on the valley floor and can be irrigated with the waters of the Lerma River that bisects the region. The ecological position of the *ejido* has not been completely positive, however. Soon after the hacienda lands were redistributed there were a series of almost yearly inundations. The hacienda managers had been able to control flooding by putting large numbers of peons to work reinforcing the banks of the river and cleaning drainage ditches. The *ejidatarios* (members of an *ejido*),[3] freed from the oppression of the hacienda, were unable to organize effectively to perform the same operations, so the dikes and other water control measures were allowed to fall into disrepair.

Another "ecological disadvantage" of this community derives, paradoxically, from its favored geographic location. Agricultural agents and other developers frequently choose the *ejido* for experimental programs. Some of these have been notably unsuccessful, in some cases because of fraudulent practices of bank and *ejidal* officials.

[3] Some of the individuals who control land in the *ejido* are women. In most cases these are widows who have inherited lands from their husbands. Although examples used later in the chapter are all males, we should make it clear that sampling and data collection included these women.

The development efforts have not been entirely detrimental, however. In the early 1960s, the *Secretaria de Recursos Hidráulicos* (a cabinet-
level department of the Mexican government that manages the country's water resources) decided to invest large amounts of money in the
area to alleviate the periodic flooding. The agency diverted a part of the
river into a new channel, improved the network of drainage and irrigation canals, and constructed two large dams to control the flow of water.
As a result of these works, the Lerma has not flooded in this area since
1969.

Thus, the *ejido* and the whole valley region now have an excellent
technological infrastructure on which to develop new economic ventures. The irrigation and flood control works give it a major advantage
over most rural areas in Mexico. On the other hand, the monocultiva-
tion of corn is rapidly depleting the soils. Although the valley once had
been a major wheat-producing area, declines in prices and recurrent
plant diseases have discouraged the planting of this crop. Wheat has
not been cultivated widely since approximately 1963, and there is no
longer any rotation of wheat and corn as had been the practice in earlier
years. New soils deposited by floods forestalled serious ecological
damage for some time, but the years of corn cultivation have diminished seriously the organic material present in the fields. Thus,
changes in the traditional techniques of cultivation as well as changes in
the traditional cultigens have become imperative.

Wealth Homogeneity or Heterogeneity?

We chose an *ejido* as the unit of study because there are many reasons
to expect that the individuals in an *ejido,* as a group, are more
homogeneous in wealth and other characteristics than the general population of the communities or the region. Some of the other members of
these communites are poorer than the *ejidatarios* because they have no
land. On the other hand, there are a few people in the area who are
wealthier than any member of the *ejido* because they engage in fairly
profitable nonagricultural economic activities (e.g., commercial ventures, marketing stone from a nearby quarry, or government employment).

There are a number of factors that should have maintained relative
socioeconomic equivalence among the members of the *ejido*. First, Mexican land reform laws require that members of the *ejidos* should have
approximately equal amounts of land.[4] Founders of the *ejido* also de-

---

[4] In fact, there is some variation in the size of landholdings. There are some families in

cided that quality of the land held by each individual should be more or less equal. (One of the more damaging side effects of this ideology was the fractionalization of lands. That is, the 2.38 ha—about 5.88 acres—owned by each *ejidatario* is composed of at least four widely separated small plots.) Another "leveling mechanism" has been the periodic flooding, which regularly wiped out major portions of the crops. The development programs have also tended to equalize people's economic positions because they have been largely unsuccessful. Finally, the region periodically has been subjected to a number of livestock illnesses. Those who had managed to accumulate some wealth in animals often lost part of their herds in these epidemics.[5]

In actuality, all of these would-be leveling mechanisms have not created a situation of "shared poverty." In spite of them, economic differences among the *ejidatarios* are easily found. They are reflected in household possessions and construction, number of animals owned, other business investments and endeavors, types and amounts of food eaten, and other features (see B. DeWalt 1979:106–121).

To examine socioeconomic differences, we collected information on household possessions and other aspects of wealth, but for purposes of this chapter, we will make use of the "levels of wealth" ratings provided by two key informants. Both informants were *ejidatarios* themselves and knew each of the other 123 farmers in the *ejido*. One of the informants was a wealthy individual and had recently been president of the ruling body of the *ejido*. The other was younger, of moderate economic means, and had recently been a *delegado* (representative of the community in the municipal government).

The technique of informant ratings used was similar to that of Silverman (1966). The key informants were given cards, each containing the name of one person, and were asked to sort these into groups on the basis of socioeconomic position. Instructions were purposely vague to allow the men to develop their own criteria for rating. They were also free to use as many categories as they wished. One of the men easily sorted people into five categories, whereas the other used only four. The correlation coefficient (Pearson's r) between the two ratings was .75, indicating substantial agreement between the two men. No underlying

---

which sons, for example, have *ejidal* land but prefer to work in Mexico City. The father then works his son's land as well as his own. There are also two brothers, both of whom have served as president of the *ejido*, who have managed to accumulate considerable land through some shady dealings. Neither brother is really interested in the land, however, and many of their fields lie fallow.

[5] The cargo system, which has been widely discussed as a "leveling mechanism" in Mesoamerican communities, is present in this community. It is of much less importance than the other factors mentioned in terms of economic leveling.

term such as *rispetto* (found by Silverman among Italians) or *categoria* found by Simon (1972) in Mexican research was encountered here as a basis for the ratings.

The ratings of the two informants were combined to yield a scale ranging from a low of 2 to a high of 9. Four quartiles of economic position, each of which contained approximately one-fourth of the *ejidatarios*, were then formed by combining some of the ranks.[6] Although this procedure obscures some of the variation present, it is a common procedure in such analyses (Cancian 1967) and greatly facilitates presentation and comparison.

The following brief descriptions of three individuals illustrate some of the range of variation in wealth among the members of this *ejido*.

Antioco is one of the poorer individuals in the community. Both key informants put him in the lowest category of economic position. Antioco does not have a team of animals nor any of the agricultural implements needed to work the land. His badly constructed one-room adobe house is unpainted, has a dirt floor, and contains only a radio, an iron, and a few other items of clothing and utensils. His corn production is usually sufficient to feed him, his wife, and three children for the entire year unless an emergency occurs and he needs to sell some grain to obtain cash. The family's diet is very poor. They eat only twice a day and rarely have meat or *frijoles* (beans that supply important parts of protein not found in corn). Still, they are better off than many families because Antioco is fairly young and can work his own lands and occasionally obtain work as a day laborer for about 12 to 15 pesos ($1–1.25) a day. His three other children (whose ages range from 13 to 17), who are now working in Mexico City, also are often able to send money to help the family get along.

Juan is one of the people in the middle range of the ratings (both informants put him in Category 2). In addition to his *ejido* land, Juan also works stone in a nearby quarry. Although he sometimes employs 3 to 6 people, Juan must depend on an intermediary to haul the stone to buyers in Mexico City. Even so, he earns about 50 to 70 pesos ($4–6) a day when he works. The family has a bed, a radio, a petroleum stove, and a bicycle. They are now planning to move into a new house where they can obtain running water. Juan owns a team of animals, and he

[6] The individuals with Scale Score 2 formed the lowest quartile, whereas those with Scale Score 5 formed the third. Scale Scores 3 and 4 were combined to form the second quartile, whereas individuals with Scale Scores 6–9 were pooled as the highest quartile. We would not claim that these quartiles have any validity as "classes," but since these data do support other theories that have used the same method (as we will see), we have elected to present them in this way.

also has many of the agricultural implements he needs to work the land. While he raises a large amount of corn, he also sells some to pay for the tractor he hires and the fertilizer he uses. The diet in Juan's household includes meat at least once a week, as well as frequent servings of beans. Although the family is fairly well off, there is no doubt that their economic position would be better still were it not for Juan's heavy drinking and playboy activities.

Both informants placed Florentino in the highest category. He is known widely as one of the richest men in the region. Although his family was poor, people report that Florentino has had good luck in life. When he was very young, he managed to scrape together some cash which he used to buy wheat from other *ejidatarios*. He later resold the wheat at a profit. Then a clothing dealer from Mexico City contacted him and asked if he would like to start selling clothes in the markets in the region. He loaned Florentino the money and clothes to begin the business, an activity in which Florentino is still involved. Later, he began breeding animals with considerable success and presently has 40 cattle, 27 goats, a few pigs, sheep, and burros, as well as the team of mules he uses to work the land. About 8 years ago, he made a capital investment of 8000 pesos ($640) to install a mill for grinding corn, and he recently bought a record player and speaker system. He now charges a small fee from people who wish to dedicate a record (which can be heard throughout the community) to wives, brothers, sisters, or other loved ones. The installation of the broadcasting system was made possible because of the fact that he headed the committee that recently brought electricity to the community. Florentino and his family live in a large house, painted inside and out, with concrete floors, and a kitchen that is larger than most houses in the community. They also have a television, radio, sewing machine, bicycle, and four beds. A doctor attended the delivery of nine of his ten children.

Of these three families, Antioco's household, the poorest, is living below or barely at the subsistence level. Juan's family has the economic advantage of the husband's stone-working skill, which allows him to earn approximately three times the daily minimum wage (about 20 pesos) established by the government for the area. Florentino's family has a potential capital of well over 100,000 pesos and, in general, lives at a level comparable to that of some of the merchants of Temascalcingo.[7]

---

[7] There has been some intradisciplinary diversity in the way in which anthropologists have used the term "peasants" (cf. Firth 1956:87; Kroeber 1948:248; Foster 1967b:6–11; Redfield 1960; Wolf 1966:12–17. Whereas there has always been some question as to whether certain communities or types of peoples (e.g., potters, fishermen) fit into the category, we are also sure that within some "peasant communities" there are individuals

Wealth and Adoption of New Technology—Chemical Fertilizer

Chemical fertilizers had not been used widely in the *ejido* nor in the general region until 5 years ago. This was due, in part, to a negative experience with a program about 20 years ago that provided credit for fertilizer and a new type of wheat seed. The yield was worse than in most normal years because the seed was not well-adapted to the area, and, as a result, the *ejidatarios* concluded (quite logically) that fertilizer was not worth the investment.

The first wide-scale use of chemical fertilizers began in 1970 when an agronomist came to the region to supply credit for fertilizers along with technical assistance for the improvement of the corn crop. About three-fourths of the farmers participated to some degree in this program. For reasons that are somewhat unclear, a dispute arose between the *ejidatarios* and the agronomist, and the farmers were prevented from doing any work on their own fields. Field hands from other communities were brought in to perform the necessary agricultural chores. Because of this, the eventual costs to the *ejidatarios* were very high. Despite increased yields of corn, many of the farmers had difficulty paying their debts. Despite the overall negative result of this experience, the people felt that the merits of fertilizer had been demonstrated, and many of them have continued to use it.

We obtained data on fertilizer use for 1970 and also for 1972 and 1973. In 1972, there was no credit available, so all costs of fertilizer were borne by the farmers at the time of purchase. In 1973, bank credit was available through the president of the *ejido* who acted as an intermediary. Farmers did not have to pay for their fertilizer until after the harvest. Some *ejidatarios* continued to pay cash for their fertilizer.

---

who do not have many or all of the characteristics defined as peasant. Certainly, the individuals in the *ejido* studied who are engaged in commercial activities (like Florentino who is described in this chapter) do not easily fit the characterization of peasant. This is, in itself, an indication of the intracultural diversity that exists in this community and probably exists in many other "typical peasant communities" that have been described in the literature.

We have decided to retain the use of the term for comparative purposes with other similar communities described as peasant and also because many of the individuals in the *ejido* do fit traditional definitions of peasants: the basic unit of production is the family; the primary goal is the provisioning of the household; production techniques are rudimentary; and they have an asymmetrical structural relationship with a dominant group as part of a larger, compound society (Firth 1956; Kroeber 1948; Redfield 1960; and Wolf 1966). We should stress, however, that there are a number of *ejidatarios* who hire others to labor on their lands, derive only a small part of their household provisioning from their agricultural efforts, sell most of their production, have begun to use tractors on their fields, and have become (or are becoming) closely linked with the political and commercial "elite" of the town center, Temascalcingo (also see DeWalt, Bee, and Pelto, 1973).

TABLE 12.1
Adoption of Fertilizer by Quartiles of Wealth

		Wealth				
		1	2	3	4	Totals
Nonadopters	Never	1	2	0	0	3
		(22%)	(44%)	(36%)	(29%)	(33%)
	Once	4	9	8	5	26
Adopters	Twice	11	6	8	6	31
		(78%)	(56%)	(64%)	(71%)	(67%)
	Three times	7	8	6	6	27

Data concerning use of fertilizers were obtained by means of structured interviews with a sample of 87-household heads.[8] Table 12.1 shows the relationship between use of fertilizer and people's economic status. Because all but three of the farmers in the sample used fertilizer at least once during the 3 years, we have defined those who used fertilizer only once or never as "nonadopters." Those who have used fertilizer in 2 or 3 years are considered "adopters."

Analysis of these data reveals a number of somewhat surprising results that deserve emphasis:

1. Although the individuals of this *ejido* are considered to be very conservative by development agents in the valley, only 3 of the total number of 87 respondents have never used chemical fertilizers.
2. Over 66% of the farmers have used fertilizer on at least parts of their fields in 2 or more of the years studied. This indicates very rapid and widespread adoption of fertilizer use.
3. The *poorest* economic group has the highest percentage (78%) of fertilizer adopters. The wealthiest people in our sample also have a high proportion (70%) of fertilizer users, whereas the two middle economic groups have the lowest percentage of adopters (56% and 64%).

Comparison of Figure 12.5, showing the relationship of wealth and adoption of fertilizer use with the diagrams of the four theoretical

[8] Interviews were conducted with a one-half random sample of households. Additional interviews were obtained from other *ejidatarios* to yield a two-thirds stratified (on the basis of informant ratings of wealth) sample of the *ejido*. The results presented here are based on this larger stratified sample, since we are not making statements about statistical significance or inference.

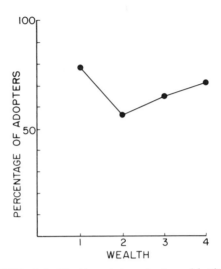

FIGURE 12.5. Wealth and the adoption of fertilizer.

positions presented earlier shows that these data closely conform to Homans' U-shaped curve. Both the upper and lower economic groups have high rates of adoption of chemical fertilizers, whereas the two middle economic groups appear to be more "conservative."

The second type of new technology available in the region arises from a promotion sponsored by Recursos Hidráulicos. Since they began their flood control works, this agency has also been experimenting with several types of clovers and other grasses. On the basis of their experimental plots in the valley, they have found a combination that grows well and provides excellent alimentation for animals.

Although the eventual aim of Recursos Hidráulicos officials is to introduce beef and dairy cattle to exploit these food sources, many *ejidatarios* in the region have seen the immediate advantage of these forage crops as a food source for the animals they currently own. There is a critical shortage of animal food, especially in the dry months of March through May when there is little wild green vegetation. With the irrigation currently available, these forage crops, *pradera*, can flourish during all seasons of the year. Some *ejidatarios* solicited technical assistance from engineers of Recursos Hidráulicos to establish their own fields of *pradera*. Others learned the technical details through the demonstration effects provided by their neighbors.

Table 12.2 presents data showing the relationship between economic position and the adoption of these *pradera*. Because these fields are easily distinguishable, we were able to determine the owner of each

TABLE 12.2
Adoption of **Pradera** in 1973 by Quartiles of Wealth

	Wealth									
	1		2		3		4		Totals	
Nonadopters	38	(100%)	28	(85%)	26	(87%)	12	(55%)	104	(85%)
Adopters	1	(0%)	5	(15%)	4	(13%)	10	(45%)	19	(15%)

piece of land planted with forage crops. Because we also obtained ratings of wealth for every *ejidatario*, data presented in Table 12.2 are for the entire 123 members of the *ejido*.

We see that there is a very different relationship between economic position and the adoption of forage crops. Not one of the poorest individuals has been able to invest in this crop. Even among the two middle quartiles of economic position less than 15% of the people have sown *pradera*. On the other hand, almost 50% of the people in the highest economic group are adopters of this new crop.

Figure 12.6, which presents the data contained in Table 12.2 in diagrammatic form, looks very similar to the Cancian model. The poorest and wealthiest individuals are in their expected positions, and the upper middle group is adopting the planting of *pradera* less frequently than the lower middle group. Although the difference in percentage of adopters is only 2%, we would expect that the upper middle group

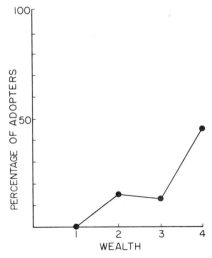

FIGURE 12.6. Wealth and the adoption of *pradera* in 1973.

should have a much higher rate of adoption based on the predictions made from the linear model, the only other model to which it is similar.

It appears that the previously posited theoretical models cannot encompass both of the instances of technological adoption found in our study. The very large range of variation that was found in wealth and in differential adoption of the two new agricultural items made the homogeneity model completely unacceptable. The linear model did not receive support from either Figure 12.5 or 12.6, although Figure 12.6 does resemble the linear model to some extent. However, those in the upper middle group were not adopting the sowing of *pradera* as often as this model would predict. Homans' middle-class conservatism model received support from Figure 12.5. The poorest and wealthiest individuals were adopting the use of fertilizers more frequently than those in the two middle economic groups. Figure 12.6, however, did not support the Homans' model because the poorest individuals have not adopted *pradera* cultivation. Cancian's modified middle-class conservatism model received support from Figure 12.6 in which the upper middle group was found to be adopting *pradera* cultivation less frequently than the lower-middle economic group. Figure 12.5 did not support the Cancian model because the poorest individuals had the highest percentage of adopters and the upper middle group had a higher percentage of adopters than did the lower middle group.

The reader should be cautioned that the relatively small number of individuals in each economic group when the data is divided into quartiles makes conclusive support for any model tenuous. Shifts of only a few individuals from the adopter category to nonadopter or vice versa because of some nonrandom variation (e.g., measurement error) in the data could change the shape of the models. However, since Figure 12.5 is based on a two-thirds stratified sample of the community, and our second model (Figure 12.6) on the total sample of 123 *ejidatarios*, we can be fairly confident that our empirical examples are descriptive of what is happening in the *ejido*.[9] Sampling error, one of the most serious problems in much research, is not a problem here. In addition, the fact that the patterning of our empirical data does coincide with other theoretical models that have received at least some empirical support in the literature indicates that something other than chance or some form of error is responsible for the particular relationships that do appear.

[9] The "universe" to which we are generalizing is the *ejido*. Because we have measurements of our variables from, in the first case, two-thirds of the individuals and, in the second case, all of the *ejidatarios*, we can be confident of generalizing our results to that "universe." We then used the empirical examples from this *ejido* as "cases" to be compared with other cases to try to determine the cross-cultural applicability of our results.

The Cancian modified middle-class conservatism model, in particular, has now been tested in a number of cross-cultural contexts and has received support (Cancian 1972:154–157). The Homans' middle-class conservatism model has received less empirical support in these studies. In any case, we are left with the somewhat puzzling finding that Figures 12.1 and 12.2 are dissimilar. To explain these findings, it will be necessary to introduce information concerning the economics of corn and *pradera* production.

Economics of Corn Production

Most studies of the adoption of new agricultural techniques assume that use of the new technology will result in significant economic benefits for the population. That is, there is usually no attempt to assess systematically future effects of the pattern of adoption of new techniques. Most important, however, is the fact that the net economic gain of the population adopting the new techniques rarely is considered. In the present case, it is imperative to consider these economic costs and profits if we are to make some sense of the patterns of adoption that were found.

As we have stated before, credit programs were available in 2 of the 3 years for which we have data on adoption of fertilizer. The credit in 1970 covered not only the cost of the fertilizer but also included credit for hybrid seeds, the use of tractors, and for pesticides and herbicides. In 1972, no credit was available, but many of the farmers were able to raise the money to buy fertilizer for at least a part of their fields. Credit was again available in 1973, but this time only for fertilizer. The possibility of adoption of fertilizers was thus enhanced by credit programs in 2 of the 3 years. The total cost of suggested applications of fertilizer for each hectare was about 550 pesos ($44). Although this is a considerable sum for a peasant, many of them acknowledge that the investment is now required. They complain that the soils no longer support the yields of corn that they did in the past, unless fertilizer is applied.

Table 12.3 presents data showing the average total corn production of each economic group. The data are for 1972 only, a year in which no credit was available.[10] The data are grouped to show differences in

[10] It is interesting to compare the pattern of adoption of fertilizer for 1972 alone, a year in which no credit was available, with that for all 3 years. The most interesting feature of these data is that when credit is not available, the pattern of adoption shifts toward greater conformity with the Cancian model and with our Figure 12.2. The upper middle group, as we see in Table 12.4, does have the characteristic (of the Cancian model) lower

TABLE 12.3
Average Total Corn Production of Fertilizer Users in 1972 by Quartiles of Wealth[a]

	Wealth								Totals	
	1		2		3		4			
Without fertilizer on any field	42.6[b]	(14)[c]	62	(16)	65.8	(15)	54.3	(7)	56.8	(52)
Using at least some fertilizer	63	(9)	77	(7)	71.8	(5)	126.4	(8)	85.4	(29)

[a] Differences in the number of cases reported in the economic groups here and in Table 12.1 are due to some missing data on production.
[b] Refers to number of *costales*, 'bags'.
[c] Refers to number of respondents.

production between individuals using fertilizer on at least part of their fields as opposed to those who used no fertilizer. Although, as a group, the *ejidatarios* who use fertilizer are obtaining higher average productions, there is a great deal of variation, and some of those not using fertilizer actually obtained higher yields than some of those using fertilizer. There are a number of other very important factors in raising corn. Plowing, harrowing, weeding, and cultivating are important, and there is little homogeneity in the performance of these tasks. The quality of the land is also a major factor.

A number of important points about the results presented in Table 12.3 should be emphasized.

1. The poorest group has a lower average total production than the

rate of adoption than the lower-middle economic group. Comparing Table 12.4 with Table 12.1, we see that the percentage of adopters declined more among those in the upper-middle economic group (from 64 to 27%) than among those in the lower-middle rank (from 56 to 32%) when data for all 3 years are compared with the year in which no credit was available. It appears that there is a greater reluctance among those in the upper middle group to use fertilizer when they have to risk their own money. That is, when credit programs are not available, the decision making of the *ejidatario* becomes more similar to the risk-taking situation described by Cancian. When credit programs are present, risk-taking is diminished because an integral part of the credit is insurance which helps pay the debt of the *ejidatario* in case of crop loss.

Despite this modicum of added support for Cancian's idea about the relative unwillingness to risk among those in the upper-middle rank, Table 12.4, when diagrammed, still looks more similar to the Homans U-shaped model than to the Cancian model. Both the poorest and wealthiest individuals have higher rates of adoption than the two middle groups. We are impressed with the fact that, even when the economic parameters are changed and the use of fertilizer means laying out cash, many of the poorest were still motivated enough to muster the necessary resources to do so. The data which we present on the economics of corn and *pradera* production help to explain this finding.

TABLE 12.4
Adoption of Fertilizer in 1972 by Quartiles of Wealth

	Wealth								Totals	
	1		2		3		4			
Nonadopters	14	(61%)	17	(68%)	16	(73%)	8	(47%)	55	(63%)
Adopters	9	(39%)	8	(32%)	6	(27%)	9	(53%)	32	(37%)

other three groups. This is true among those who use fertilizer as well as among those who do not.

2. On the average, those using fertilizer on their land have higher total productions than those in the same economic group not using fertilizer.
3. Those in the top economic group who use fertilizer have an average total production almost 100% higher than those in the other economic groups who also use fertilizer. This emphasizes the importance of other agricultural techniques.

As another way of assessing the benefits gained from using fertilizer, we collected information about production of fields in which fertilizer was used versus those in which it was not. The mean production per hectare of the 132.8 ha which did not receive the benefits of fertilizer was 27.5 *costales*, 'bags,' or about 1375 kilograms. Of the 35.1 ha of land on which fertilizer was used, there was an average production of 46.2 *costales* or 2310 kg. While the almost doubled production of fields with fertilizer seems impressive, we need to consider what this means in terms of cash value. Basing computations on the cash value prevailing in the region in early 1973 (90 Mexican cents per kilo), each hectare of corn with fertilizer was worth an average of 854.50 pesos (about $67) more than the average hectare of corn which was not fertilized. Subtracting the costs of the fertilizer (550 pesos) leaves the *ejidatario* with an average net gain of slightly over 300 pesos—the price of 12 kilos of meat, a cheap ratio, or slightly more than half the needed cash to buy fertilizer for 1 ha of corn the following year.

Thus, on the basis of the average increases in corn yield per hectare, it seems clear that no one is significantly improving their economic position from use of fertilizer in the production of corn. Only those who have more than one allotment of *ejidal* land, or who are obtaining considerably higher than average yields, would be able to earn enough money from corn production to result in any measurable increase in

their economic well-being. Some economic benefits are probably accru-
ing to the wealthiest who are producing much higher than average
yields of corn (cf., Table 12.3). In general, among the *ejidatarios* themselves, corn is not seen as a viable cash crop. Repeatedly, we were told
by informants that *maiz no es negocio*, 'corn is not business'.

Corn may not be "business" for these Mexican peasants, but it is the
staple of their diet. Thus, in order to understand the pattern of adoption
of fertilizer, it may be necessary to look at its subsistence value. Sahlins
(1972:84) has made a useful distinction between what he calls the "use
value" and the "exchange value" of a commodity. Figures 12.1–12.6
consider corn in terms of its exchange value—that is, the price that
farmers would receive if they sold their corn to local merchants or to
CONASUPO, the Mexican government agency that buys corn. Basing
his argument partially on Sahlins' distinction, Chibnik (1978) has argued recently that it may be more appropriate in peasant societies to
value subsistence production at or near its retail value. In other words,
we should value corn used for subsistence purposes at the *retail price*
that farmers would have to pay if they were not able to produce enough
corn for their own needs.

This distinction may tell us something more about the patterning of
fertilizer adoption data. Table 12.3, for example, shows that there is a
fair amount of consistency in corn production for people in economic
Group 1 who use fertilizer and for people in Groups 2 and 3 who do or
do not use fertilizer. These five groups all produced an average of
between 62 and 77 *costales* of corn in 1972. It may be that about 65
costales of corn is the "minimum subsistence standard of living [Whar-
ton 1971:161]"; that is about what an average family needs for its own
consumption in any given year. If most people in the middle economic
groups are able to produce this amount of corn without using fertilizer,
this may account for the relatively low proportion of fertilizer adopters
in these groups. Note that the additional average amount of corn produced by people in these middle groups who use fertilizer is not
substantially more than that produced by nonadopters. Thus, we might
speculate that especially among people in the middle economic groups,
the decision to adopt or not adopt fertilizer hinges on whether their
land is already producing enough to meet their subsistence needs. Once
they produce this amount, there will be little incentive for them to
increase their corn production, and they will invest their resources
elsewhere.

For poor individuals, the constraints are somewhat different. Many
individuals are in Group 1 partially because they have less land and
land of poorer quality than other people. Those who can afford to invest

in fertilizer do so, and their average gains put them into the corn production range of people in the middle two economic groups. This is probably why the lowest economic group has a higher proportion of adopters than any of the other groups (see Table 12.1, p. 301). Poor individuals who do not use, or who probably cannot afford, fertilizer produce substantially less corn, creating problems for them to meet their consumption needs. Having to buy corn is especially difficult for those who have no source of steady income.

The wealthiest individuals operate on a different scale than most of the members of the other groups. They can afford to invest in producing more corn and still maintain other business or agricultural interests. Most of the wealthy individuals hire others to do their agricultural work, and thus invest little besides money in their fields. The average total production among the wealthy who use fertilizer (126.4 *costales*) seems sufficient to earn a profit and justify their investment. Wealthy individuals can also afford to not produce the minimum subsistence standard of corn. Many of them are already increasingly substituting wheat products (bread) for corn tortillas in their diet. Thus, wealthy individuals who do not use fertilizer have the lowest average corn production (54.3 *costales*) of any group except the poor who do not use fertilizer.

Following Chibnik's suggestion to value subsistence production at its retail rather than market price allows us to see why individuals in at least the three lowest economic groups might want to use fertilizer on their fields. Let us look, for example, at the difference between a total average production of 42.6 *costales* (obtained by poor farmers without using fertilizer) and the 65 *costales* we suggested as the minimum subsistence standard of living. The retail price of corn in the Temascalcingo region in 1973 ranged from about $1.20 (pesos) per kilo just after the November harvest to $2.10 pesos per kilo in the months before the following harvest. Most people try to make it through the year without buying corn, and much of the amount purchased is bought at higher prices late in the year. Valuing corn at approximately 2 pesos per kilo and figuring about 22 *costales* (50 kilos per *costal*) as the "average shortfall" from the subsistence standard yields a cash difference of 2200 pesos (176 dollars). This amount of cash would make especially the poorest and middle economic level farmers interested in using fertilizer, even when the cost of the additional input is subtracted. Thus, although the exchange value of corn is relatively low and does not seem to provide adequate justification for utilizing fertilizer, its use value does make it understandable why many peasants from the Temascalcingo region are interested in adopting fertilizer.

The Economics of Pradera Production

The economics of sowing forage crops provide a strong contrast to the production of corn. Although *pradera* needs to be reseeded only once every 6 or 7 years and requires less fertilizer than corn, a large capital investment is needed to establish these fields. The land must be plowed with a tractor, seed and fertilizer must be purchased, and the fields must be fenced with barbed wire to prevent animals from entering. Unfortunately, credit is available only to groups of farmers, and there is great reluctance to form cooperatives because of past negative experiences. In the past, cooperatives in the *ejido* have failed because, according to many of my informants, a few of the powerful men wanted all of the returns for themselves and were unwilling to do any of the work. The *ejidatarios* also have noted the events taking place in a nearby *ejido*, which did form a cooperative to sow *pradera*. A few of the *ejidal* leaders there have been jailed for corrupt practices and, despite impressive-looking fields of *pradera*, the cooperative has yet to yield a profit for its members. Thus, those *ejidatarios* who now wish to sow these forage crops must do so with their own resources.

Given these conditions, it is not surprising that we find that only the wealthiest have been able to sow this crop with any degree of consistency. Those in the lower economic groups have difficulty raising the necessary cash. In the three lower quartiles of economic position, only 9 of the 101 individuals have been able to sow *pradera*. This is true even though the great majority of *ejidatarios* with whom we talked expressed very keen interest in beginning the cultivation of this crop.

The interest in *pradera* is well-founded. Those who have sown it report that they are reaping good profits from the crop, either by selling the *pradera* to others or by using it to fatten their own livestock. It is difficult to determine production with regard to forage crops, since they are left either as pasture for animals or cut as needed. However, from a farmer in another *ejido* in the region, we obtained the following relative estimate of corn versus *pradera* as a cash crop. This individual has a large block of private land in addition to his *ejido* land and has been to a number of agricultural schools. As a result, he uses many of the recommended techniques for maximization of corn production, including the use of tractors, fertilizers, and insecticides. He said that the best output he can obtain from a hectare of land sown with corn is 5000 kg of grain. (This is considerably higher than the average yields reported in Table 12.3.) Discounting his costs, he estimates a profit of less than 3000 pesos per hectare with corn. We should note that his costs include only that portion of the labor that he hires. With *pradera*, all of which he sells, he reported a profit of 4000 pesos per hectare per year. That is, the difference in profit per hectare is at least 1000 pesos more if forage crops are sown instead of corn. Although production and profits in the *ejido* we studied may not be comparable to these figures, there is little doubt that those individuals who are sowing *pradera* are obtaining better profits than those sowing corn.[11]

Thus, these two opportunities to adopt new agricultural techniques differed in a number of significant aspects. Adoption of chemical fertilizers is a new technique requiring relatively little economic investment, and credit has been available (in some years) to aid the farmers. It is not a technique that involves much risk because it is used on the traditionally grown crop—corn—and the most that could be lost in case of crop loss is the relatively small sum required for purchase of the fertilizer. Those who obtain credit also receive crop insurance that offers some minimization of risk in case of loss. Used on the traditional subsistence crop, fertilizer offers little hope of significant economic returns because corn is not (nor is it viewed by the people as) a viable cash crop. Given the situation of low risk coupled with relatively meager returns, it is not surprising that use of fertilizer is heaviest among those individuals (the poorest) who require better yields of corn simply to bring them to the minimum subsistence standard. Those in the middle groups, many of whom are already meeting their subsistence requirements of corn (many of whom also have other economic means of support), are not very interested in taking even a minimal risk in investing in improved corn production because it will not significantly improve their economic standing. Many of the rich, with sufficient capital for fertilizer as well as for other important operations in corn cultivation, invest heavily in fertilizers and, as we have seen, seem to be reaping sufficient yields to justify their investments. While this pattern of adoption (cf., Figure 12.5) supports Homans' model, we should note that the explanation for this pattern which is being advanced here is quite different from Homans' views of low status people as nonconformists. For these peasants, adoption of new agricultural techniques is not simply behavioral idiosyncrasy: it may be a matter of economic survival.

[11] Based purely on our own observations, we would guess that there is considerably more variation in corn production than in *pradera* production. Those who sow the forage crops seem to follow recommended patterns of cultivation and fertilization to protect their considerable investment. All of the fields of these individuals appeared to be producing well. It is very easy, on the other hand, to see which fields will produce good corn harvests and which will produce very poor harvests. There are visible differences in height, density, and color of the plants, depending on whether fertilizer has been applied and on the cultivation practices of the owner.

The adoption of sowing *pradera* conforms more closely to the risk-taking model to which Cancian originally applied his ideas. Planting of forage crops requires a considerable initial capital investment and involves some risks. It is a new crop unknown in the region and requires skills and knowledge quite different from those involved in the peasants' traditional cultigens. However, concomitant with the risk, there is also the promise of considerable economic gain because *pradera* has a ready market in the region, and is thus suitable as a cash crop as well as for feeding the peasants' own animals. Many of the wealthier people, who have the capital to be relatively unconcerned about the risks involved, have sown the crop. And, as predicted by Cancian's model, the lower-middle economic group has a higher rate of adoption than would be expected on the basis of wealth alone, whereas the upper middle group has a lower rate than that predicted. On the basis of Cancian's evidence and the data presented here, it seems that there is more of a willingness to risk among lower middle-class individuals hoping to improve their relative economic standing, and less of a willingness to risk among those in the upper-middle group who do not wish to jeopardize their relatively high status in a venture that might cause their decline from that position.

Wealth and Pradera Adoption in 1977

In 1977, we were able to make another trip to the Temascalcingo region. This provided us with the opportunity to collect data to test another of Cancian's predictions. He found that in later years, after a new piece of technology has proven itself, the upper-middle group makes a strong comeback and adopts at a higher rate than any other group except the wealthiest (1972:156–157). In particular, Cancian has considerable supporting evidence from many areas of the world showing that the upper-middle group adoption rate is greater than the rate of the lower-middle group in later stages of the adoption process (1977:6–2 to 6–3F). Some indications that this might be happening were already present in 1973. A number of individuals from the upper-middle economic group (and a few other men as well) were talking about possibly combining efforts to sow a portion of their land with *pradera*. They hoped to save money by pooling resources to buy seed in quantity and by sowing a large tract where they all had land.

In 1977, as part of a follow-up analysis of the effect of choice of agricultural strategies on dietary patterns, we were able to collect information on further adoption of *pradera*. During the intervening 4

years, several individuals had died, some people who had been allowing others to cultivate their lands had begun to work it themselves, and
one or two had given up agriculture for a different full-time pursuit. To
make the data as comparable as possible, we limited our investigation
to the same 123 people as in the 1973 sample, but with replacements.
That is, we replaced the landowner from the first time period with the
1977 owner. This was necessary because wealth ratings from 1973 were
used. We believe it is justfied, however, because in every case land
ownership was transferred to a close relative (i.e., wife, son, or brother)
whose economic status would be expected to be comparable to that of
the previous owner.

Data presented in Table 12.5 relate economic position to adoption of
pradera. In 4 years, the planting of this crop in the *ejido* had more than
doubled. The primary reason for this, in addition to the demonstrated
viability of *pradera* as a cultigen and as a money-maker, is that a
cooperative had been formed to establish a dairy herd. The first step
was to sow a large amount of land with forage crops. The leaders of the
cooperative selected one area of the *ejido* for the field of *pradera*. Some
people with land in that area who were not members of the cooperative
and refused to trade their plots had their land plowed and planted with
pradera anyway. Although it was unclear whether the cooperative
would ever acquire dairy cows (see DeWalt 1979:272–273), many people
had fields of forage crops, either in the cooperative or that they had
sown themselves.

Table 12.5 demonstrates that, even though some people did not really
"adopt" *pradera* of their own volition, Cancian's hypothesis about the
"later stage" of the adoption process is supported by the pattern. Seven
more people in the upper-middle group, as opposed to only four more
in the lower-middle group have adopted *pradera*. The adoption rate of
the upper-middle group has increased in the second stage of the adoption process and thus overall, the percentage of this group with *pradera*
has surpassed the percentage of the lower group planting this crop.
Combined with the five new adopters in the lowest economic group

TABLE 12.5
Adoption of Pradera in 1977 by Quartiles of Wealth

	Wealth									
	1		2		3		4		Totals	
Nonadopters	33	(87%)	24	(73%)	19	(63%)	6	(27%)	82	(67%)
Adopters	5	(13%)	9	(27%)	11	(37%)	16	(73%)	41	(33%)

and the six additional adopters in the highest group, the distribution now most closely approximates the linear model.

Summary and Conclusions

Empirical evidence on the adoption of two innovations in a Mexican *ejido* demonstrated that the relationship of economic position and adoption of new technology is not a simple one. Three of the four earlier theoretical models gained some support from the information presented here. None of the models, however, was able to account for all of the patterns of adoption found among Mexican peasants.

The only theoretical model that did not receive any support was the homogeneity model. It was demonstrated that peasant homogeneity in the form of conservatism or in economic wealth was not an accurate depiction of this community of peasants. Considerable variation was found in wealth such that some individuals were existing very close to or below the subsistence level, whereas others had capital and goods totaling well over 100,000 pesos. We also found considerable intracommunity variation in terms of conservatism (and nearly every other kind of cultural behavior). Many individuals (over 65%, in fact) had adopted the use of chemical fertilizers on their fields, thus dispelling the general myth of peasant conservatism. Only 15% of the *ejidatarios* had adopted the sowing of *pradera* in 1973.

The relationship between economic position and adoption of fertilizer was similar in patterning to the model of middle-class conservatism proposed by Homans. His view of conformity was not used in our explanation of this pattern of adoption. Instead we interpreted this information in light of the importance of corn as a subsistence crop in the community and posited that the goal of all but the wealthiest members is to produce enough corn for consumption needs.

The relationship between economic position and adoption of *pradera* was found to conform closely to the modified middle-class conservatism model for which Cancian found considerable empirical support. The upper middle-class conservatism, which this model proposes, was explained in terms of this group's unwillingness to risk investment in *pradera* cultivation, given their relatively secure economic position and the large amount of cash required.

The linear model was found to apply to the data on adoption of *pradera* as of 1977. The "catching-up" that Cancian hypothesizes for the upper middle group, once an innovation has proved itself, transformed the patterning of *pradera* adoption into a very good approximation of the linear model. We should point out that this pattern is also a part of Cancian's theoretical model. Thus, the data on adoption of forage crops very clearly offers further support to what Cancian has already found for other areas of the world. Nevertheless, the data linking economic position and fertilizer adoption does not fit his model.

The variability of the patterning of these data suggests that (a) there may be some validity to hypotheses about middle-class or upper middle-class conservatism; and (b) the relationship between economic position and adoption of new technology is not one which can be adequately understood on the basis of measurements of these two variables alone; more contextual information needs to be provided to make the models understandable. In other words, we need to be able to identify more clearly the parameters or the types of situations in which one or the other of the models will apply. For example, in early stages of the adoption of an expensive crop where people are used to practicing subsistence cultivation, Cancian's Stage 1 model may apply. But later in the adoption process, his Stage 2 model will apply. Some of the other information that needs to be provided has been suggested in the examples considered earlier in the chapter. These include (a) the subsistence base and minimal subsistence standard of living; (b) the amount of investment required; (c) the availability of credit; (d) the possible "exchange value" benefits of the new crop; (e) other nonagricultural job prospects; and (f) the amount of risk involved in adopting the new techniques. In other situations, many other variables may become important. The ultimate message is that our models have to become more complex if we are to begin to understand processes of agricultural decision making (also see B. DeWalt 1979:248–260).

It seems reasonable to us that eventually we may discover that there are only two or three different types of situations of economic change, each of which takes on a different model of the relationship between economic position and adoption of new technology. That is, looking for patterns of sociocultural change may require a sorting out of several different theoretical models that reflect significant differences in the types of technology being introduced and the kinds of socioeconomic constraints and incentives present.

Acknowledgment

Thanks are due to Kenneth Hadden and Gretel Pelto for comments and suggestions. Special appreciation is expressed to Pertti Pelto for his assistance. The support of fieldwork by the National Institute of Mental Health through grant number 1 F01 MH54604–01 is gratefully acknowledged.

References

Banfield, Edward
  1958 The Moral Basis of a Backward Society. New York: Free Press.
Cancian, Frank
  1967 Stratification and Risk-taking: A Theory Tested on Agricultural Innovation. American Sociological Review 32:912–927.
  1972 Change and Uncertainty in a Peasant Economy. Stanford: Stanford University Press.
  1977 The Innovator's Situation: Upper Middle Class Conservatism in Agricultural Communities. Social Sciences Working Paper 132. University of California, Irvine.
Chibnik, Michael
  1978 The Value of Subsistence Production. Journal of Anthropological Research 34(4):551–576.
DeWalt, Billie
  1979 Modernization in a Mexican Ejido: A Study in Economic Adaptation. Cambridge and New York: Cambridge University Press.
DeWalt, Billie, Robert Bee, and Pertti Pelto
  1973 The People of Temascalcingo: A Regional Study of Modernization. Storrs: University of Connecticut. Mimeo.
Firth, Raymond
  1956 Elements of Social Organization. London: Watts.
Foster, George
  1967a Peasant Society and the Image of Limited Good. In Peasant Society: A Reader, J. Potter, M. Diaz, and G. Foster, eds. pp. 300–323. Boston: Little, Brown.
  1967b Tzintzuntzan: Mexican Peasants in a Changing World. Boston: Little, Brown.
  1972 The Anatomy of Envy: A Study in Symbolic Behavior. Current Anthropology 13:165–202.
Homans, George C.
  1961 Social Behavior: Its Elementary Forms. New York: Harcourt Brace and World.
Huizer, Gerrit
  1970 "Resistance to Change" and Radical Peasant Mobilization: Foster and Erasmus Reconsidered. Human Organization 29:303–312.
Iwanska, Alicja
  1971 Purgatory and Utopia: A Mazahua Indian Village of Mexico. Cambridge: Schenkman.
Kahl, Joseph
  1968 The Measurement of Modernization: A Study of Values in Brazil and Mexico. Austin: University of Texas Press.
Kroeber, Alfred L.
  1948 Anthropology. New York: Harcourt.
Lerner, Daniel
  1958 The Passing of Traditional Society. New York: Free Press.
McClelland, David C.
  1961 The Achieving Society. New York: Van Nostrand.
Nash, Manning
  1961 The Social Context of Economic Choice in a Small Community. Man 61:186–191.
  1971 Market and Indian Peasant Economies. In Peasants and Peasant Societies. Teodor Shanin, ed. pp. 161–177. Baltimore, Md.: Penguin.

Pelto, Pertti
    1973   The Snowmobile Revolution: Technology and Social Change in the Arctic. Menlo
           Park, Calif: Cummings.
Pelto, Pertti, and Gretel Pelto
    1975   Intra-cultural Diversity: Some Theoretical Issues. American Ethnologist 1975:1–
           18.
Redfield, Robert
    1960   Peasant Society and Culture. Chicago: University of Chicago Press.
Rogers, Everett
    1969   Modernization among Peasants: The Impact of Communication. New York:
           Holt, Rinehart and Winston.
    1971   Communication of Innovations: A Cross-Cultural Approach. Toronto: Collier–
           Macmillan.
Sahlins, Marshall
    1972   Stone Age Economics. Chicago: Aldine.
Silverman, Sydel
    1966   An Ethnographic Approach to Social Stratification: Prestige in a Central Italian
           Community. American Anthropologist 68:899–921.
Simon, Barbara D.
    1972   Power, Privilege, and Prestige in a Mexican Town: The Impact of Industry on
           Social Stratification. Unpublished Doctoral dissertation, University of Min-
           nesota, Minneapolis.
Wharton, Clifford
    1971   Risk, Uncertainty, and the Subsistence Farmer: Technological Innovation and
           Resistance to Change in the Context of Survival. *In* Studies in Economic An-
           thropology. G. Dalton, ed. pp. 152–178 Anthropological Studies, No. 7. Wash-
           ington: American Anthropological Association.
Wolf, Eric R.
    1955   Type of Latin American Peasantry: A Preliminary Discussion. American An-
           thropologist 57:452–70.
    1966   Peasants. Englewood Cliffs, N.J.: Prentice–Hall.

PART III

AGRICULTURAL DEVELOPMENT POLICY AND PROGRAMS

Chapter 13

Decision Making and Policymaking in Rural Development

SARA S. BERRY

How useful are studies of agricultural decision making for the design of rural development policies? In the following chapter, I will use "studies of agricultural decision making" to refer to those that seek to explain farmers' economic behavior in terms of constrained utility maximization models. On one level, since the literature on agricultural decision making is itself inconclusive concerning the particular content and method of poor farmers' decisions, it can be used to justify conflicting approaches to effecting rural development, and may therefore be said to have contributed little to the cumulative improvement of rural development strategies. However, a review of some of the methodological and substantive limitations of the decision-making approach to analyzing agricultural performance may help to clarify issues to which policymakers ought to address themselves. I shall try to show in this chapter that it is sometimes useful to point out what we do not know.

Studies of farmers as constrained utility maximizers generally postulate that poor farmers organize their agricultural activities (e.g., level of output, crop mix, methods of cultivation, investment choices, innovation) to maximize the utility they derive from the outcomes of these activities, subject to existing constraints on their access to productive resources, technical knowledge, and market opportunities. Since most agricultural decisions must be made in the absence of perfect information about the actual result of any particular act, the "outcome" of an act

321

AGRICULTURAL DECISION MAKING:
Anthropological Contributions to
Rural Development

is usually expressed as a probability distribution of possible outcomes. Utility is postulated to be a function of the moments of the probability distribution, and decision makers are thought to act to maximize subjective expected utility (Anderson, *et al.* 1977; Balch *et al.* 1974; Savage 1954). In other words, utility depends both on farmers' expectations (or forecasts) of the results of their actions and on their preferences or attitudes toward the expected results. Within this analytical framework, the principal tasks of empirical research are (*a*) to specify correctly the constraints that farmers face, and the determinants of the utility they expect to derive from possible outcomes; and (*b*) to collect accurate data with which to estimate the parameters of the hypothesized relationships between expected outcomes and constraints. Predicted results are compared with observed behavior to test the ability of alternative models to explain or predict actual events, and models that perform well are then recommended to policymakers as useful guides to the effects on rural development of proposed government programs.

Decision-making models have been used in most disciplines in the social sciences. Scholars in different disciplines often assign different weights to, e.g., "economic" or "cultural" variables in their models, or generate better data on variables normally stressed within their own fields than on those they conventionally hold constant, but the basic postulates and empirical methods of decision-making studies are the same across disciplines. In studies of farmers in underdeveloped economies, for example, economists often recognize in principle the influence of social and cultural variables on both utility and constraints, just as anthropologists understand that prices, technology, and demand make up part of the system of social relationships within which individuals function. If economists have tended in the past to ignore sociocultural factors in practice, this gap in the literature is now being filled by economic anthropologists and others who attempt to take a wider range of independent variables into account in their models.

Specific conclusions that have been drawn about the determinants of farmers' decisions vary within as well as among disciplines. One can, for example, cite both economists and anthropologists who argue that poor farmers allocate resources in a manner consistent with profit maximization—and others in both fields who deny that this is so (see, e.g., Schultz 1964 versus Lipton 1968; Schneider 1974 versus Dalton 1971). Indeed, one of the clearest "findings" of the whole corpus of literature on agricultural decision making in underdeveloped economies is that there is no evidence that poor farmers' goals or decision-making processes are consistently different from other people's (Roumasset 1976). The debate joined over a decade ago between Schultz and Lipton over what poor farmers seek to maximize has spawned a host of increasingly sophisticated examples and counterexamples, but no consensus.

The inconclusiveness of this literature with respect to the distinctive nature of poor farmers' economic performance derives in part from certain conceptual features of the decision-making approach to analyzing social behavior. In particular, decision-making analysis (*a*) assumes that individuals' attitudes or preferences influence their behavior independently of the circumstances in which they act; and (*b*) tends to assume that individuals or groups of homogeneous individuals base their actions on independent assessments of given sets of circumstances. It thus treats or explains social processes as sums of individual acts rather than as complex processes of interaction among both individuals and groups. In the following pages, I will discuss each of these points and comment on their implications for rural development policymaking.

Do Attitudes Make a Difference?

The premise, common to most decision-making studies, that individuals' preferences exert an independent influence on their behavior is inherently plausible, but difficult to test. Because decisions can never be taken under conditions of certainty about the actual outcomes of particular acts, they must be based on estimates of future outcomes. This is, of course, the basic reason for expressing outcomes as probability distributions rather than as single-valued variables. But this means that decisions are based on both expectations about future outcomes and on attitudes toward those expectations. A farmer's choice among, for example, different crop combinations depends both on how the expected net value of the output is estimated and on how the farmer feels about it. But, since both expectations and attitudes are subjective and hence not susceptible to direct observation, efforts to test alternative decision-making models against observed behavior face a serious identification problem. To cite a common example: if we observe that a particular group of farmers chooses a crop combination whose yield we estimate to have a lower mean but also a lower variance than a more profitable but riskier alternative, we cannot conclude that these farmers prefer lower risk to higher profits. It may be, instead, that they estimate the likely values of the outcome differently from the way we have estimated them.

Various methods have been used to resolve this identification problem, none of them completely successful. Some studies deal with it in effect by assuming it away. In studies of short-term resource allocation (choice of input levels, combinations or crop mix), for example, it is customary to assume that farmers estimate future outcomes in a specific way and then seek to infer the shape of their utility functions from their behavior (as in, for example, Moscardi and de Janvry 1977). Conversely, in the literature on supply response, it is usually taken for granted that farmers seek to maximize expected profit and that behavior varies under similar circumstances because of variations in forecasting procedure. Such assumptions are convenient, but do little to advance our understanding of the decision-making processes involved.

Alternatively, some researchers have tried to elicit direct evidence on farmers' expectations or attitudes through simulated decision games (Dillon and Scandizzo, 1978). Farmers are asked to evaluate or choose between sets of hypothetical payoffs designed to approximate the range of actual options available to them. Such studies are open to the objection that experimental situations do not, in fact, replicate the uncertainty or the actual constraints of real decision problems and hence cannot induce people to reveal either their preferences or their estimating procedures in a meaningful way. Like the larger body of research by experimental psychologists on choice and decision processes from which they derive (Slovic et al. 1977), these studies have raised doubts about the predictive value of expected utility maximization models, but have so far produced no clearly satisfactory alternative.

Another approach to analyzing the effects of subjective processes on social behavior, which is closer to conventional social science methodology, is to use observable characteristics or behavior patterns of the decision makers as proxies for expectations or attitudes. Cancian's work on risk (Chapter 7 in this volume) is an interesting example. He suggests that people are rank-seeking creatures, whose economic and social acts vary systematically with their relative socioeconomic status. Thus, he argues that although the ability to bear risk (which, following Knight, he defines as "a known probability that investment in [a] new seed . . . will result in losses during the coming year") increases with income and/or assets, upper middle-class farmers will be more averse to uncertainty (unknown probabilities of loss) than lower middle-class farmers because they have more to lose in terms of rank. During the early stages of the introduction of an agricultural innovation to a given community, when local experience with the innovation is limited or nonexistent, and therefore uncertainty is high, lower middle-class farmers are expected to adopt it more readily than upper middle-class farmers because they have less to lose should the innovation fail.

As time passes, and farmers gain experience with the new crop or cultivation method, however, the rate of adoption among upper middle-class farmers should increase and even surpass that of lower middle-class farmers whose resources are more limited. Evidence from a number of empirical studies is said to be consistent with these predictions, although Cancian does not discuss how sensitive the results of these studies are to the authors' choices of reference groups within which to measure individuals' relative status.

In effect, what Cancian is doing is to use rank as a proxy for preferences, and time elapsed after the introduction of an innovation as a proxy for farmers' estimating procedures. As farmers gain experience with the innovation, they estimate a lower probability of loss due to ignorance about proper use of the new technique under local conditions, and therefore attach a higher expected value to the innovation. However, one might argue with equal plausibility that upper middle-class farmers are relatively slow to adopt innovations not because they are afraid of losing rank, but because they can afford to wait for more information (and hence lower costs), whereas lower income farmers may be so pressed to try to earn extra income (to meet household consumption needs, and avoid debt) that they feel they must try anything that comes along. In other words, because they are somewhat better endowed with resources to meet current needs, upper middle-class farmers are able to shift the risks of initial experimentation onto their poorer neighbors.[1] Cancian's results could be interpreted as evidence of profit-maximizing (rather than uncertainty-averting) responses to a situation in which imperfections in factor markets have given rise to differential access to resources, which in turn means that the cost of innovation is different for farmers in different income or landholding classes.

I will come back to the issues of market imperfections and the incidence of risk in a moment; for now, the point I want to make is that we have no way of choosing between my interpretation and that of Cancian on the basis of his evidence. The problem, as Lipton (1979) has pointed

[1] This interpretation of Cancian's results does not depend on Knight's questionable definitions of risk and uncertainty as "known" and "unknown" probabilities of loss, respectively. Probablities of future events are never known with complete certainty: there simply is no such thing as certain risk and hence, no possibility of eliminating the need to forecast if one wants to make decisions in terms of future outcomes of present acts. Even studies arguing that farmers make no explicit attempt to forecast future prices may still analyze their behavior in terms of some rough assumption about future outcomes—cocoa or coffee will be worthwhile in the long run and therefore farmers plant it, subject to current resource availabilities and competing needs [Berry 1976; Ortiz 1973].

out, is not that social behavior may be irrational and hence not susceptible to rationalization, but that it is overdetermined and hence susceptible to multiple rationalizations. Humans may be universally rank-
seeking, profit-maximizing, or both—we will probably never know.

The assumption that preferences exert an independent influence on
behavior has influenced not only descriptive research on farmers' behavior, but also the design of rural development strategies. If differences in behavior are attributable to differences in attitude, it may be
argued that development strategy should seek either to change attitudes
(through some form of education—informational and/or persuasive), or
to channel incentives and resources to those individuals whose attitudes
are favorable to desired changes. If, for example, one believes (as many
studies of agricultural decision making imply) that subjective aversion
to risk leads farmers to act in a nonprofit-maximizing fashion (Berry
1977), one may believe that they should be induced to change their
ways through programs designed to improve their forecasts and/or allay
their fears by providing more and better information about the likely
costs and benefits of alternative production plans. Such reasoning has
been used to justify policies such as disseminating information about
prices and cultivation practices, agricultural extension services to demonstrate new techniques, or price stabilization and crop insurance
schemes.

Although there is certainly nothing wrong with educational approaches to improving economic performance, such programs are often
disappointing in their results. The problem is that whereas education
may be a necessary condition for social change, it is by no means a
sufficient one. Improving the flow of information to a decision maker
does not necessarily increase the capacity to act on it. A poor farmer
may know about improved seed, fertilizer, high urban prices, insurance companies, and the stock exchange, without being able to gain
access to them to increase output or income. Agricultural extension
services often have little impact on poor farmers' incomes and productivity precisely because they tend to address themselves to the problem
of supplying information, without providing farmers the means to
make use of it. Similarly, to the extent that price stabilization schemes
actually serve to destabilize farmers' incomes, by eliminating supply-
determined changes in price that tend to offset shifts in output, they
tend to reduce farmers' ability to bear risk rather than to increase it.

If educational policies designed to reduce the risks of profitable
production plans fail to induce risk-averse farmers to change their
behavior, it is often argued, the way to increase agricultural output is to
channel available resources to those "who are best able to use them"—

to farmers whose preferences lead them to invest, innovate, and allocate resources in an efficient manner. This may be done deliberately, by providing loans, extension services, or improved inputs to farmers who have already demonstrated their capacity to innovate or invest, or indirectly through a structure of price supports, subsidies, taxes, and public expenditures favoring large-scale, commercial and/or capital-intensive farming methods (Griffin 1974; Johnston and Cownie 1969). In either case, such programs are often justified on the grounds that they maximize returns to scarce resources in the short run, thus enlarging the total output available for further investment or income redistribution later on.

However, insofar as successful farmers are not those who are willing to expand output, but those who can afford to do so, programs which subsidize "progressive" farmers tend to achieve agricultural growth at the cost of increased rural inequality (Kilby and Johnston 1975; Lele 1974). Development for the few is, in turn, not only questionable on ethical grounds but probably also economically counterproductive. As with import-substituting industrialization, rural development policies that promote the use of complex, capital- and/or import-intensive methods of agricultural production cannot be said to contribute to the efficient use of available resources. And, by raising the incomes and strengthening the ability of rural elites to control private resources and public policies, such inegalitarian development strategies make it more difficult to redistribute output and assets in the future. In short, programs which seek to subsidize progressive attitudes often serve to intensify the constraints on increasing production and income for a majority of the rural population.

Poor Farmers Profit When They Can

Fortunately for social analysis, our ability to predict behavior does not depend entirely on our success in measuring attitudes. Indeed, one of the major contributions empirical decision-making studies have made to the literature on rural development has been to show that often one need not postulate irrationality or subjective resistance to change to explain poor farmers' behavior. When actual costs and returns to alternative agricultural activities are fully and accurately measured, it often turns out that poor farmers prefer, for example, subsistence to commercial production, or mixed to monocropping, or existing cultivation methods to new ones, because it pays them to do so. Such choices frequently lead to higher incomes than would the supposedly more productive alternatives, given the constraints under which poor farmers produce, sell and consume.

Correctly identifying and measuring the relevant constraints is often a difficult task, for which decision-making theory itself offers few guidelines. It is in the quality of empirical decision-making research that, for example, anthropologists have made some of their most important contributions by taking the trouble to collect full and accurate information on individual farmers' opportunity costs. In her meticulous study of agricultural decision making by Indian farmers in a Colombian village, for example, Ortiz (1973) shows that by carefully specifying differences in the conditions under which farmers make decisions about different crops, it is possible to reconcile apparent anomalies in their decision processes. Her informants plant both coffee and food crops, but, whereas resources are allocated to coffee cultivation on the basis of careful long-range planning, decisions about food crop production appear haphazard and disorganized.

The difference, it turns out, is a matter of timing. Coffee trees take several years to mature and bear for a long time thereafter. Once planted, a plot of coffee trees will absorb part of the farmer's land and labor, and yield him some income for many years. The decision to commit resources to a plot of coffee is therefore considered very carefully and timed to avoid conflicting with other foreseeable demands on the farmer's resources. The fact that income from a plot of coffee trees fluctuates unpredictably from year to year only reinforces the long-term nature of the decision. The farmer knows he is investing in a "life-time" income rather than in a steady flow and makes his calculations accordingly.

In the case of food crops, however, both consumption needs and opportunity costs vary continuously, and a farmer would plant food crops "until he had enough, at which point he would stop and do something else. It became clear that he could not express his need in terms of specific number of plants or specific acreage because *enough* was a relative measure" depending both on his family's needs and "on what he had to forego in order to plant that much. . . . Decisions regarding subsistence production and occasional wage labor were made in the course of action, whereas decisions regarding cash crops were usually made beforehand [Ortiz 1973:7–8]." Consequently, farmers often changed their plans between the time they cleared land for food crops and the time they sowed the seed. Far from being haphazard, decisions about food crop production were adjusted continuously to changing circumstances.

Similarly, in studies purporting to show that poor farmers fail to respond to opportunities to earn higher incomes, through changes in crop combinations, cultivation practices, or capital formation, or that they allocate resources differently from more prosperous farmers, the results often depend on the way in which the author(s) measure constraints. For example, in the case of "conventional" versus Chayanovain cost–benefit calculations discussed by Barlett, Chapter 6, this volume, Chayanov's claim that capitalist firms and family farms employ different decision-making processes amounts to little more than the assertion that they use different methods of imputing labor costs and the value of nonmarketed output. Similarly, farmers' failure to innovate or to increase production for the market often reflects their inability rather than their unwillingness to do so. The potential conflict between the goals of profit and security has been overstressed in some of the literature; in fact, more profitable options do not always entail greater variance of expected income and, even when they do, are not necessarily riskier, since increased profits enhance the decision maker's capacity to bear risk (Berry 1977; Roumasset 1976). In brief, farmers' readiness to take advantage of new income earning opportunities often depends more on their assets than on their attitudes (cf., Lipton, in press; Moscardi and de Janvry 1977).

Also, as we have seen, poor farmers have not only proved generally responsive to feasible opportunities to increase their income, but also their responses are often more "appropriate" to the factor endowments and institutional structures of underdeveloped economies than are those of large-scale agricultural producers. Numerous studies have shown that small-scale, land- and labor-intensive methods are more economically efficient, given prevailing conditions, than large-scale mechanized ones (Lipton 1974; Kilby and Johnston 1975; Yotopoulos 1968). These findings suggest, in turn, that if resources could be redistributed from large-scale farmers to smaller ones, the result might well be faster growth as well as more equally distributed incomes.

The difficulty is that redistribution is not easy to effect: plans to help poor farmers help themselves rarely succeed in reaching the "bottom 40%" and, when they do, poor farmers often fail to participate, either by not showing up or by diverting subsidized credit, fertilizer, and other needs to "nonproductive" uses. To understand why, for example, poor farmers buy subsidized inputs only to resell them to more prosperous neighbors and use the cash for bridewealth, school fees, funeral expenses, bribes, or sacrifices, it does not really help to reiterate that such expenditures increase farmers' utility which, in turn, they try to

maximize. Rather, these things call for an explanation of the circumstances in which poor farmers find themselves, which economists'
decision-making analysis does not provide.

Rural Poverty as Social Process

Part of the reason for the inadequacy of decision-making analysis as a
guide both to explaining and to effecting social change is that it considers individuals in isolation, whereas in fact both the "opportunities"
and the "constraints" to which individuals respond at any point in time
arise out of interactions among individuals and social groups. Every
beginning student of economics is taught that market supply is the sum
of individual firms' marginal cost curves, or that individual profit
maximization leads to socially efficient allocation of resources, *only*
under conditions of perfect competition. But the rarity of those conditions is seldom acknowledged in studies of decision making and the
difficulty of drawing social conclusions from studies of individual behavior is all too often ignored.

Studies of agricultural decision making, for example, often assume
that the consequences of an act fall on the decision maker alone, and
hence that individuals make decisions without reference to the actions
of their neighbors. In fact, this assumption is not warranted in any
situation in which some people have the power to influence the terms
on which they acquire, use, or dispose of resources, since such people
are in a position to shift some of the burdens or costs of their activities
onto others.

Shops supplying seed, fertilizer, and implements, moneylenders, and
traders who purchase farm products at or near the "farm gate" not only
increase their profits at poor farmers' expense, but also shift the risks of
their own businesses onto poor farmers, often at a premium. For example, several studies have argued that it is impossible to explain the wide
range of interest rates charged to rural borrowers in underdeveloped
economies solely in terms of the higher costs of administering small
loans to low-income farmers. (AID 1973; Lele 1974; Masson 1972;
Roumasset 1976.) Poor farmers are not more likely to default on loans
than are wealthier ones—on the contrary—lending to them is no riskier
than lending to the well-to-do. Nonetheless, they are often forced to pay
interest rates of 100% or more per annum, whereas better-off or better-
connected farmers pay the legal maximum of 10–15% (Lipton 1979;
Long 1968; Roumasset 1976). In other words, poor farmers' "failure" to
invest may reflect not so much their own relative aversion to risk as

successful risk aversion by moneylenders (and/or landlords) which leads to discriminatory rural interest rate structures and hence to under-investment in small farms.

Similarly poor farmers' reluctance to specialize in production of cash crops and/or to sell foodstuffs may be profit-maximizing responses to the structure of market prices, as well as behavior designed to stabilize income. In many rural communities with poor transportation and mar-keting facilities, farmers who sell their crops and rely on the market for their own consumption needs may eventually pay higher prices for staple foodstuffs than they receive for selling the same commodities. Thus it is more *economical* to produce foodstuffs for their own consump-tion than to grow them for sale (Ortiz 1973, 1979). Seasonal fluctuations in food prices may also raise the cost of meeting household consump-tion needs by purchase, especially in years of poor harvests, and lead farmers to grow part or all of the food they consume to maximize their annual real incomes. Also, high storage costs may prevent poor farmers from accumulating their own buffer stocks in good years to cover household needs in years of poor harvests. Hence resources may be used for subsistence production year after year, which might have been available for cash crop production if storage were cheaper. And, finally, the effects of public expenditure programs or technical charge on indi-viduals' output, employment, and incomes may vary within a given community according to differences in individuals' ability to control resources and/or gain access to new opportunities (Hart and Sisler 1978).

The fact that some people's actions constitute other people's con-straints means that one cannot study individuals' behavior in isolation—in the language of normal economic analysis, one cannot ignore externalities. But there is more to it than that. People's behavior depends not only on what other people are doing, but also on the form and quality of the social relationships among them. Farmers use agricul-tural loans for ceremonial expenditures in part because such expendi-tures serve to affirm or enhance their commitment to principles of social interaction (among kinsmen, neighbors). These principles, in turn, shape farmers' access to productive resources.

Parkin (1972) has shown, for example, how Giriama farmers in south-eastern Kenya used traditional authority structures to facilitate capital accumulation, and hence acquired a vested interest in maintaining those structures, even though they did not themselves hold positions of traditional authority. When copra production became commercially important after World War II, the younger Giriama men who competed most actively with one another in planting and acquiring palm trees relied on the locally respected testimony of elders when their claims to ownership of trees were challenged in court. Consequently, although the younger farmers were becoming the wealthiest group in the community, they upheld the principle of gerontocracy in local political affairs, including the tradition of holding elaborate and expensive funerals for deceased elders. Thus, the amounts spent on funerals tended to *rise* with the accumulation of agricultural capital, in part because the two forms of resource use were mutually reinforcing.

Economic performance often depends on the quality of social relationships, as well as on their form. Economists have long recognized, for example, that most firms (and farms) probably do not operate on their marginal cost curves at all, but at points inside the curve which are determined by the ways in which people interact within the productive enterprise and which, accordingly, cannot be predicted in terms of constrained utility maximization (Leibenstein 1976). Similarly, the theory of "imperfect competition," which seeks to explain the effects of market power on economic performance, can give a determinate answer to this question only in the case of "pure monopoly"—in which interactions among sellers are eliminated by assumption. In both of these cases, economic performance depends not only on objective information about market prices and techniques, but on the quality of relationships among the actors—for example, acts and feelings of solidarity, mutuality, or alienation between owners (or managers) and workers within an enterprise, or of cooperation or conflict among buyers and/or sellers in a market.

The success, for example, of Parkin's rural capitalists depends partly on their ability to command the loyalty and support of creditable witnesses—an ability that depends in turn on complex political and cultural processes that cannot be reduced meaningfully to a scale of payoffs. Funeral expenses are not payments for supportive testimony, but rather one indicator that may lead us to examine the social processes involved but does not measure, much less explain them. Similarly, Lipton (1968) has described how, in an Indian village, farmers did not adopt a yield-increasing innovation (contour ploughing) although they understood its value, because for purposes of inheritance, hillside land was held and cultivated in vertical rather than horizontal strips. The inheritance system served, as Lipton showed, to spread risks among inheriting sons. To attribute the historical persistence of such inheritance patterns and their influence on people's responses to changing opportunities simply to an (immutable?) wish to avert risk for one's children seems to be both an unverifiable and unnecessary simplification.

Policy Implications

If economic performance cannot be understood as the sum of individuals' utility-maximizing responses to given circumstances, then it cannot be changed effectively simply by manipulating the prices at which people acquire inputs and dispose of output. Policymaking, like production, is a social process, whose outcome depends on changing relationships among people, as well as on the availability of resources, and on the ability of individuals to rationalize their use. Failure to grasp this point lies at the root of the liberal belief that governments can, through legislation and law enforcement, bring about a more socially desirable distribution of income and resources that individual decision makers will then use in an economically efficient manner to increase aggregate welfare. Policy simply does not work that way. Social and economic benefits do not "trickle down" from the rich to the poor; redistributive land reform does not prevent rural capitalists from growing rich at the expense of their neighbors; legislators do not pass loophole-free tax reform laws or subsidize the poor and powerless at the expense of their own supporters. Nor do farmers allocate resources to maximize profit (or minimize risk) if, by doing so, they would alienate people through whom they may be able to acquire resources or access to opportunities in the future. Just as decision-making analysis cannot explain how people interact over time to create the opportunities and constraints that confront the individual at any moment, so policies based on decision-making analysis have not succeeded in eliminating rural poverty and stagnation because they do not address the sources of power and patronage that shape resource allocation. Decision-making analysis tells us that better cropping patterns, input subsidies, price supports, water control, crop insurance, agricultural loans, and investment in rural marketing and transport facilities will all tend to increase the output and incomes of farmers who have access to them. It does not help us to find out who those farmers are, how they arrived at their present position, or how they are likely to act to extend or deny such access to others. Policymakers, like all decision makers, need more than information and money to effect social progress.

References

AID
   1973   Spring Review of Small Farmer Credit. Washington, D.C.
Anderson, J. R., J. L. Dillon, and J. B. Hardaker
   1977   Agricultural Decision Analysis. Ames, Iowa: Iowa State University Press.

Balch, M., D. McFadden, and S. Wu, eds.
  1974   Essays on Economic Behavior under Uncertainty. Amsterdam: North-Holland.
Berry, Sara S.
  1976   Supply Response Reconsidered: Cocoa in Western Nigeria. Journal of Develop-
         ment Studies 13:4–17.
  1977   Risk and the Poor Farmer. Washington, D.C.: AID.
Dalton, George, ed.
  1971   Economic Development and Social Change. Garden City, N.Y.: The Natural
         History Press.
Dillon, J. L., and P. L. Scandizzo
  1978   Risk Attitudes of Subsistence Farmers in Northeast Brazil. American Journal of
         Agricultural Economics 60:425–435.
Griffin, Keith B.
  1974   The Political Economy of Agrarian Change. London: Macmillan.
Hart, Gillian, and D. Sisler
  1978   Aspects of Rural Labor Market Operation: A Javanese Case Study. American
         Journal of Agricultural Economics 60:821–826.
Johnston, Bruce F., and John R. Cownie
  1969   The Seed–Fertilizer Revolution and Labor Force Absorption. American Eco-
         nomic Review 59:569–582.
Kilby, Peter, and Bruce F. Johnston
  1975   Agriculture and Structural Transformation. New York: Oxford University Press.
Leibenstein, Harvey B.
  1976   Beyond Economic Man. Cambridge, Mass.: Harvard University Press.
Lele, Uma J.
  1974   The Roles of Credit and Marketing in Agricultural Development. In Agricultural
         Policy in Developing Countries. Nurul Islam, ed. pp. 413–441 London: Macmil-
         lan.
Lipton, Michael
  1968   The Theory of the Optimising Peasant. Journal of Development Studies 4:327–
         351.
  1974   Towards a Theory of Land Reform. In Peasants, Landlords and Governments. D.
         Lehmann, ed. pp. 269–315 New York: Holmes and Meier.
  1979   Agricultural Risk, Rural Credit, and the Inefficiency of Inequality. In Risk,
         Uncertainty and Agricultural Development. J. H. Boussard and J. A. Roumasset,
         eds. New York: Agricultural Development Council.
Long, Millard
  1968   Why Do Peasant Farmers Borrow? American Journal of Agricultural Economics
         50:240–248.
Masson, R. T.
  1972   The Creation of Risk Aversion by Imperfect Capital Markets. American Eco-
         nomic Review 62:77–86.
Moscardi, E., and Alain de Janvry
  1977   Attitudes Towards Risk Among Peasants. American Journal of Agricultural Eco-
         nomics 59:710–716.
Ortiz, Sutti R. de
  1973   Uncertainties in Peasant Farming. New York: Humanities Press.
  1979   The Effect of Risk Aversion Strategies in Subsistence & Cash Crop Decisions. In
         Risk, Uncertainty, and Agricultural Development. J. H. Boussard and J. A.
         Roumasset, eds. New York: Agricultural Development Council.

Parkin, D. J.
   1972   Palms, Wine and Witnesses. San Francisco: Chandler.
Roumasset, James A.
   1976   Rice and Risk: Decision Making Among Low-Income Farmers. Amsterdam: North-Holland.
Savage, L. J.
   1954   The Foundations of Statistics. New York: Wiley.
Schneider, Harold D.
   1974   Economic Man: The Anthropology of Economics. New York: Free Press.
Schultz, Theodore W.
   1964   Transforming Traditional Agriculture. New Haven: Yale University Press.
Slovic, P., B. Fischhoff, and S. Lichtenstein
   1977   Behavioral Decision Theory. Annual Review of Psychology 28:1–40.
Yotopoulos, Pan A.
   1968   On the Efficiency of Resource Allocation in Subsistence Agriculture. Stanford Food Research Institute Studies 8:125–136.

Chapter 14

Agricultural Decision Making in Foreign Assistance: An Anthropological Analysis[1]

ALLAN HOBEN

Introduction

It is essential to remember that bureaucrats are as rational as peasants, and that their behavior, like that of peasants, must be understood in terms of the institutional contexts in which they work.

The Puzzle

There is a gap between the rhetoric and policy guidelines of development agencies, the allocation of their resources, and the outcomes of their projects. Although in recent years academics and policy planners have come to recognize that farmers' decisions are rational responses to local conditions, this change in perspective is not carried through in the selection of projects.[2] In this analysis, bureaucrats are seen as rational decision makers too, and the discrepancy can be understood only in the context of the institutions in which they work.

[1] I am grateful to the Ford Foundation and to the Overseas Development Council for their support in the preparation of this chapter.

[2] The assertion that individual farmers are optimizers does not imply, of course, that the pattern of agricultural behavior resulting from individual choices is optimal from an environmental, ecological, economic, or a developmental perspective, nor does it imply that farmers calculate probabilities in their heads (Quinn 1978).

337

AGRICULTURAL DECISION MAKING:
Anthropological Contributions to
Rural Development

Official guidelines call for the "fine tuning" of projects to local ecological, social, and economic conditions through improved data collection and analysis, the participation of project beneficiaries in decision making, and an ongoing two-way exchange of information with the beneficiaries through extension. Although some progress has been made in implementing these guidelines, information about farmer decision making, and the participation of the farmers themselves, still have very little effect on resource allocation decisions in agricultural development projects.

Problems occur at each stage. The quality of analyses included in project design documents is very uneven. Many are merely descriptive rather than analytical and do not identify key issues relevant to the project for which they were prepared (McPherson 1978; Perrett 1978). Readily available information often is omitted. When social scientists are brought in to the design process, they often are selected with little regard for their geographic or substantive backgrounds or for their language competence. Even when high quality, readable, and relevant analyses are prepared by consultants, they often are ignored or emasculated through editing. They seldom influence crucial decisions concerning project inputs, budget, staffing, site selection criteria, or indicators by which project impact is to be assessed. Moreover, regardless of the sensitivity to local conditions evinced in project design, it may be lost in the process of project implementation. And finally, little use is made of ecological, social, or microeconomic analyses in monitoring or evaluation. As a result, there is little learning from project experience in these areas, and the same kinds of projects are designed and approved repeatedly, despite mounting evidence of poor performance.

The purpose of this chapter is to explain the difficulty large development agencies have in using the findings of anthropologists and other field-oriented social scientists and to suggest ways that these can be overcome. My central thesis is that bureaucrats are as rational as peasants, and that the ways they use and do not use information in their work can be understood by examining the institutional contexts in which, and the processes through which, they make policy, program, and project level decisions about agriculture.

To understand this difficulty and the discrepancy between policy and resource allocation, I will discuss the background of the problem, provide a brief overview of the organizational setting and decision-making process in one development agency, the United States Agency for International Development (AID), and examine AID's response to the new legislation requiring it to recognize peasant rationality.

Much of the material on which the analysis is based concerns AID, where I worked from 1976 to 1979 as Senior Anthropologist for Policy under an Intergovernmental Personnel Act agreement with my university. It was gathered using a participant–observer approach, supplemented by extensive interviews with colleagues and analysis of documents. Whereas my analysis is, therefore, most directly pertinent to AID, there is considerable evidence that it is applicable to other large bilateral and multilateral donors.

Conclusions

This chapter presents a number of conclusions that will be of interest to scholars and planners concerned with the institutional constraints on the implementation of current, people-oriented, agricultural development policies. Some of these conclusions are not surprising. Others confront assumptions current in academic and development circles. All of them will be developed and documented at greater length in my forthcoming book on decision making for development.

The most important of these are:

1. Pressures arising from the external institutional environment and from the internal dynamics of development agencies create organizational objectives and incentives that often conflict with officially stated policy and with individual employees' professional judgment and personal values.

2. The most enduring and pervasive institutional objective in development agencies and banks is to have a program, that is, to obtain funds and to obligate those funds through loans or grants. This objective frequently overrides all others, including the rational allocation of "scarce" resources among competing host country developmental needs.

3. Other things being equal, decision makers have institutional incentives to seek out and use information about local farming systems (or any other type of information) only to the extent that, in their experience, it will contribute to the objective of "moving money" in a timely and efficient manner.[3]

4. Donor organizations are not monolithic, nor are their decision-making processes unitary. Structural incentives for using, ignoring, or distorting information about local farming conditions must therefore be examined in relation to a variety of decision-making contexts. The most

[3] It should be stressed that by institutional incentives I refer to only those incentives that arise because of an employee's role in a developmental organization and not to his or her personal motives.

important ways that these decision-making contexts will be differentiated for the purposes of this discussion are according to the "location" of the decision makers in the organization and the "location of the decision" in the standardized series of stages through which policy is translated into budget and development allocation decisions.

5. Incentives for advocating careful farm-level analysis and feasibility studies are greatest in central offices responsible for policy formulation and weakest in operational, field-oriented units responsible for designing and implementing projects. This pattern exists because the central policy office needs documentation that will convince external constituencies and critics, who can affect the agency's funding levels, that all of its activities are sound. The policy office does not, however, have to conduct the analyses.

The operational units, by contrast, are constrained by the need to obligate amounts of money that often are determined rather arbitrarily and may fluctuate unexpectedly to satisfy or at least to placate a variety of host country interest groups, to take account of international political relationships, or to conform to complex project design, documentation, and approval procedures. It is hardly surprising that the operational units are often unenthusiastic about centrally mandated requirements for time-consuming analyses that may raise host country sensitivities and may restrict the options for spending.

6. Decisions concerning project location, commodities, technical inputs, and host country implementing institutions are generally made on an ad hoc basis to take advantage of what donor agency staff perceive to be "targets of opportunity."

7. In making these kinds of major decisions in the early stages of project design, decision makers do not normally make probabilistic judgments in which they assign weights to alternative courses of action. Instead, they use a simplifying heuristic entailing a series of conditions that must be met if they are to move ahead with the project.

8. To avoid the risk of innovative failure, and because of the complexity of project design requirements, new projects and their major components are usually modeled on earlier projects with which the donor agency's staff or contractors are familiar.

9. Detailed economic, financial, environmental, and social analyses are then carried out, in large measure, to justify the decisions already made in documentation required by the donor to demonstrate that the project conforms to policy and is sound.

10. Since the initial decisions raise expectations among host country counterparts and represent a commitment of scarce staff time, major

changes in project design tend to be viewed as costly and are resisted, even if analyses indicate that the initial decisions are suboptimal.

11. Project documents are advocacy documents designed to obtain approval and do not necessarily reflect either the information available to project designers or the actual decision-making process. Whereas cost–benefit and other types of economic analyses are used to lend elegant authority to project documents, economic analyses generally play no more role in project-level decision making than do insights from other sources.

Despite these institutional constraints, the recognition of small farmer rationality at the policy level is beginning to increase sensitivity to local conditions and needs in agricultural development programs. Social scientists who want to work directly or indirectly with development agencies should pay more attention to the organization and decision-making processes within the agencies themselves to identify the best points to introduce new ideas and information.

The Ideological Setting

The increased attention currently being given by donors to the rationality of low-income producers' decision-making processes and to the dynamics of local farming systems in agricultural development policy, reflects a change in the understanding of small farmers' behavior as well as changes of emphasis in concessionary foreign assistance.

How Did the Peasant Become Rational?

Until recently, development planners and a majority of scholars concerned with development assumed that the agricultural practices of low-income rural people are governed by tradition, change only slowly, and are often poorly adapted to local conditions. Moreover, it was assumed that traditional rural societies were more or less static, and that their institutions must be broken down or greatly modified because they were constraints on more rational development.

Today, by contrast, leading scholars in diverse disciplines, including agricultural and developmental economics, anthropology, economic history, human geography, and rural sociology recognize that low-income producers' behavior must be understood as the result of recurrent decisions about the use of productive assets, the organization of labor, marketing, saving, and investment;[4] that experimentation with new crops and crop mixes is commonplace and attempts to introduce major technological innovations not unusual, even in communities beyond the reach of extension services;[5] and that many indigenous small-scale farming systems are sensitively adjusted to local ecological, economic, and political conditions—and their fluctuations.

Finally, development planners, as well as social scientists, are coming to understand that agricultural development programs are unlikely to reach low-income intended beneficiaries unless they take account of the strengths of existing ecological, social–cultural, economic, and political institutions, which persist because they meet real needs.[6] New technologies and organizational forms will be accepted only if they meet these needs more effectively.

Why Are Donors Concerned with Peasants?

During the past decade, there also has been an increasing emphasis in developmental policy on agriculture and, in particular, on the involvement of low-income people in agricultural and rural development. During the 1960s, development assistance was largely concentrated on the urban, industrial sector on the assumption that this would stimulate such higher savings and growth rates that the benefits would eventually "trickle down." Technical assistance and capital assistance to agriculture was used primarily to strengthen agricultural research, education, and extension at the national level. Other types of capital and physical infrastructure projects, such as transportation and fertilizer plants, also indirectly affected agriculture.

By the end of the decade, it became clear that, even in developing countries that were achieving high growth rates in their gross national product, the trickle-down effects were not working. The poor were as

[4] Recent major case studies and comparative works that illustrate this conclusion include Barlett 1978; Brush 1977; Chibnik 1974; Epstein 1962; Greenwood 1973; Johnson 1971; Just 1975; Lipton 1968; Norman 1971, 1974; Ortiz 1973, 1976; Schluter 1976; Schultz 1964; Scott 1976; Shanin 1971; Wharton 1971.

[5] Detailed studies demonstrating that peasants regularly experiment with technological innovations have been carried out in all geographic regions. Excellent works of this type include Barlett 1977; Berry 1975; Cancian 1972, 1979; Cole and Wolf 1974; Collier 1975; De Walt 1979b; Greenwood 1976; McLoughlin 1970; Moerman 1968; Netting 1968, 1969, 1974, 1978.

[6] Studies illustrating this point include American Society for Agronomy 1976; Bandong 1977; Binswanger, Krantz, and Virmani 1976; Binswanger 1977; CATIE 1978; Harwood 1975; Hildebrand 1976; IBRD 1974, 1978, 1979; Kass 1978; Lele 1975; NRC 1977a, 1977b; SAREC 1979; Tourte 1974; USAID 1978a, 1978b, 1978c.

badly off as ever in terms of income, underemployment, infant mortality, and nutrition. It also was evident that the increasing economic dualism between the "modern" and traditional sectors was being exacerbated by policies that subsidized capital-intensive industrial growth at the expense of agriculture.

Within agriculture, investment was also skewed. The emphasis was on capital-intensive high technology farming for the production of cash, export-oriented crops. Agricultural research, education, extension, credit, and other investments tended to benefit high-income farmers, and, in some instances, to further imbalance the distribution of land against the small holder.

In response to these growing concerns, most major donors have begun to give greater attention, at least in their policy and rhetoric, to income distribution and employment, to agriculture and rural development, to food crops, and to the use of more labor-intensive appropriate technologies in agricultural development. The major elements of this new approach to agricultural development were incorporated into the United States Foreign Assistance Act through amendments in 1973 and in 1975, mandating AID to shift the focus of its funding toward agriculture and toward low-income farmers.

Recognition that small farmers are reasonable decision makers made the task of development planning in AID and other agencies more difficult than it had been when macroeconomic theories held sway. The variability of the ecological, social, microeconomic, and political contexts in which peasants make decisions adds great complexity to the planning process.

The Organizational Setting

What Does AID Do?

The Agency for International Development serves as the major conduit for United States bilateral foreign assistance. In fiscal year 1978, AID's program budget for developmental assistance was approximately $1 billion, of which just under $600 million was in the functional account designated as "Agriculture, Rural Development and Nutrition." An overview of activities and trends within this account, which is also referred to somewhat optimistically as "Food and Nutrition" within AID, are shown on Table 14.1. AID also administered approximately $2¼ billion of economic assistance appropriated under the Security Supporting Act. Some of this assistance is also used for agriculture

TABLE 14.1

A.I.D. Program Levels—Agriculture and Rural Development ($ Millions)

	FY 1975 $	FY 1975 %	FY 1976[a] $	FY 1976[a] %	FY 1977 $	FY 1977 %	FY 1978[b] $	FY 1978[b] %	FY 1979[b] $	FY 1979[b] %
1. Asset distribution and access	6.9	1.1	16.3	2.4	22.2	4.3	35.0	6.0	54.5	7.4
a. Land tenure	(.4)	(.1)	(2.6)	(.4)	(2.6)	(.5)	(5.8)	(1.0)	(5.9)	(.8)
b. Local participatory institutions	(6.5)	(1.0)	(13.7)	(2.0)	(19.6)	(3.8)	(29.2)	(5.0)	(48.6)	(6.6)
2. Planning and policy analysis	9.8	1.6	15.1	2.2	17.1	3.3	32.4	5.5	65.4	8.9
3. Development & diffusion of new technology	53.2	8.7	89.0	13.0	77.1	15.1	131.5	22.3	141.9	19.2
a. Centrally funded research	(3.5)	(.6)	(6.0)	(.9)	(9.5)	(1.9)	(17.5)	(3.0)	(24.4)	(3.3)
b. International centers	(10.5)	(1.7)	(15.7)	(2.3)	(20.6)	(4.0)	(24.0)	(4.1)	(26.6)	(3.6)
c. Bilaterally funded research	(15.4)	(2.5)	(18.0)	(2.6)	(22.0)	(4.3)	(48.6)	(8.2)	(52.9)	(7.2)
d. Education and extension	(23.8)	(3.9)	(49.3)	(7.2)	(25.0)	(4.9)	(41.5)	(7.0)	(38.0)	(5.1)
4. Rural infrastructure	238.8	39.0	205.0	30.0	233.4	45.7	246.0	41.7	311.8	42.3
a. Land and water development and conservation	(137.0)	(22.4)	(146.3)	(21.4)	(134.9)	(26.4)	(115.3)	(19.6)	(163.3)	(22.1)
b. Energy, including rural electrification	(25.6)	(4.2)	(.3)	—	(36.1)	(7.1)	(57.3)	(9.7)	(78.0)	(10.6)
c. Rural roads	(76.2)	(12.4)	(58.4)	(8.6)	(62.4)	(12.2)	(73.4)	(12.4)	(70.5)	(9.6)
5. Marketing and storage, input supply, rural industry, and credit	303.4	49.6	357.0	52.3	161.9	31.6	144.7	24.5	164.1	22.2
a. Marketing and storage	(32.9)	(5.4)	(38.4)	(5.6)	(40.3)	(7.9)	(20.2)	(3.4)	(39.3)	(5.3)
b. Input supply	(200.0)	(32.7)	(234.5)	(34.4)	(86.6)	(16.9)	(71.3)	(12.1)	(80.8)	(11.0)
c. Rural industry	(23.2)	(3.8)	(12.5)	(1.8)	(25.7)	(5.0)	(15.0)	(2.5)	(2.4)	(.3)
d. Credit	(47.3)	(7.7)	(71.6)	(10.5)	(9.3)	(1.8)	(38.2)	(6.5)	(41.6)	(5.6)
Total	612.1	100.0	682.4	100.0	511.7	100.0	589.6	100.0	737.8	100.0

SOURCE: Figures based on Development Assistance appropriation, Food and Nutrition account, excluding nutrition and program development funds.

[a] FY 1976 figures include transitional quarter.

[b] FY 1978 and FY 1979 figures include Sahel programs. Columns may not add due to rounding.

and rural development, but much of it is still directed to urban-oriented, large-scale capital projects, commodities, and budget support.

AID currently (Fall 1979) has Development Assistance programs in over 60 countries. Programs are concentrated in low-income countries, though some middle-income countries still have small programs. Security Supporting Assistance is concentrated in Egypt, Israel, the Philippines, and other nations in which the administration and the Congress believe the United States has strategic interests. While AID's organization, procedures, and decision-making processes are very similar in both types of programs, the present discussion is primarily concerned with the less overtly political Development Assistance programs.

Most AID Development Assistance resources are presently transferred to host countries through the vehicle of "targeted" projects, rather than through technical assistance to a ministry, a program, or general budgetary support. A project is a discrete effort, usually lasting from 3 to 5 years. It requires extensive documentation in accordance with standardized regulations. It must be approved individually for funding by the Congress and, once approved, must be implemented by a contractor working with a host country organization. Resources normally provided through AID programs include: technical assistance by United States experts; short- or long-term participant training for host country nationals; commodities, such as fertilizer or contraceptives; equipment; and construction costs. AID also plays a central role in allocating PL 480 food aid for relief and rehabilitation, for Food-for-Work projects, and, more recently, in support of host country policy changes in areas such as agricultural price policy. AID's formal organization and programming procedures reflect its official task of translating congressional guidance and budgetary appropriations into projects.

How Is AID Organized?

The major units in AID's formal organization are nine *bureaus,* headed by *assistant administrators* who report to the appointed head of the agency, the *administrator.* Four of these are *regional bureaus* with direct line responsibility for programs in the countries under their jurisdiction: *Africa, Asia, Latin America and the Caribbean,* and the *Near East.*

The other bureaus provide support and liaison functions. They include: the *Program and Policy Coordination Bureau,* which is responsible for translating legislation into policy guidelines and making sure that they are reflected in AID's annual budget submission; the *Bureau for Intergovernmental and International Affairs,* responsible for liaison with other donors, United States agencies, and for analyzing international aspects of economic development; the *Bureau for Program Management and Services,* which provides "housekeeping" services; the *Bureau for Development Support,* which funds research and provides technical support to the regional bureaus and the "field"; and the *Bureau for Private and Development Cooperation,* with responsibility for miscellaneous programs, such as Food for Peace, and support for private and voluntary organizations.

Each bureau is divided into *offices,* many of which are further subdivided into *divisions.* There are also nine other specialized offices that report directly to the administrator, including the *Office of the Administrator,* the *Office of the Executive Secretary,* the *Office of Public Affairs,* the *Office of the General Counsel, Personnel Management,* the *Office of Equal Opportunity Programs,* the *Office of Financial Management,* and other minor miscellaneous units.

Finally, there are overseas *missions* in most of the developing countries in which AID has a Development Assistance program. In principle, mission directors report to the head of their regional bureau, although, as political appointees, some have influential support in Congress or in other domestic constituencies that enables them to exercise a degree of autonomy from AID–Washington.

AID's hierarchical structure is generally four to five levels in depth, and its internal organization is characterized by recurring cleavages between "Washington" and "the field," "function" (e.g., agriculture, health, population) and geographic region, as well as the usual line–staff and technical–managerial distinctions.

The Agency's organization also exhibits much functional redundancy. There are, for example, units with overlapping responsibilities for agricultural policy and programs in the Policy Bureau, the Development Support Bureau, in all of the regional bureaus, and in each of the overseas missions. Because of their overlapping interests and jurisdictions, personnel in units of this type spend a considerable proportion of their work time in liaison, in disputing one another's opinions, and in reviewing each other's project, program, policy, and research documents.

This situation is not helped by the fact that many employees assigned to functionally specialized units and positions do not, in fact, have appropriate professional qualifications. This is not because AID employees are inherently poorly qualified; indeed, as a group they are well-educated, with more than one-third holding advanced degrees; but because AID's personnel system simply reclassifies people to meet changing demands without giving them additional training. An AID agricultural officer is, therefore, not necessarily trained or experienced in agriculture.

In all, AID has approximately 2000 professional (nonclerical) employees, of whom about one-half are assigned to overseas missions at any time. This represents a much greater emphasis on field staff than is found in most other donors and is considered, rightly or not, to give the agency a comparative advantage in knowledge of host country conditions and in the delivery of technical (as opposed to purely financial) assistance.

AID's organization is not so much the result of rational planning and efficient functional differentiation as it is of history. The birth and death, rise and fall, and transfer of particular organizational units must be understood against a background of Byzantine bureaucratic maneuvers played out as successive waves of reorganization have swept over the agency.

It is hardly surprising, then, that many less formal types of organization and association have arisen within the agency that influence the ways decisions actually are made. Structured conflicts of interest and difficulties in communication generated by competition for staff, budget, and jurisdiction are counterbalanced by the continual formation of ad hoc problem-oriented working groups and task forces, by informal crosscutting ties between employees, based on previous association, by common interests in development or common professional or regional background, and by quasi-groups centering around influential individuals.

The situation in regard to employees' role expectations is similar. Indeed, the formal rules and procedures governing vital activities—such as the recruitment, promotion, and overseas assignment of agency personnel; project design and approval; and the selection of contractors—are so cumbersome that little could be accomplished without such an informal set of understandings.

These informal forms of association and behavioral expectations influence styles of leadership and decision-making processes. In this environment, a successful AID bureaucrat is one who knows how to "work the informal system" in order to mobilize the personnel, contractors, and fiscal resources required by the task at hand, despite the impediments of AID structure and regulations. A powerful bureaucrat is one who has these operational skills, who, like a Yako elder, is a member of the task-oriented, crosscutting working groups, possesses a well-developed information network, can defend his bureaucratic "turf," who is adept in the use of the current idiom and symbols of the development subculture, and who can use all these skills to initiate new programs that obligate significant funds.

These complex patterns of organization and leadership, combined with functional redundancy of units, geographic dispersion of decision makers, and the reclassification of employees without serious regard for their qualifications all contribute to a rather decentralized and diffuse process of decision making. Virtually all resource allocation decisions are made by committees and reviewed by still other committees, usually made up of representatives of several organizational units. Under these circumstances, decision-making processes involve strategies of cooptation and trade-offs. The outcome of this process of decision making is a low degree of individual accountability, and a Gresham's law of information use, as decision makers are forced to find a lowest common denominator of shared knowledge and agreement about complex issues in distant lands.

How Does AID Allocate Resources?

In AID, as in other major donors, the decision-making process through which resources are allocated involves the preparation of three types of documents: policy guidance, country program analyses, and project-specific papers. Policy guidance states the Agency's goals and objectives, based on the Foreign Assistance Act. Country program documents analyze the development assistance requirements of individual countries in light of AID policy. Project-level documents specify in great detail how resources are to be used to solve particular problems.

The preparation and review of policy and program-budget documents is the central activity for AID's line organizational units, the regional bureaus, and the missions. It occupies much of the time of their employees and takes precedence over all other activities. It affects their career incentives by rewarding them for procedural and tactical knowledge, for becoming experts at moving money. It is also the basis of most interaction, complementarity, and tension between different parts of AID's organization, and provides the main decision-making arenas for resource allocation in all sectors. Virtually all AID project documents are prepared collaboratively by a team of AID and contract personnel. Resource allocation decisions, explicit or implicit in project documents, therefore represent the result of a complex decision-making process, rather than the analytical thought of an individual. Often these "committee"-type decisions seem to be equal to less than the sum of their parts for reasons that will become clear in the following sections.

Because document preparation and review is closely tied to the Agency's annual budget cycle, it governs the nature and pace of work during the course of the year, much as the agricultural cycle governs the life of the farmer. At the same time, the cyclical nature of program budget preparation affects employees' incentives for gathering and analyzing data on local conditions, and places arbitrary and stringent limitations on the time frame during which information of this type can affect allocation choices.

Formally this process begins in the spring with the passage by Congress of AID's annually updated *Authorization Bill,* which spells out the purposes for which the agency is to use its budget. The Bureau for Policy Planning and Coordination is then responsible for interpreting the legislation, for incorporating it into various documents to guide the preparation of country programs and projects.

Each AID overseas mission is required to prepare a document, currently termed the *Country Development Strategy Statement* (CDSS) by the end of the following January, in accordance with the guidelines from the policy bureau. The purpose of the CDSS is to describe and analyze the host country's development problems, to assess the commitment and capability of the host country government to deal with these problems, and, taking into account other donors' plans, to develop a 5-year strategy that adapts AID policy to the country's needs. The CDSS does not specify the particular projects through which the mission's strategy is to be carried out, but it indicates whether the projected *Indicative Budget* levels for the period, which have been set for that country in Washington, according to a simple formula, are adequate, too high, or too low.

The mission next prepares a brief *Project Identification Document* (PID) for each of the projects it proposes to develop. The PID is sent to Washington, where it must be reviewed, modified, and approved by the relevant regional bureau and the Policy Bureau, with the participation of technical experts from the Development Support Bureau. If the PID is approved, the project it adumbrates is incorporated into AID's *Annual Budget Submission,* prepared during the summer and submitted in the fall to the Office of Management and Budget for inclusion in the president's executive budget. If it survives these hurdles, the project is presented to Congress in the administration's *AID Appropriation Bill* for funding the following spring. Thus, policy to project approval requires a minimum of 2 years.

As soon as a project PID is approved in AID–Washington, the mission, with Washington support, can begin work on the *Project Paper* (PP). This is a much longer planning document that describes and analyzes the project in detail, states its expected purpose and goal as well as its inputs and outputs, contains a detailed budget, an implementation plan, and economic, financial, environmental, and social feasibility analyses. The project design represents the fine tuning of AID policy and country strategy to a particular setting.

It is only after the PP has been approved that AID can issue a *Request for Proposals* to solicit bids for project implementation from prospective contractors and negotiate a *Program Agreement* with the host country. During the period of project implementation the mission must submit a *Project Evaluation Statement* annually and an indepth *Evaluation* at a scheduled date. In principle, these documents help determine whether project funding should be continued or extended. In all, it takes at least 2 years for a project to pass through the program and budget cycle from the PID stage to the beginning of implementation—usually considerably longer. For changes in the legislation to be reflected in completed projects, it takes at least 5 years.

How Does AID's Institutional Environment Affect Organizational Objectives and Individual Incentives?

AID's institutional objectives and individual incentives have been profoundly affected by the ambiguity of its goals, perceived uncertainty of its institutional environment, and its vulnerability to criticism. This analysis of the way AID goals have been displaced by the need to move money draws heavily upon Tendler's (1975) brilliant discussion of the topic.

Throughout its "temporary" existence, the Agency has operated with a sense of uncertainty in regard to levels and continuity of funding, scope of operations, geographic areas of concentration, context, and emphasis, due to congressional vagaries in funding and the administration's foreign policy imperatives. At the same time, the Agency has been open to criticism from Congress, other government agencies, and other domestic interest groups which are not necessarily well informed or interested in development. Generally, this criticism has been based on GAO audit reports and inquiries into inefficiency, poor auditing, or the misappropriation of funds; it seldom focused on the effects of AID programs on development or on those who were to benefit from it. Significantly, criticism from United States and host country academics, and professionals with technical competence in development, has had little impact on the agency's institutional environment, and virtually no provision has been made until very recently for the feedback of positive or negative information from low-income peoples affected by AID programs.

Under these external institutional pressures, the Agency's legislated goal of fostering development has been displaced by both its means (obligating funds, having projects, increasing its personnel complement, getting larger appropriations) and, at times, by the goals of other agencies and interest groups (for example, the Treasury, USDA, the State Department, the Department of Defense, and domestic grain and oilseed producers).

Whereas the more political types of goal displacement have received more attention from AID's critics, the more enduring pressure to move money, to have programs, has more profoundly affected the agency. This pressure exists and affects incentives in any money-spending bureaucracy. In AID, however, organizational output has been defined largely in terms of its ability to obligate funds. This pressure has had a pervasive effect on four areas of AID functioning. One is the way AID classifies, reclassifies, and rotates its employees with little serious regard for their professional qualifications, experience, or country knowledge. Its system of project classification and record keeping ensure institutional amnesia in regard to what was tried in the past and whether or not it worked. Emphasis is placed on designing projects, rather than on implementing or evaluating them, and on the formal and informal standards used to judge achievement. For example, a regional bureau is regularly judged by its obligation rate, a mission director by the annual increment in his program, and efficiency is normally defined by the ratio of funds obligated to operating budget.

The Decision-Making Process

In principle, the process of program and project development entails a hierarchical, sequenced series of choices concerning the allocation of resources. Choices made early in the sequence involve a wider range of alternatives—for example, between countries, sectors, and regions—and require rather general types of data. Choices made later in the sequence involve a more restricted set of alternatives—for example, between crop varieties, techniques for extension, or user cost rate structures—and require more specific kinds of data.

Ideally, the appropriate type of information should be integrated with the decision-making process at each stage of program development. In practice, this rarely happens because a project is not, in fact, the product of a series of decisions in which options are progressively eliminated.

In reality, many projects originate with a "target of opportunity," which may be a host country request, the politically determined or ad hoc selection of a region, a commodity, a technology, or a host country ministry. Under these circumstances, higher-order alternatives are precluded from the outset. Nevertheless, the documentation requirements of AID's program and budget cycle require that much effort be devoted during the project design to rationalizing, post hoc choices that, in fact, were never considered. Anthropologists, economists, and other analysts brought into project design under these circumstances are likely to find themselves in an adversary role with their advice unheeded.

Analyses included in a PP to justify choices already made to protect a project during review are referred to, in-house, as "boiler-plate." A good amount of cutting and pasting goes into their production, with choice material occasionally even borrowed from previous project papers that have withstood the test of Washington scrutiny.

Major mission-level project identification decisions generally are not based on probabilistic judgments of all relevant variables in alternative courses of action but rather on simplifying heuristics.[7] At the project identification stage, the heuristic takes the form of an implicit list of questions, all of which should be answered in the affirmative if the mission is to invest time in preparing a PID and submit it to Washington. Since these questions reflect the unique constellation of constraints facing a particular mission director at a particular point in time, the exact content of the heuristic is variable. Characteristically, however, the mission director and his key aides must ask:

1. Is the proposed project consistent with effective AID policy, that is, the policy embodied in Washington project approval decisions, rather than in policy papers?

---

[7] The critical role of simplifying "cultural paradigms" or cognitive models in decision-making processes is now widely recognized in many disciplines. Slovic, Fischhoff, and Lichtenstein (1977) note that recent psychological research indicates that when people perceive, process, and evaluate the probabilities of uncertain events they are not "intuitive statisticians" but instead use simplifying heuristic models and may persist in using them despite major information-processing deficiencies. Allison (1971) makes a similar point about decision making in foreign policy. Gladwin, an economist studying the process by which farmers decide whether to adopt new technology, notes that people do not make complex calculations of the overall utility of each alternative but use procedures that simplify their decision-making calculations (Gladwin 1976). The same point has been made by Barlett (1977) and Quinn (1975, 1978).

2. Is it consistent with the mission's analysis (in its Country Development Strategy Statement) of the way that AID policy should be adapted to host country conditions?
3. Is it acceptable to host country political leadership?
4. Is it acceptable to a host country ministry or agency that will be responsible for implementing it? (Donors often engage in wishful thinking in answering this question, only to find out much later that the implementing agency selected has a resounding lack of interest in the project and consequently lets it languish in the backwaters of the bureaucracy. Tendler (1975:103) notes the tendency of donors to exclude the host government from the project identification and design process in order to assure themselves a reliable supply of projects.)
5. Will the project complement or balance the mission's "portfolio" of projects? For example, a mission that has a strong program in agriculture and health-care delivery may desire projects in population or education. This desire for a balanced or at least a mixed portfolio is in part a reflection of AID's congressionally mandated "functional accounts" and partly a risk-aversion strategy on the part of the mission director who does not want to put all his eggs in one sectoral basket.
6. Is the cost of the project consistent with the mission's budgetary levels or aspirations?
7. Does the mission have a sufficient work force with appropriate skills to manage the labor-intensive process of project design, and what are the opportunity costs of this use of staff time?
8. Are there likely to be any special objections to the project raised by the United States ambassador or particular members of Congress?
9. Has AID designed and implemented similar projects previously?

At no point in the decision-making process is the question of what the farmers want and need necessarily given high priority.

A Project Identification Document usually contains only a brief and generalized statement of a problem and a proposed solution. If approved, it constitutes a claim on AID's fiscal resources and a commitment to a particular host government agency or to political officials to deliver more or less well-specified resources. Because of this commitment, and because scarce mission staff time has been invested in PID preparation, it is very difficult to stop the process of project design and implementation after the initial PID is approved (though it is possible to alter it considerably), no matter how suboptimal the proposed activity turns out to be upon further analysis.

The task of producing a Project Paper is enormously complex and time consuming because of the specialized supporting analyses required, which inevitably exceed the time and talent of mission personnel, and the degree of logistic and financial information that must be included. Moreover, much of the design work is done by poorly coordinated visiting teams of experts from AID–Washington or universities and must be completed to meet arbitrary deadlines imposed by the program budget cycle.

To cope with this situation, project designers tend to use past projects or parts of projects and the documents associated with them as models. Much of the decision making in project design is consequently a matter of choosing between alternative models brought to the attention of the project design team leader and the AID project design manager.

In part, past projects are used as technical, logistic, and financial guides. Thus, for example, it is possible to plug in a seed multiplication station, an extension unit, a credit component, or a farmer training center into an agricultural project simply by lifting a section from another PP and adjusting the scale of operations and costs to local conditions.

AID projects are also both implicitly and explicitly posited on theories of cause and effect and hence also serve as conceptual guides. Regardless of whether they are based on experimental evidence, disciplinary dogma, past experience, or merely professional folklore, the theories inherent in past projects have an important cognitive, evaluative, and expressive role in the world of the developer.[8] For example, some agricultural specialists believe that rural poverty derives mainly from the low productivity of labor and therefore seek means of raising that productivity. The productivity of land, the supply of labor, or the distribution of income do not enter into their analysis of the situation.

These paradigms of and for development have provided the personnel of donor agencies with shared ways of thinking and talking about what they are doing and of explaining why they believe it will work to those upon whom they depend for funding. Like other models, they not only provide criteria for choosing between alternatives, but they define these alternatives and hence the kinds of information that are considered relevant. In this way, they generate their own categories of data, which lend to their adherents a comforting aura of concreteness. For example, the "model farmer" paradigm, which held sway

[8] Whereas anthropological concepts and methods are helpful in analyzing the role, incentives, and strategies of decision makers, the unique contribution of anthropology to research on bureaucracy, and of research on bureaucracy to anthropology, is in analyzing the ways that implicitly held "cultural paradigms" structure decisions and affect information processing.

recently, rested on the self-fulfilling assumption that progressive farmers have larger landholdings because they are progressive, whereas, on the other hand, small holders are inherently more traditional. Aid, therefore, should be given to those who have the attribute of being progressive. Alternative hypotheses concerning the political–economic bases of wealth were not explored, nor were data generally gathered that could have tested them.

Like other long-used conceptual paradigms, they are not challenged easily by merely factual evidence of failure, for like the Trobrianders' magical beliefs, they provide a rationale for explaining away their apparent lack of success and for displacing the blame onto others. For example, since it is often assumed that pastoralists are not price responsive, their failure to sell livestock in a marketing project is taken, *prima facie*, as evidence of their traditional values, and more rational explanations are not sought. Moreover, in the world of development, with its many conflicting constituencies, the voices of low-income intended beneficiaries are not clearly heard. In short, there has been little feedback from project experience that could effectively challenge the validity of past project models. Past projects, or parts of them, thus continue to be used as models for future projects, as long as they do not create undue problems in implementation, or bring public censure upon the donor.

In a study of the design of AID pastoral livestock projects in Africa, for example, I was able to determine that most livestock projects involving livestock-dependent populations in arid zones fail to achieve their objectives, or even exacerbate the problems they address, because they are based on incorrect assumptions about the nature of pastoral systems—about the range, water, and herd arrangement strategies of pastoralists, and about the relationship of pastoralism to the wider environmental, political, and economic context in which it is found. I was able to trace this recurrent problem to the fact that AID classifies all of its pastoral zone projects in the livestock subsector of agriculture. Because of the professional background and experience of host country persons, donor specialists, consultants, and contractors working in this subsector, livestock production and land use management, rather than the nutrition, health, security, or income of pastoralists, become project objectives. In the words of one senior AID official, "Cattle, rather than people, are treated as the target population."

Furthermore, because of the experience of those involved, the primary focus of livestock projects is almost invariably on cattle, rather than sheep or goats, and on beef production, rather than dairy products (including ghee and cheese) or hides. This is true regardless of pastoralists' actual preproject income sources and strategies, the risks to which they are exposed, or the ways they cope with them (Hoben 1979). Despite widespread agreement among observers that pastoral livestock projects do not succeed in meeting their environmental, production, or income-generation objectives, more than $600 million have been spent on them in sub-Saharan Africa by all donors combined.

The Challenge of the New Directions Congressional Mandate

The Legislation

In 1973 and 1975, the Congress introduced major changes into AID's authorization bill that reflected the emerging perspectives on the rationality of "peasants" and their role in agricultural and economic development discussed in the previous sections. These changes, generally referred to in the agency as the "New Directions" or the "congressional mandate," emphasize income distribution as well as growth, the selection of labor-intensive "appropriate technologies" and other means of employment generation, the participation of low-income intended beneficiaries in decision making, and the need to adapt programs to local ecological, social, and cultural conditions. More recent amendments to the legislation add emphasis to helping people meet their "basic needs," including adequate nutrition, shelter, clothing, health service, and education.

The law is quite explicit. Section 102 of the Act, as amended, states that:

> the principal purpose of United States bilateral development assistance is to help the poor majority of people in developing countries to participate in a process of equitable growth through productive work and to influence decisions that shape their lives, with the goal of increasing their incomes and their access to public services which will enable them to satisfy their basic needs and lead lives of decency, dignity, and hope. Activities shall be emphasized that effectively involve the poor in development by expanding their access to the economy through services and institutions at the local level, increasing labor-intensive production and the use of appropriate technology, expanding productive investment and services out from major cities to small towns and rural areas, and otherwise providing opportunities for the poor to improve their lives through their own efforts [Foreign Assistance Act of 1979, Section 102 (a)].

A special section on agricultural research adds that:

> Agricultural research carried out under this Act shall (1) take account of the special needs of small farmers in the determination of research

priorities, (2) include research on the interrelationships among technology, institutions, and economic, social, environmental, and cultural factors affecting small farm agriculture, and (3) make extensive use of field testing to adapt basic research to local conditions [Foreign Assistance Act of 1979, Section 103 (a)].

The incorporation of these changes into the Foreign Assistance Act was the result of a well-coordinated effort by a comparatively small group of congressmen and staffers who were able to mobilize political support from rather diverse interest groups. Foreign aid has never been popular with the electorate and has required careful shepherding through Congress each year by the administration for which aid is, among other things, a useful instrument of foreign policy. In the post-Vietnam neoisolationist mood of the period, there was more than the usual public resentment against spending tax dollars on programs that seemed to benefit only foreign elites, and the aid bill was in jeopardy of being eliminated altogether. The New Directions legislation's rhetorical emphasis on strengthening independent family farms and local democratic institutions, harking back to midAmerican agrarian utopian ideals, made the aid bill politically safer to vote for. At the same time, it appealed to some more conservative elements in the Congress and the administration who were alarmed at the political destabilization that seemed to be resulting from the rapid but imbalanced urbanization and industrial growth of the 1960s.

A brief examination of the ways that AID decision makers have responded to the congressional mandate serves at once (a) to explain the apparent contradiction between AID policy and practice in regard to beneficiary participation in decision making and to matching projects to local, social, cultural, and ecological conditions; (b) to illustrate the way that different decision-making contexts influence the use of information in AID; and (c) to indicate the ways in which AID is, in fact, beginning to take greater account of information about local farming systems and decision-making processes.

The Program and Policy Coordination Bureau Response

The passage of the New Directions legislation put The Program and Policy Coordination Bureau (PPC) in the difficult position of trying to ensure that the work for which it could be held directly accountable by congressional watchdogs was in at least apparent conformity with the law without interfering with the agency's internal imperative of obligating funds or its external role as a United States State Department instrument of the administration's foreign policy. Under these conflicting pressures, it is not surprising that PPC has been more successful in introducing the mandate into its guidelines than in enforcing its provisions through its program and project review and budgetary control functions.

The new approach called for by the mandate was not widely understood, desired, or even accepted by AID personnel. Many, especially those in managerial positions, regarded the New Directions as just another "barnacle" on the already encrusted Foreign Assistance Act—a nuisance that would increase the workload per dollar obligated still more, and hence would have to be dealt with as expeditiously as possible.

AID's role as an instrument of foreign policy also made the introduction of the New Directions more difficult. The mandate's enthusiasm for the popular participation of the poor majority in decision making is not always shared by host government leaders, nor is it always consistent with short-term United States foreign policy objectives. The allocation of aid to repressive regimes that do not tolerate, much less encourage, popular participation indicates to AID employees as well as to outside observers that the rhetoric of official policy guidelines from PPC are not necessarily effective policy.

The Program and Policy Coordination Bureau, then, faced a number of problems in attempting to comply with the mandate. The most general problem was the paradox of introducing "bottom up" and participatory decision making into the agency's programing by means of "top down" centralized administrative control. This problem was compounded by the fact that AID's missions must work collaboratively with and through host country government elites.

To make matters worse, the normal bureaucratic resentment against PPC because it occupied a position "upstream" on policy and budget had been exacerbated by the AID administrator's attempt to introduce into the Agency a complex, centralized management and data-processing system under PPC's supervision. The Program and Policy Coordination Bureau was thus already seen as the source of a steady stream of bothersome new paperwork requirements. Moreover, for historical reasons, the Agency had few field-oriented social scientists or area studies experts on its rolls, and its promotion and assignment procedures had effectively discouraged most employees from acquiring indepth knowledge of a country or region.

The Program and Policy Coordination Bureau had a variety of well-known bureaucratic means for responding to the New Directions legislation. The most important of these were: issuing general policy statements; establishing substantive and procedural guidelines for the preparation of project and country program documents; reviewing documents prepared by other bureaus for conformity with the guidelines; enforcing conformity with the substance of the guidelines by withholding budgetary approval; recruiting staff with requisite skills; and sponsoring research. Although all of these means were used to some extent, the relative emphasis on each was consistent with PPC's need to take account of the constraints already discussed. In purely strategic terms, regardless of individual employees' personal commitments, the bureau had to (a) show evidence of compliance; (b) make sure that all AID documents used to support resource-allocation decisions make reference to New Directions objectives; (c) devolve the responsibility for compliance with the legislation, and most of the work it entails, onto the regional bureaus and overseas missions; and (d) ensure that the implementation of the new policies did not prevent the agency from obligating appropriated funds.

From PPC's standpoint, general policy statements are a very attractive means for responding to new legislation. They are effective for demonstrating compliance, since they reach an audience much broader than AID, and yet, because of their generality, they seldom place rigid restrictions on what AID missions can do in particular cases. As they are unrestrictive in application, they are unrestricted by pragmatic and political considerations. Usually they are well written, well researched, and up-to-date on the state-of-the-art literature in relevant disciplines. They provide a sensitive discussion of complex issues and generally reflect the spirit and the letter of the law quite closely. AID's current Agricultural Development Policy Paper exemplifies all of these characteristics and gives a ringing endorsement to the major provisions of the New Directions legislation.

Guidelines requiring that special analyses be included in project documents are another widely used way of responding to new, externally imposed policy initiatives. Issuing such guidelines is not only a sign of compliance in itself, but it also ensures that all project papers will be in at least apparent conformity and, at the same time, it devolves the responsibility for preparing the analyses to missions and regional bureaus, for in effect it requires *them* to certify that their projects conform to policy.

Requiring additional analyses makes the process of project design more labor intensive, and this, in turn, forces missions to develop larger projects, for labor, and not financial resources, are perceived as the scarce factor in the production of AID programs. Paradoxically, requirements for analyses that will fine tune projects to local conditions may, in this way, result in larger projects that are, in fact, less responsive to local needs and local participation. However, the imposition of new project design requirements does not usually threaten obligation rates by "stopping" projects, so long as primary emphasis in review is placed on what the analysis says, rather than on whether it is truthful.

In September 1975, in response to the New Directions mandate, PPC introduced a requirement for social soundness analysis into AID's handbook on project design. Specifically it calls for a wealth of information about the identity of local groups, the way they are organized, the way they allocate their time, their motivation, whether they will all be able to participate in the proposed project, and the obstacles to project implementation.

It also asks for an analysis of the ways that sociocultural factors—particularly those that relate to the communication, leadership, authority, and patterns of mobility—will affect the diffusion of innovations introduced by the project. Finally, it calls for analysis of the way that all of these factors will affect the access of each group concerned to productive assets, employment, and other benefits.

In October 1978, PPC introduced guidelines for a much revised country program analysis, the Country Development Strategy Statement (CDSS). They require that AID missions provide an "analytical description" of the poor, including: who they are, in economic, social, cultural, and locational terms; in what ways they are deprived; how they support themselves; why they are poor, in terms of macroeconomic factors; social stratification; and the political, administrative, and institutional structure (USAID 1978a:8–10).

In principle, PPC enforces its policy guidelines by reviewing documents prepared by the missions. In practice, this task is very difficult if the criteria of compliance are not explicit and quantifiable, if review is time consuming, if analysis requires special qualifications, or if criticism threatens to lower the Agency's obligation rate. Because of these difficulties, there is a tendency in PPC, as in policy units in other bureaucracies, to accept analyses on their face value without questioning their truthfulness or whether their conclusions are reflected in project design.

The Program and Policy Coordination Bureau initially had neither the capacity nor the incentive to insist upon full compliance with its new guidelines' requirements for the analysis of local social, cultural, and ecological conditions. Criteria of adequacy were not clear. Employees of PPC had little social science and area studies expertise, nor was there sufficient time to conduct a literature search for each project document to be reviewed. There was great reluctance to ask outsiders with country knowledge to participate in the review, since it was feared that they would not "understand" the Agency's special needs and might create problems by attacking the Agency in public.

Even in instances where the issues could be clearly drawn, the bureau was reluctant to insist on compliance. Project and program documents have much less visibility outside the Agency than do guidelines and policy statements. Moreover, substantive, as opposed to procedural, criticisms of analyses are resented as "second-guessing" the missions, who are thought to be better informed by virtue of their presence in the host country. Above all, raising basic problems that cannot be dealt with through editing threatens to slow down or stop projects to which many people are already committed, and ultimately it threatens the higher-order common AID goal of obligating funds.

The Response from the Field

The New Directions policy statements and requirements for new types of analysis were not well received in AID's overseas missions or regional bureaus. The policy's central thrust made programing more difficult by eliminating many of the kinds of capital-intensive and high technology projects with which mission leadership had experience, and which host countries had come to expect from AID. The requirements for social soundness in project papers and for an analytical description of the poor in the CDSS were also resented because they created an additional workload for mission staff, required skills and information that most missions did not have, slowed down the project design process, raised complex issues that threatened project approval in Washington, and were generally seen as unnecessary, sometimes as politically offensive, by host government officials.

All new policies undergo a subtle but fundamental transformation when they are introduced into AID programs because their role in the project level decision-making process is different from and more restricted than it is assumed to be by their drafters. Policy statements from PPC set long-term goals and discuss the ways that intermediate objectives and AID assistance can contribute to the attainment of these goals. They do not normally specify the particular means, the types of projects, or input, to be used in all instances, but rather attempt to spell out the analytical criteria by which AID missions should select the means that will be most appropriate to host country conditions.

For the mission, by contrast, the means are the ends, for projects must be designed and funded regardless of the nature of the host country's problems or policies. To be sure, the mission must evoke long-term goals to add legitimacy to program and project documents, but goals are of little significance in the decision-making process, which tends to begin, rather than to end, with projects. From the mission's perspective, viable projects, rather than fiscal resources to fund them, are usually in short supply. When the mission's personnel turn to policy, it is generally to find out whether a project that has presented itself as a target of opportunity through a host country request or another donor's efforts will be approved in Washington.

In this way, through a process of trial and error, new policy statements are transformed from criteria for addressing long-term developmental problems to a typology of projects that are considered appropriate for funding. Since the resulting typologies of projects are based, in large part, on precedent from project review, they tend to vary somewhat from one regional bureau to another and even from mission to mission. Some of the "folk" policy that emerges has no basis in written policy. In one regional bureau's missions, for example, the belief that rural roads are not New Directions has persisted for several years, despite repeated assurances to the contrary from the highest authority in PPC. Through the same process, the analytical criteria intended to help the missions select projects become guidelines for project justification.

The missions' response to the New Directions policies illustrates these processes. Faced with the usual program pressures and constraints and the additional need to develop a balanced New Directions portfolio of projects, most mission directors could not afford to undertake costly and time-consuming studies or to design large numbers of new and unprecedented types of projects.[9] Instead, they sought to (a) identify types of projects that should not be considered because they are not New Directions; (b) identify types of projects with which they were already familiar that seemed to be consistent with the New Directions; (c) "package" projects and inputs with which they were familiar and comfortable in ways that would appear to be more consistent with the New Directions; and (d) to identify or design "show" projects that would embody the essence of New Directions.

The new requirement for social soundness analysis merely exacerbated the problem of finding and funding acceptable projects, for the more carefully it was carried out, the more likely it was that well-

---

[9] It is probably not accidental that the most thorough analytical work for program development I encountered while in AID was being done in Afghanistan and Ethiopia, which for political reasons had little new project activity. Both programs have subsequently been effectively discontinued. Two other missions noted for their attention to analysis of host country conditions were frequently criticized within AID for the small size of their programs in relation to the size of their staff.

precedented, ready-made project designs would be found unsuitable to local conditions. Not surprisingly, many missions complained that the social scientists they brought in for a few weeks were too negative and did not understand their problems.

The great majority of the analyses included in project papers during the first 2 years of the requirement are ethnographic, descriptive, discursive, and have little direct bearing on the proposed project. The most consistent theme is the assertion that the local people are poor enough to qualify for assistance, and that they lack the skills, inputs, or services to be supplied by AID. Seldom is there an adequate discussion of unequal access to productive assets, political power, or government services within the intended beneficiary or, in AID terminology, target group. Nor is there normally a discussion of the way people perceive the problem that AID proposes to alleviate, or of the way individuals and groups presently deal with the problem. Most importantly, there is almost never an adequate discussion and assessment of previous attempts by other donors or by the host country to implement a similar project.

Although there was considerable variation in the quality of the missions' response to the new CDSS requirement for an analytical description of the poor, it, too, was shaped by strategic bureaucratic concerns. To ensure higher funding levels and to keep a wide range of project options open, many missions assert that their country contains a large group of people living in a rather undifferentiated state of poverty and that almost anything AID may do will help. The host government, the reader is assured, has recently adopted much improved policies consistent with the New Directions but lacks the capacity to implement them.

As in the social soundness analyses, the political economy, asset distribution, the role of elites, and past rural development experience are generally underdeveloped themes. AID employees are well aware of the Catch-22 nature of the CDSS analytical requirements. Indeed the program officer in one of AID's African missions told me that he feared his candid treatment of the way urban elites were amassing landholdings along the developing road network in his first CDSS had adversely affected his mission's budget. He intended to modify his analysis for the next CDSS.

The Beginning of Change

Though the New Directions legislation had little initial effect on the extent to which most AID agricultural projects took account of peasants' farming strategies, it has fostered changes that have increased the

Agency's capacity to do so and that are beginning to affect agricultural decision making in AID's programs. Certain types of feasibility studies, project activities, and research topics already favored by some AID employees have gained legitimacy.

The greatest impact has been made by a change in the composition and training of AID's staff. By June of 1977, the number of anthropologists working full time with AID under a variety of contractual arrangements had risen from 1 to 22. Subsequently it has risen still further. There was also a marked increase in the number of agricultural economists and newly recruited foreign service officers with Peace Corps or other overseas grass roots working experience. In addition to the social scientists who work with AID full time, scores of others have served as consultants.

Paradoxically, though the main bureaucratic justification for bringing more field-oriented social scientists into the Agency was to handle the new requirement for social soundness analysis in project papers, their major contributions have been in other areas. Short-term consultants, for example, have often exerted more influence on mission decisions through informal discussions of possible future activities than through their formal reports on projects under design. Anthropologists assigned to missions on a full-time basis have been most effective when involved on a day-to-day basis in all stages of program development, project design, implementation, and evaluation. In this role they are able to understand the decision-making context to the mission and to raise problems and suggest alternatives in a timely fashion.

The New Directions perspective on small-farmer decision making has also been stressed in many of AID's in-service training programs. About 80 to 100 midlevel employees a year complete a 3-month, team-taught intensive course in development studies.

Whereas formal training may have modified stereotyped conceptions of the "tradition-bound" peasant, the most significant learning has occurred when AID employees have had a positive experience working with individual social scientists solving AID problems in an AID context. Such "converts," particularly from middle and higher level management, who have found that social analysts can make AID tasks easier as well as better, have been the most effective agents of change within the Agency.

The New Directions legislation also has encouraged administrative entrepreneurs to create new organizational units with a bureaucratic interest in having the Agency make more and better use of the findings of development-oriented social research. Outstanding units of this type include the Rural Development Office of the Development Support

Bureau, which has committed more than $20 million for applied social science research and consulting, and the Social Analysis division of the Near East Bureau, which reviews all incoming project documents to determine whether they raise social issues that require further work. This division, staffed entirely by social scientists with area studies backgrounds, is represented at all top-level meetings in the bureau.

The new paradigm of peasant agricultural decision making will become institutionalized in the Agency's programs only to the extent that it is exemplified in new types of projects, unless AID moves to non-project modes of assistance. While most projects are still unaffected by this new view of the beneficiaries, AID currently is funding a number of projects that are attempting to incorporate information about local farmers' needs and strategies through ongoing research and monitoring.

The Office of Evaluation in PPC is currently undertaking a comparative study of AID projects incorporating farming systems approaches to find out whether they are more successful than orthodox projects and whether they can be replicated at a reasonable cost.

Though all these changes are incremental, together they begin the increase in the agency's sensitivity to local conditions and decision-making processes. At the same time, there are structural problems with which AID and other donors have made little progress. As long as success is judged according to the ability to transfer resources—to obligate funds—in a timely and efficient manner, rather than according to the developmental impact of their efforts, donors will continue to have a powerful incentive for establishing and maintaining resource allocation routines that eliminate disruptive inputs from host country decision makers, including farmers.

Only a fundamental reorientation of organizational goals, criteria of success, and individual incentives would enable donor agencies to take serious account of the distinctive features of local farming systems in planning programs, to encourage wide participation in decision making, and to adjust the scale and pace of activities in response to changing local conditions.

References

Allison, Graham
    1971   Essence of Decision: Explaining the Cuban Missile Crisis. Boston: Little, Brown and Company.
American Society of Agronomy
    1976   Multiple Cropping. Special Publication No. 27. Madison: American Society of Agronomy.

Bandong, J. P., et al.
  1977 Determining the Appropriate Component Technology for More Productive
        Cropping Patterns in Pangasinan 1976–1977. Los Baños: IRRI Saturday Sem-
        inar.
Barlett, Peggy F.
  1977 The Structure of Decision Making in Paso. American Ethnologist 4(2):285–308.
  1978 What Shall We Grow?: A Critical Review of the Literature of Small Farmer Land
        Use Decision Making. Washington, D.C.: USAID–PPC.
Berry, Sara S.
  1975 Cocoa, Custom, and Socio-Economic Change in Rural Western Nigeria. New
        York: Oxford University Press.
Binswanger, Hans P., Bertil Allen Krantz, and S. M. Virmani
  1976 The Role of ICRISAT in Farming Systems Research, Begumpet: ICRISAT.
Binswanger, Hans P., et al.
  1977 Approach and Hypotheses for the Village Level Studies of ICRISAT
        Occasional Paper 15, Economics Program Begumpet: ICRISAT.
Brush, Stephen B.
  1977 The Myth of the Idle Peasant: Employment in a Subsistence Economy. In Peasant
        Livelihood. R. Halperin, and J. Dow, eds. pp. 60–78. New York: St. Martins.
Bryant, Coralie
  1979 Organizational Impediments to Making Participation a Reality: "Swimming Up-
        stream" in AID. Written for Rural Development Participation Review, Spring
        1979.
Cancian, Frank
  1972 Change and Uncertainty in a Peasant Economy. Stanford: Stanford University
        Press.
  1979 The Innovator's Situation: Upper Middle Class Conservatism in Agricultural
        Communities. Stanford: Stanford University Press.
CATIE (Centro Agronómico Tropical de Investigación y Enseñanza)
  1978 A Farming System Research Approach for Small Farms of Central America.
        Turrialba: CATIE.
Chibnik, Michael
  1974 Economic Strategy of Small Farmers in Stann Creek District, British Honduras.
        Unpublished doctoral dissertation, Anthropology, Columbia University.
CODERIA–DSRC Conference
  1978 Social Science, Research, and National Development in Africa. Conference Out-
        line. Khartoum.
Cole, John, and Eric Wolf
  1974 The Hidden Frontier. New York: Academic Press.
Collier, George A.
  1975 "Are Marginal Farmlands Marginal to Their Farmers? In Formal Methods in
        Economic Anthropology. Stuart Platter, ed. pp. 149–185. American Anthropolog-
        ical Association, Special Publication, No. 4.
Collinson, Michael P.
  1972 Farm Management in Peasant Agriculture: A Handbook for Rural Development
        Planning in Africa. New York: Praeger.
De Walt, Billie R.
  1979 Modernization in a Mexican Ejido: A Study in Economic Adaptation. New York:
        Cambridge University Press. (b)

Epstein, T. Scarlett
  1962 Economic Development and Social Change in South India. Manchester: Manchester University Press.
Foreign Assistance Act
  1979 Legislation on Foreign Relations through 1978: Current Legislation and Related Executive Orders (Vol. 1). Washington, D.C.: United States Senate–House of Representatives Joint Committee.
Gladwin, Christina
  1976 A View of the Plan Puebla: An Application of Hierarchical Decision Models. American Journal of Agricultural Economics 58(5):881–87.
Greenwood, Davydd J.
  1973 The Political Economy of Peasant Family Farming: Some Anthropological Perspectives on Rationality and Adaptation. Rural Development Occasional Paper No. 2. Ithaca: Cornell University, Center for International Studies.
  1976 Unrewarding Wealth: The Commercialization and Collapse of Agriculture in a Spanish Basque Town. Cambridge University Press.
  1978 Community-Level Research, Local–Regional–Governmental Interactions, and Development Planning. Washington, D.C.: USAID–PPC.
Harwood, Richard R.
  1975 Farmer-oriented Research Aimed at Crop Intensification. In Proceedings of the Cropping Systems Workshop. pp. 12–31. Los Baños: IRRI.
Hildebrand, Peter E.
  1976 A Multidisciplinary Methodology for Developing New Cropping Systems Technology for Traditional Agriculture. Paper prepared for the Conference on Economic Development in Agricultural Areas, Bellagio, August 4–6. Guatemala City: ICTA.
Hoben, Allan
  1979 Lessons From a Critical Examination of Livestock Projects in Africa. Office of Evaluation, Studies Division, Working Paper No. 26. Washington, D.C.: USAID.
IBRD
  1974 The Social Sciences and Development. Papers Presented at a Conference in Bellagio, Italy, on the Financing of Social Science Research for Development. Washington, D.C.: World Bank.
  1978 Farming Systems Research at the International Agricultural Research Centers. Consultative Group on International Agricultural Research, Technical Advisory Committee. Washington, D.C.: World Bank.
  1979 Project Implementation and Supervision (for official use only), April 30, 1979. Washington, D.C.: World Bank.
Johnson, Allen
  1971 Sharecroppers of the Sertao. Stanford: Stanford University Press.
Just, Richard E.
  1975 Risk Aversion Under Profit Maximization. American Journal of Agricultural Economics 57(2):347–352.
Kass, D. C. L.
  1978 Polycultural Cropping Systems: Review and Analysis. Cornell International Agricultural Bulletin No. 32. Ithaca: Cornell University.
Lele, Uma
  1975 The Design of Rural Development: Lessons from Africa. Baltimore: Johns Hopkins Press.

Lipton, Michael
    1968 The Theory of the Optimizing Peasant. Journal of Development Studies
          4(3):327–351.
McLoughlin, Peter M. F., Ed.
    1970 African Food Production Systems. Baltimore: Johns Hopkins Press.
McPherson, Laura, Ed.
    1978 The Role of Anthropology in the Agency for International Development: Work-
          shop Report of the Institute for Development Anthropology. Binghamton, N.Y.:
          IDA
Moerman, Michael
    1968 Agricultural Change and Peasant Choice in a Thai Village. Berkeley: University
          of California Press.
Netting, Robert McC.
    1968 Hill Farmers of Nigeria. Seattle: University of Washington Press.
    1969 Ecosystems in Process: A Comparative Study of Change in Two West African
          Societies. In Ecological Essays. D. Damas, ed. pp. 102–112. National Museum of
          Canada, Bulletin No. 230.
    1974 Agrarian Ecology. In Annual Review of Anthropology. B. Seigel, ed. pp. 21–56.
          Palo Alto: Annual Reviews.
Netting, Robert McC. et al.
    1978 The Conditions of Agricultural Intensification in the West African Savannah.
          Washington, D.C.: USAID–PPC.
Norman, David W.
    1971 Initiating Change in Traditional Agriculture. Agricultural Economics Bulletin for
          Africa 13:31–52.
    1974 Rationalizing Mixed Cropping Under Indigenous Conditions: The Example of
          Northern Nigeria. Journal of Development Studies 11(1):3–21.
NRC (National Research Council)
    1977a World Food and Nutrition Study: The Potential Contributions of Research.
           Washington, D.C.: National Academy of Sciences.
    1977b A Methodology for Farming Systems Research. In Supporting Papers: World
           Food and Nutrition Study. Washington, D.C.: National Academy of Sciences.
Ortiz, Sutti
    1973 Uncertainties in Peasant Farming. London: Athlone Press.
    1976 The Effect of Risk Aversion Strategies on Subsistence and Cash Crop Decisions.
          Conference on Uncertainty and Agricultural Development. Mexico: Agricultural
          Development Council.
Perrett, Heli E.
    1978 Social Analysis and Project Design in the Agency for International Development.
          Washington, D.C.: USAID.
Quinn, Naomi
    1975 Decision Models of Social Structure. American Ethnologist 2(1):19–45.
    1978 Do Mfantse Fish Sellers Estimate Probabilities in Their Heads? American
          Ethnologist 5(2):204–226.
SAREC (Swedish Agency for Research Cooperation to Developing Countries)
    1979 SAREC's Third Year (Annual Report 1977–1978). Stockholm: SIDA.
Schluter, Michael G. G., and Timothy D. Mount
    1976 Some Management Objectives of the Peasant Farmer: An Analysis of Risk Aver-
          sion in the Choice of Cropping Pattern, Surate District, India. Journal of De-
          velopment Studies 12(3):246–261.

Schultz, Theodore W.
1964   Transforming Traditional Agriculture. New Haven: Yale University Press.
Scott, James C.
1976   The Moral Economy of the Peasant. New Haven: Yale University Press.
Shanin, Teodor, ed.
1971   Peasants and Peasant Societies. London: Penguin.
Slovic, Paul, Baruch Fischhoff, and Sarah Lichtenstein
1977   Behavioral Decision Theory. *In* Annual Review of Psychology 28:1–39.
Tendler, Judith
1975   Inside Foreign Aid. Baltimore: Johns Hopkins Press.
Tourte, R.
1974   IRAT Approach to Development of Intensive Systems in Peasant Agriculture: A
       Case Study in Senegal. International Workshop on Farming Systems. Begumpet:
       ICRISAT.
USAID
1978a   Guidance for the Country Development Strategy Statement (CDSS). Washing-
        ton, D.C.: USAID.
1978b   A Strategy for a More Effective Bilateral Development Assistance Program.
        Washington, D.C.: USAID.
1978c   Agricultural Development Policy Paper. Washington, D.C.: USAID.
Weiss, Charles, Jr.
1979   Mobilizing Technology for Developing Countries. Science 203:1083–1089.
Wharton, Clifton R.
1971   Risk, Uncertainty, and the Subsistence Farmer: Technological Innovation and
       Resistance to Change in the Context of Survival. *In* Studies in Economic An-
       thropology. George Dalton, ed. pp. 152–179. Washington, D.C.: American An-
       thropological Association, Anthropological Studies, No. 7.

Subject Index

STUDIES IN ANTHROPOLOGY

Under the Consulting Editorship of E. A. Hammel,
UNIVERSITY OF CALIFORNIA, BERKELEY

in preparation

Eric B. Ross (Ed.), BEYOND THE MYTHS OF CULTURE: A Reader in
    Cultural Materialism

Thayer Scudder and Elizabeth Colson, SECONDARY EDUCATION AND
    THE FORMATION OF AN ELITE: The Impact of Education
    on Gwembe District, Zambia